Compara...

Comparative studies ... more important as a means for und... and evolution of mammals. Primates have ... ps and diverse ecologies, and represent a large s... This book draws together a wide range of experts from fields as diverse as reproductive biology and foraging energetics to place recent field research into a synthetic perspective. The chapters tackle controversial issues in primate biology and behaviour, including the role of brain expansion and infanticide in the evolution of primate behavioural strategies. The book also presents an overview of comparative methodologies as applied to recent primate research that will provide new approaches to comparative research. It will be of particular interest to primatologists, behavioural ecologists and those interested in the evolution of human social behaviour.

P.C. Lee is a lecturer in Biological Anthropology at the University of Cambridge and Fellow of Downing College. She began field work on baboons in 1975 and has maintained an interest in the socioecology and behaviour of primates and other large mammals ever since. She has written numerous papers and has co-edited three previous volumes on primates – *Primate Evolution*, *Primate Ecology and Conservation* and *Primate Ontogeny, Cognition and Behaviour* (all 1986) – and has co-authored *The Threatened Primates of Africa* (1988).

Cambridge Studies in Biological and Evolutionary Anthropology

Series Editors

HUMAN ECOLOGY
C. G. Nicholas Mascie-Taylor, University of Cambridge
Michael A. Little, State University of New York, Binghamton
GENETICS
Kenneth M. Weiss, Pennsylvania State University
HUMAN EVOLUTION
Robert A. Foley, University of Cambridge
Nina G. Jablonski, California Academy of Science
PRIMATOLOGY
Karen B. Strier, University of Wisconsin, Madison

Consulting Editors
Emeritus Professor Derek F. Roberts
Emeritus Professor Gabriel W. Lasker

Comparative
Primate Socioecology

EDITED BY P.C. LEE

CAMBRIDGE
UNIVERSITY PRESS

PUBLISHED BY THE PRESS SYNDICATE OF THE UNIVERSITY OF CAMBRIDGE
The Pitt Building, Trumpington Street, Cambridge, United Kingdom

CAMBRIDGE UNIVERSITY PRESS
The Edinburgh Building, Cambridge CB2 2RU, UK
40 West 20th Street, New York NY 10011-4211, USA
10 Stamford Road, Oakleigh, VIC 3166, Australia
Ruiz de Alarcón 13, 28014 Madrid, Spain
Dock House, The Waterfront, Cape Town 8001, South Africa

http://www.cambridge.org

First published 1999
First paperback edition 2001

Printed in the United Kingdom at the University Press, Cambridge

Typeset in Times 10/12.5pt [VN]

A catalogue record for this book is available from the British Library

Library of Congress Cataloguing in Publication data

Comparative primate socioecology / edited by P.C. Lee
    p.   cm. – (Cambridge studies in biological anthropology)
Includes index.
ISBN 0 521 59336 0 (hb)
1. Primates — Behavior. 2. Primates – Ecology. 3. Primates – Evolution. 4. Social
    evolution in animals. I. Lee, Phyllis C. II. Series.
QL737.P9C5784   1999
599.8–dc21   98–36457   CIP

ISBN 0 521 59336 0 hardback
ISBN 0 521 00424 1 paperback

Contents

Contributors

ROBERT BARTON
Department of Anthropology, University of Durham, Old Elvet, Durham DH1 3HN, UK

ALLISON BEAN
Department of Biological Anthropology, University of Cambridge, Downing Street, Cambridge CB2 3DZ, UK

NICHOLAS BLURTON JONES
Graduate School of Education and Department of Anthropology, Moore Hall, University of California at Los Angeles, Los Angeles, CA 90024, USA

ROBIN DUNBAR
Population Biology Research Group, School of Biological Sciences, Nicholson Building, University of Liverpool, Liverpool L69 3BX, UK

ROBERT A. FOLEY
Department of Biological Anthropology, University of Cambridge, Downing Street, Cambridge CB2 3DZ, UK

KRISTEN HAWKES
Department of Anthropology, University of Utah, Salt Lake City, UT 84112, USA

CLARE HOLDEN
Department of Anthropology, University College London, Gower Street, London WC1E 6BT, UK

KATE E. JONES
Roehampton Institute, School of Life Sciences, West Hill, London SW15 3SN, UK

PETER M. KAPPELER
AG Verhaltensforschung/Okologie, Deutsche Primatenzentrum, Kellnerweg 4, 37007 Gottingen, Germany

PHYLLIS C. LEE
Department of Biological Anthropology, University of Cambridge, Downing Street, Cambridge CB2 3DZ, UK

RUTH MACE
Department of Anthropology, University College London, Gower Street, London WC1E 6BT, UK

ANN MACLARNON
Roehampton Institute, School of Life Sciences, West Hill, London SW15 3SN, UK

MARIA A. VAN NOORDWIJK
Department of Biological Anthropology and Anatomy, Duke University, P.O. Box 90383, Durham, NC 27708-0383, USA

CHARLES L. NUNN
3323 Ridge Road, Durham, NC 27705-5535, USA

JAMES F. O'CONNELL
Department of Anthropology, University of Utah, Salt Lake City, UT 84112, USA

J. MICHAEL PLAVCAN
Department of Anatomy, New York College of Osteopathic Medicine, Old Westbury, NY 11568, USA

ANDREW PURVIS
Department of Biology, Imperial College at Silwood Park, Ascot, Berkshire SL5 7PY, UK

KATE ROBSON-BROWN
Centre for Human Evolutionary Research, Department of Archaeology, University of Bristol, Clifton, Bristol, UK, and Department of Biological Anthropology, University of Cambridge, Downing Street, Cambridge CB2 3DZ, UK

CAROLINE ROSS
Roehampton Institute, School of Life Sciences, West Hill, London SW15 3SN, UK

CAREL P. VAN SCHAIK
Department of Biological Anthropology and Anatomy, Duke University, P.O. Box 90383, Durham, NC 27708-0383, USA

KAREN B. STRIER
Department of Anthropology, University of Wisconsin, 5240 Social Science Building, 1180 Observatory Drive, Madison, WI 53706-1393, USA

ANDREA J. WEBSTER
Department of Biology, Imperial College at Silwood Park, Ascot, Berkshire SL5 7PY, UK

DAISY K. WILLIAMSON
Hominid Palaeontology Research Group, Department of Human Anatomy and Cell Biology, University of Liverpool, Ashton Street, Liverpool L69 3GE, UK

Preface

Comparative studies have become both more frequent and more important as a means for understanding the biology, behaviour and evolution of mammals. Historically, studies of primate socioecology have been in the forefront of the field and many interesting methodological developments in comparative socioecology have emerged from earlier work. This is not to say that other animals have not been examined – for example, there are excellent studies of seals, carnivores and ungulates, not to mention extensive work on birds.

But primates are particularly interesting in that they have complex social relationships and diverse ecologies, as well as representing a large radiation of morphologies. Socioecology, as used here, is taken to represent the interactions between characteristics of the resource base, its mode of exploitation, reproductive biology and life history, and the observed social system. In this sense, primates can be considered as a test case for hypotheses that the solutions to ecological problems have a social root. Thus, the chapters in this book seek to explore the diverse relations between sociality and resources, mating systems, energetics and reproduction. Questions of biological or physiological constraints on sociality are also examined.

Since the 1987 publication of *Primate Societies* by Smuts *et al.*, field researchers have added greatly to our knowledge of primate social systems and ecological variation, and this book attempts to synthesise some recent work. It is perhaps notable that the socioecology of the primates is not approached with a taxonomic structure here. Rather, this book tries to cover less well-known species that have been the focus of recent field studies, and specific issues that are of current theoretical interest for primates as diverse as lemurs and humans.

Part 1
Comparative methods

Editor's introduction

Our ability to analyse variation within and between taxonomic groups has been enhanced by the development of techniques for the statistical manipulation of comparative data, but we have yet to reach a consensus on which techniques are appropriate for specific analyses. Thus, several possible approaches are presented. A comprehensive overview of the pros and cons, as well as how to carry out different comparative techniques can be found in Harvey and Pagel (1991).

It should be noted that there are two separate issues involved in phylogenetic analyses. The first of these is fundamentally *statistical*. Although it has long been recognised that the use of 'species' data in comparative analyses on closely related taxa may violate statistical assumptions of independence of data points (e.g. Crook, 1965), this was elaborated in relation to phylogenetic similarity in allometry by Felsenstein (1985). Stated simply, closely related taxa may share traits derived through that genealogical relationship rather than as a result of selection, and species as such are not independent within lineages. This issue had been at least partially explored in earlier socioecological and life history research on primates through data reduction techniques – the use of mean values for different taxonomic levels – the 'higher node' approach (e.g. genera: Clutton-Brock and Harvey, 1977; subfamily: Harvey, Martin and Clutton-Brock, 1987).

But there is a second, more interesting, question raised by comparative analyses, that of the *evolutionary* similarity within and between related taxa (Purvis and Harvey, 1995), and it is in this context that the value of phylogenetically controlled comparisons is most apparent. One of the most common and accessible techniques, Comparative Analysis by Independent Contrasts (CAIC), is presented by Purvis and Webster in Chapter 3. The value of CAIC lies in its simplicity and in the detailed primate phylogeny derived by Purvis. Some problems with the method are also considered.

The fundamental question, however, remains whether the comparative study seeks to determine if evolutionary change in traits has occurred, or

1

whether it seeks to identify variation between species or groups of species in an attempt to determine causality in this observed variation. Often, a comparison of the results obtained from several different analytical techniques may allow for more robust interpretations. This procedure is used in a number of the chapters in subsequent parts of the book. Another technique for exploring evolutionary variation is that of nested analysis of variance. Originally devised to determine which taxonomic level explained the observed variance in a trait, and thus to limit comparisons to that 'independent' level, it has a further utility in partitioning variance between these taxonomic levels and thus provoking evolutionary explanations. Methods such as correcting for degrees of freedom in nested ANOVAs also address the problem of statistical dependence (see Smith, 1994). Interestingly, there may be times when different taxonomic levels explain variation for distinct variables, suggesting that it would be difficult, if not impossible, to 'control' for phylogeny by selecting a single independent higher taxonomic node for analysis. For example, among primates, variance in adult body weight is greatest at the level of the subfamily, whereas that of density is greatest at the population level (Vella, 1995).

If two species share traits, is this the result of evolutionary convergence or simply due to sharing ancestral traits between closely related descendants? If we are exploring evolution within and between lineages, then obviously the lineages themselves are part of the data we are examining. It becomes critical to know both the phylogenetic relationships and to tease apart the ancestor–descendant traits, as noted by Purvis and Webster. The potential to determine separate evolutionary events by cladistic analysis is outlined by Robson-Brown (Chapter 2). Such techniques are far more accessible with current programs, but users need to be aware of the debates about homology and analogy explored by Robson-Brown.

Other techniques, which rely on 'species' data but allow for an assessment of the effects of phylogeny on the observed patterns, are potentially available; for example the use of maximum likelihood estimators for co-evolution in discrete traits (e.g. Pagel, 1994; Mace and Holden, Chapter 15), or multidimensional scaling of traits which can produce visible clusters among close phylogenetic relatives (e.g. Bean, Chapter 13). MacLarnon, Chivers and Martin (1986) produced evidence for phylogenetic similarity in gut areas among primates using multidimensional scaling, with a consistent cluster of colobines in analytical space, despite observed differences in diets from fruits, through seeds to mature leaf (Davies and Oates, 1996). The power of such analyses lies in their ability to explore patterns explicitly due to shared descent. Other possible means for incorporating phylogeny that do not rely on phylogenetic subtraction, and thus the assumption that the mean of nodes reconstructs a single ancestral state (e.g. Pagel and Harvey, 1988; Stearns, 1992), could lie in non-linear modelling, in nested analysis of covariance, or in principle components data reduction techniques. Consensus on the 'most' appropriate technique is still to be found.

The point of providing several different techniques and perspectives in this book is to focus researchers on making explicit the hypothesis being tested. Is it an evolutionary explanation, a mechanical or physiological one, or a functional relationship? These issues are presented by Mac-Larnon in a general overview of methodology (Chapter 1). When and why should species be expected to vary? How do rates of evolution within lineages vary? What are the effects on traits? Are predictive trends the aims of the analysis or are we seeking mechanisms in evolution? The technique used, or combinations of methods, needs to be tailored to suit the questions. Even after 30 years of debate, no single method can yet be considered sufficient or even the most appropriate, and it is the question not the methodology that should drive the exploration.

References

Clutton-Brock, T.H. and Harvey, P.H. (1977). Primate ecology and social organisation. *Journal of Zoology, London* **183**, 1–39.

Crook, J.H. (1965). The adaptive significance of avian social behaviour. *Symposium of the Zoological Society of London* **14**, 181–218.

Davies, A.G. and Oates, J.F. (1996). *Colobine Monkeys*. Cambridge: Cambridge University Press.

Felsenstein, J. (1985). Phylogenies and the comparative method. *American Naturalist* **125**, 1–15.

Harvey, P.H., Martin, R.D. and Clutton-Brock, T.H. (1987). Life histories in comparative perspective. In *Primate Societies*, ed. B.B. Smuts, D.L. Cheney, R.M. Seyfarth, R.W. Wrangham and T.T. Struhsaker, pp. 181–96. Chicago: University of Chicago Press.

Harvey, P.H. and Pagel, M. (1991). *The Comparative Method in Evolutionary Biology*. Oxford: Oxford University Press.

MacLarnon, A.M., Chivers, D.J. and Martin, R.D. (1986). Gastrointestinal allometry in primates and other mammals, including new species. In *Primate Ecology and Conservation*, ed. J.G. Else and P.C. Lee, pp. 75–85. Cambridge: Cambridge University Press.

Pagel, M. (1994). Detecting correlated evolution on phylogenies: a new method for analysis of discrete categorical data. *Proceedings of the Royal Society, Series B* **255**, 37–45.

Pagel, M. and Harvey, P.H. (1988). Recent developments in the analysis of comparative data. *Quarterly Review of Biology* **63**, 413–40.

Purvis, A. and Harvey, P.H. (1995). Mammal life history evolution: a comparative test of Charnov's model. *Journal of Zoology, London* **237**, 259–83.

Smith, R.J. (1994). Degrees of freedom in interspecific allometry: an adjustment for the effects of phylogenetic constraint. *American Journal of Physical Anthropology* **93**, 95–107.

Stearns, S. (1992). *The Evolution of Life Histories.* Oxford: Oxford University Press.

Vella, A. (1995). Primate Population Biology and Conservation. PhD thesis, University of Cambridge.

1 *The comparative method: principles and illustrations from primate socioecology*

ANN MACLARNON

Introduction

There are two major means of investigation across a wide range of sciences, both natural and social. These are the experimental method and the comparative method. In the so-called hard sciences – physics and chemistry – and also the 'harder' end of biology, investigation is more commonly by experimental manipulation. Other biological questions, notably those concerning evolutionary history and adaptation, are more or less inaccessible to experimentation, as are other aspects of the natural world, such as astronomical phemonena. Exploration of these areas and the development of explanations are undertaken largely by the comparative method, whereby common patterns and principles of variability are sought out, providing the basis for possible interpretation in terms of causes and effects. Similarly, in the social sciences, comparisons can be made across space and time of different societies, divisions or aspects of societies, with the aim of uncovering the origins and explanations of present features and past changes.

The comparative method has its origins in the realisation of the Enlightenment that the natural world can be understood and explained in terms of common principles and predictable variation. It involves testing the generality of suggested explanations for characteristics or phenomena, in contrast to *ad hoc*, one-off explanations that may merely reflect coincidence rather than causal connection. Predictions can be made from proposed general principles, and tested on further species, societies, stars or galaxies, and if borne out, they provide increased support for the validity of a principle.

The fundamentals of the comparative method were first expounded in the mid-nineteenth century by John Stuart Mill in his book *A System of Logic* (1872, 1967) in the chapter 'Of four methods of experimental inquiry'. These four methods essentially describe the basic principles of logical

5

deduction used in scientific inquiry today, including the comparative method. Despite the fact that Stuart Mill's examples mostly come from the physical rather than the living world, the applications of the methods as outlined, their difficulties and limitations, are entirely pertinent to the comparative method in biology, including socioecology. The four methods are as follows:

1. Method of Agreement. 'If two or more instances of the phenomenon under investigation have only one circumstance in common, the circumstance in which alone all the instances agree is the cause (or effect) of the given phenomenon.' (1967, p. 255).

2. Method of Disagreement. 'If an instance in which the phenomenon under investigation occurs, and an instance in which it does not occur, have every circumstance in common save one, that one occurring only in the former; the circumstance in which alone the two instances differ is the effect, or the cause, or an indispensable part of the cause, of the phenomenon.' (1967, p. 256).

   These two methods can be combined in the Joint Method of Agreement and Difference:
   'If two or more instances in which the phenomenon occurs have only one circumstance in common, while two or more instances in which it does not occur have nothing in common save the absence of that circumstance, the circumstance in which alone the two sets of instances differ is the effect, or the cause, or an indispensable part of the cause, of the phenomenon.' (1967, p. 259).

3. Method of Residues. 'Subduct from any phenomenon such part as is known by previous inductions to be the effect of certain antecedents, and the residue of the phenomenon is the effect of the remaining antecedents.' (1967, p. 260).
   (Note: By 'antecedent' Stuart Mill is referring to conditions rather than ancestors.)

4. Method of Concomitant Variation. 'Whatever phenomenon varies in any manner whenever another phenomenon varies in some particular manner, is either a cause or an effect of that phenomenon, or is connected with it through some fact of causation.' (1967, p. 263).

The main principles of scientific inquiry are established in the first two Methods, while the third and fourth can be seen as special cases of the Method of Difference. The Method of Difference describes a basic principle of good experimental design whereby all factors bar one are the same for all samples, and thus any difference in findings between the samples is related to the one differing factor (see also Chapter 3). As Stuart Mill discusses, this is a better form of experimental design than the Method of Agreement, because it is easier to ensure that virtually all circumstances are identical, as the Method of Difference requires, than to be certain that one, and only one circumstance of relevance is the same, for the Method of Agreement. However, the strict conditions required by the Method of Difference can rarely be found in the natural world, where observations of similarities and differences between natural phenomena are the only available sources of data and where experimental manipulation is not possible. The requirements of the pure Method of Difference are highly unlikely to be met if the experimental design cannot be controlled. As Stuart Mill puts it, 'In the spontaneous operations of nature there is generally such complication and such obscurity . . . and [these operations are] therefore so seldom exactly alike in any two cases, that a spontaneous experiment, of the kind required by the Method of Difference, is commonly not to be found' (1967, p.257). Hence, when using the comparative method rather than experimental inquiry, the Method of Agreement is generally more appropriate.

For the Method of Agreement, the circumstances in common must be the only ones that could possibly have a cause-and-effect relationship with the phenomenon of interest. In practice, it is difficult to be certain that this requirement is met, even when the experimental design can be controlled. Hence, if the experimental design is controllable, the Method of Difference is preferable. However, the Method of Agreement can still be useful even when the absolute exclusion of other possible relevant common circumstances is not possible. At the very least, even if phenomenon *a* is only found when circumstance *A* is in place, then circumstance *A* may be a *condition* for the existence of *a*, though they are not necessarily related as cause and effect. This is where the Joint Method of Agreement and Difference enables closer approximation to the determination of a cause-and-effect relationship, and in fact many of the applications of the comparative method in evolutionary biology essentially use the Joint Method of Agreement and Difference. By comparing different circumstances under which a phenomenon occurs and does not occur, it can be deduced which of the different circumstances are at least *conditions* for the presence or absence of a phenomenon, even if a causal link cannot be established with certainty. It is always possible that another, unidentified, third factor actually *causes* both the variation in circumstances and the presence or absence of the phenomenon. Put into modern terms, 'correlation does not mean causation'. However, establishing a conditional link is a useful step that other information or comparisons may help to make firmer.

The Method of Residues can be seen as the Method of Difference under special circumstances, in which a phenomenon is caused by several factors. If a case in which all but one of these factors is present is compared with another case in which all the factors occur, the difference in the size of the phenomenon between the first and second instances is related to the differing factor. As with the Method of Difference, the difficulty in using this method for deduction from observations of the natural world comes in making the assumption that all circumstances are similar save one. Nevertheless, the Method of Residues as described by Stuart Mill forms an interesting parallel with the method of phylogenetic contrasts recently developed for comparative biology (see Chapter 3). This method utilises subtraction (or calculation of residues) between the sizes of a phenomenon or feature in pairs of closely related taxa, which share many factors because of common ancestry. It investigates whether such subtracted differences are associated with differences between the paired taxa in some other feature of biological or adaptive concern.

Stuart Mill's Method of Concomitant Variation is applicable in cases in which phenomena are always present to a greater or lesser extent, and hence it is not possible to compare the effects of their presence and absence. In such cases, comparison can be made between the size of two phenomena in different contexts, and rules can be deduced about the relationship between changes in one phenomenon and changes in the other, which may reflect a cause-and-effect relationship. This method therefore applies to phenomena or variables that are continuous as opposed to categorical variables. Like the Method of Residues, the Method of Concomitant Variation is really a special case of the Method of Difference, and it is widely used in comparative biology, including applications of phylogenetic contrasts to continuous variables.

Throughout his explanation of the four methods, Stuart Mill emphasises the difficulty of determining which of two related phenomena is the cause, and which the effect. In an extended example investigating the cause of dew formation, the problem is resolved by recognising the primacy of basic physical properties such as the heat conduction of different materials, and these are therefore identified as the causal factors. In comparative biology, reference to basic biological laws and knowledge is similarly useful.

The use of the comparative method in evolutionary biology essentially follows the methods outlined by Stuart Mill and encounters the difficulties he describes. The resulting logical deductions resemble those possible from experimental results, but with the handicap that research design played no part in determining the combination of phenomena and circumstances, variables and subjects in each 'natural experiment'. There are therefore inevitable gaps in a simple line of deduction, and the comparative method involves making best use of whatever 'natural experiments' are available, for example the species (populations, higher level taxa, etc.) that exist. Good biological knowledge of the factors and features that vary between sample species, and of fundamental biological laws and processes is essential. Detailed methodological issues are also important where they affect results. The chapters of this book provide rich and varied examples of the use of the comparative method in socioecology.

The first step: investigation of patterning

The first step in comparative analyses is to describe the patterns of distribution of the characteristics of interest across a chosen sample, in order to establish whether the conditions for one of the four methods exist. This involves investigating whether particular variants of one categorical variable are associated with particular variants of another, or whether continuous variables are correlated across the sample. Essentially, this is a similar process whether the variables concerned are categorical or continuous, and many features can be described either way. For example, dietary variation can be categorised according to the predominant food, as insectivory, frugivory etc., or measured according to the proportion of a particular food type in the diet, such as percentage fruit (e.g see Chapter 13). Where there is a choice of either a categorical or continuous measure, factors such as the nature and quality of available data, and the question under investigation are important. For example, duration of lactation is a measure used in both Lee and van Schaik *et al.*'s chapters (Chapters 5 and 8). Lee's study focuses on variation in the length of the lactation period itself and the degree of correlation with other continuous life history variables. However, for van Schaik *et al.*, the feature of interest is categorical: whether the lactation period is longer or shorter than the gestation period. If it is shorter, post-partum oestrus, enabling reconception immediately following a birth, would be a viable evolutionary option, given that energetically the mother must wean one infant before having to feed a second. It should be borne in mind, however, that where it is possible to choose between categorical and continuous versions of a variable, this could affect the results, particularly levels of significance.

In comparing different species, variation in overall body size is commonly an important factor. The question of interest may be how a variable, say brain size, correlates with body size, or attention may be focused on residual variation from scaling relationships, such as relative brain size. Both types of investigation are utilised in this volume (e.g. Chapters 4 and 5). Allometric methods of analysis are commonly necessary in cross-species comparisons as many features do not scale to body size, or to each other, in a linear fashion. Rather, they are related through power functions. Thus, simple ratios, for example of neonate weight to maternal weight, should only be used with great care. The intention may be to 'remove size' from the comparison of species, but such ratios will not be 'size free' unless the variables concerned scale linearly with one another – that is, with the same exponent in relation to body size. For example, Charnov's recent life history theory (see Chapters 4 and 6) predicts that several life history characteristics will scale to body size with the same exponent (0.25). Hence, ratios of these variables are expected to be constant across species. However, instances of size-free ratios are rare in comparative biology.

The analyses presented by van Schaik *et al.* (Chapter 8) illustrate the centrality of investigating patterning to the comparative method. Data were collected for primate species for a wide range of features such as the incidence of infanticide, infant care styles, whether lactation is longer or shorter than gestation, mating patterns during pregnancy, the presence or absence of post-partum oestrus, the development of sex skin, and whether females produce calls related to mating. These data were examined to determine whether the pattern of distribution across species of variants of one feature is associated with that of another – the basic requirement if variables are causally or functionally related. For example, the species in which the mother alone carries the infant do not have post-partum oestrus, whereas most of those which park their infants, or in which carrying is shared with other individuals, do. In Ross and Jones' study (see Chapter 4), patterning is similarly fundamental, but here the variables are largely quantitative, such as maternal weight, age at first reproduction, mortality rates and interbirth intervals. The first step was to investigate the pattern of correlation between the variables. For example, across primates, taking variation in body weight into account, levels of mortality among infants and juveniles are correlated with birth rates, and species with higher pre-reproductive mortality reproduce faster. However, adult mortality rates are not correlated negatively with age of first reproduction for females, as was predicted; species with higher adult mortality do not start reproducing earlier (at least not in the small sample available).

The question of homology

A fundamental requirement of the comparative method is that the features compared across a sample should be homologous. Robson-Brown (Chapter 2) provides an overview of attempts to define homology and their application in socioecology. Some definitions require that a common feature in two species must derive from a common ancestor in order to be homologous ('taxic' definitions), whereas others require only that features are indistinguishable ('operational' definitions). For the types of analyses contained in this book, the former definition is not only extremely restrictive, in that the necessary work on the detailed development, structure and evolutionary history of the features concerned has rarely been done, but it would also render most of the analyses invalid. Many working in comparative biology agree that only when the presence of similar features in different taxa results from separate evolutionary events can these taxa provide independent data points for the testing of informative correlation or association (see Chapter 3). Clearly, this requirement is mutually exclusive with the more restrictive definition of homology, and acceptance of both would rule out the use of the comparative method. Indeed, the separate evolution of closely similar features in several species or taxa can provide a useful basis for the deduction of the adaptive or functional cause of their evolution, as many examples in this book illustrate. This is a version of the Method of Agreement.

Van Schaik *et al.* (Chapter 8) present data showing that male infanticide has not been observed in primate species in which communal infant care is well developed. The forms of communal care involved are not identical in all the sample species; for example, the caretakers other than the mother vary. Also, this type of behaviour must have evolved more than once, given its distribution across the primate phylogeny. Therefore, 'communal care' does not meet the conditions for the stricter, 'taxic' definition of homology. However, the basis of the use of the comparative method in van Schaik *et al.*'s study is that similar behaviour evolves for similar functional reasons, and it is not necessary for either factor to be absolutely identical across the sample for the results to enhance evolutionary understanding. The Method of Agreement is a flexible tool that requires reasonable, but not rigid, application.

Nevertheless, care must be taken in deciding whether or not such co-categorisation of non-identical features is reasonable. Kappeler (Chapter 10) provides an extensive discussion of whether lemur social structures are similar enough to those of anthropoids for their comparison to provide a useful test of the generality of theories originally developed for anthropoids. He examines whether lemur species can be divided into similar categories to anthropoids on the basis of variation in four major aspects of female–female relationships: philopatry, nepotism, tolerance and despotism. Female–female relationships in lemurs, he concludes, display a number of different features, and different combinations of features from those of anthropoids. Reasons for this lack of comparability are suggested, including fundamental grade differences between the groups such as the lesser visual acuity of lemurs. This is clearly an example of too many concurrent dissimilarities alongside other similarities, rendering the comparative method unusable, and further theoretical groundwork is needed first. In other words, the conditions are not met for the Method of Agreement, the Method of Differences, or the Joint Method.

The assumption of adaptational equilibrium

It is a basic assumption of the use of the comparative method in evolutionary biology that species are in adaptational equilibrium. That is, their combinations of features, morphological, behavioural and ecological, have evolved to be co-adapted – they are in an evolutionary steady state. This is normally a reasonable assumption given that evolution is generally viewed either as occurring gradually over a very long timescale, making significant perturbations or disequilibria unlikely, or as occurring rapidly in small populations interspersed by much longer periods of stasis, or equilibrium, such that the latter are much more likely to be sampled. However, within primates, there are two taxa for which there is good evidence that the assumption of equilibrium is not valid, at least for certain features. The first is humans, for whom recent, extraordinarily rapid cultural and technological changes may have occurred too fast for any necessary balancing biological changes to keep up, although Foley (Chapter 14) argues that humans may well be adapted at least to the changes brought about with the advent of agriculture. However, to overcome the possible problem of human disequilibrium, Blurton Jones *et al.* (Chapter 6) use data from remaining hunter–gatherer groups, our presumed long-standing state, to compare reproductive life history strategies with those of great apes. Kappeler (Chapter 10) refers to the second probable disequilibrium that results from the relatively recent extinction of numbers of lemur species. Some extant species may have taken advantage of vacated niches, for example by switching from nocturnal to diurnal activity patterns, or aspects of their environment may have altered, such as the array of sympatric species or predator pressure. Such species may not yet be fully adapted to their new situation and, hence, at least some of their features may not yet be in equilibrium with current conditions. Lack of agreement between factors present for these lemur species compared with other species, or lack of consistent patterns of differences, may therefore not yet provide the basis for conclusions about selective causes and adaptive effects.

Hypothesis testing and variable selection

All the chapters in this book demonstrate the utility of the comparative method for hypothesis testing. In most cases, this is overt and the hypothesis is stated at the outset, although in some, a wide range of variables is first tested for correlational patterns leading to the formation of hypotheses. For example, Bean (Chapter 13) investigates a wide range of possible causal factors for variation in great ape foraging strategies, including the effects of sexual dimorphism, producing hypotheses about the relative importance of climatic, ecological and energetic factors. However, even in such cases, the selection of variables for testing can be seen as implicitly hypothesis related. For example, Plavcan (Chapter 9) sets out to test the hypothesis that sexual dimorphism in anthropoid primates is the result of sexual selection produced by male competition for mates. His first task is to select suitable variables to measure sexual dimorphism and male competition. The former is apparently more straightforward, but as his own analyses show (Table 9.1, p. 247), different measures of sexual dimorphism do not necessarily give identical results concerning whether or not a particular difference in male competition levels has a significant effect on dimorphism. Male competition is even harder to measure, and Plavcan discusses this in some detail. He prefers a categorisation of species according to the intensity and frequency of male–male competition, or the use of species' operational sex ratios (i.e. the ratio of males in a group to available breeding females) to the more frequently used categorisation by mating systems. The former two measures are better surrogates for the likely reproductive consequences of male competition, ultimately the measure of evolutionary interest. The selection of variables is also addressed by Williamson and Dunbar (Chapter 12) in connection with modelling the relationship between group size and habitat in baboons and chimpanzees. Climatic variables are used to measure food resource availability, and Williamson and Dunbar provide detailed support for their choice of climatic measures and indices. In view of the differences that variable selection can make to results, these must always be given careful consideration, and the same hypotheses tested using different variables can produce contradictory results.

The use of outgroups

In using the comparative method to test possible causes and functional reasons for the evolution of adaptations, it is important to establish what the ancestral state of a feature was, and hence which species or groups of species have undergone adaptational change requiring explanation. Also, theories or hypotheses proposed to explain variation in features can be shown to be more robust, or refuted, by testing them on more than one sample. Both of these aspects of comparative methodology involve the use of outgroups, such as other orders of mammals in addition to primates, or strepsirhines as well as haplorhines. This use of the Method of Agreement is an important aspect of analyses presented here in a number of the chapters.

Based on information on female relationships and dispersal patterns in New World monkeys, Strier (Chapter 11) questions the common assumption that female kin-bonding is the primitive state in primates. If, as she suggests, female kin-bonding, which is common in Old World monkeys, is a derived adaptation rather than the evolutionary starting point for primates, its explanation will need to be reconsidered and to encompass a broader range of factors than are usually included. Strier uses New World monkeys as an outgroup to test the food resource and predation hypotheses for the maintenance of female kin-bonding, which were largely derived from data on Old World monkeys, and concludes that the hypotheses are insufficient as general explanations.

Kappeler (Chapter 10), as described above, demonstrates how the attempt to use outgroups can show up weaknesses, or the local specificity of apparently general theories. Until the problems he identifies in characterising or categorising lemur social groups in a homologous fashion to anthropoid groups are solved, socioecological theories about the causal or functional factors that shape the social structures of all primate groups, both strepsirhines and haplorhines, cannot be successfully formulated and tested.

In both these cases, outgroups are used to test whether the patterns of similarity and difference noted for more restricted samples are local coincidences or are more likely to indicate causal connections. In other words, outgroups provide an important means of checking whether the presumed conditions for the Methods of Agreement or Difference are really present. In particular, the Method of Agreement requires that only one factor of relevance is shared in common, and a factor that is shared coincidentally could be mistaken for one of causal importance. A broader, more varied sample reduces the chances of such an error.

Outgroups are used in comparative biology to provide an evolutionary context for the features of interest, and to provide a check on the robusticity or generality of hypotheses. In some senses, the difficulties encountered in trying to uncover the adaptational reasons for rare characters are at the opposite end of the methodological spectrum. If a single species or taxon has evolved a unique variant of some feature, then there is only a sample of one available for the testing of possible explanatory hypotheses. This is insufficient for the Method of Agreement, leaving only the more difficult Method of Difference. Put in other terms, in most chapters in this volume, statistical methods are used to test whether feature x is significantly associated with situation y, or at least the frequency of their occurring together is investigated. However, for Blurton Jones *et al.* (Chapter 6), such an approach is not possible as there is only one human species, with its more or less unique set of reproductive life history features, such as a very extended juvenile period from weaning to first reproduction, high fertility rates during the female reproductive period, a long female post-reproductive lifespan, and low adult mortality. The authors were able to identify clearly which features of the reproductive lifespan of humans are unusual, and which fit common mammalian, or at least great ape, patterns using Charnov's predictions for mammals and comparative data on great apes. The first step of the comparative method, the identification of patterning, does, therefore, serve a useful purpose in such extreme circumstances, enabling the authors to focus on the features requiring explanation specifically for humans. In this second stage, comparison is also useful, as features which are different require explanatory factors which also differ. This is the Method of Difference. The unique degree of help available to reproductively active human females from their post-reproductive mothers (the grandmother effect) is Blurton Jones *et al.*'s suggested explanatory link between the suite of unique life history features. In combination, these features actually fit with Charnov's model of trade-offs between growth and reproduction, although the combination is unique. This is a form of the Method of Agreement.

Foley (Chapter 14) describes an alternative means of using the comparative method to examine the socioecological adaptations of humans. The 15 or so extinct hominid species provide a comparative framework for extant humans, albeit that evidence from the fossil record is inevitably patchy and limited. However, as Foley points out, analyses using extant species only are also using samples made patchy by the uneven effects of extinction. These 'terminal twigs' of the surviving branches of evolutionary history are themselves a time-limited snapshot of past adaptive radiations. Palaeobiological evidence provides access to the conditions under which adaptations actually evolved, including those no longer represented in the modern-day survivors. As well as some disadvantages, the fossil record therefore also has advantages as a source of data for socioecological investigations.

Within-species analyses

Most chapters and analyses in this book concentrate on comparisons of species, or higher level taxa or species groupings. However, a number also demonstrate the use of comparisons within species to test hypotheses (e.g. Lee, Chapter 5; Mace and Holden, Chapter 15). This also serves as a reminder to beware of treating species as invariable, and of the potential dangers of combining species-specific data from different individuals or populations. Barton (Chapter 7) discusses one aspect of this, the so-called Economos' problem. Lee (Chapter 5) uses intraspecific data from rhesus macaques to show that target weaning weights in relation to maternal weights appear to exist within species, as well as there being species-typical values, so strengthening support for the importance of metabolic constraints on this life stage. However, individual mothers can covary the relative rate of infant growth and duration of lactation required to reach the target weight, hence providing the mechanism for differential responses to local ecological and individual social factors. Strier (Chapter 11) shows that intraspecific variation among *Saimiri* in the defensibility of food resources does correlate with predictions from theories developed for Old World monkeys, that female philopatry and kin-bonding are causally linked with resource defence, so providing one example in which New World monkeys are similar in these respects to Old World monkeys, amongst many others in which New World monkeys differ. Williamson and Dunbar (Chapter 12) use interpopulation variation in climatic measures, activity budgets and group sizes to model aspects of the behavioural ecology of *Papio* baboons, gelada and chimpanzees. The comparative methodology used in such analyses is not different from that for interspecific comparisons, it just focuses on patterns of correlated variation among population groups.

Mace and Holden's study (Chapter 15) is specifically limited to humans and involves treating sub-Saharan African societies as separate data points to examine the possible adaptive value of different cultural inheritance patterns, particularly matrilineal and patrilineal descent systems. The authors defend the use of the comparative method and an evolutionary approach to the study of human cultural traits against their rejection by many contemporary social anthropologists. In doing so, they provide useful parallels for the value of the method for studies of social features in non-human primates, even though the complexity and means of transmission of such behaviours may differ. Their results indicate that patrilineal descent tends to evolve following a shift to pastoralism, and that matrilineal descent, coupled with a 'roving male' strategy, can be adaptive for both females and males when inherited resources are of little intrinsic value. However, such a system may be unsustainable once land becomes valuable in itself or hard to protect.

The necessity of biological knowledge

Good biological knowledge is essential to the successful use of the comparative method in evolutionary biology in a number of ways. Firstly, in order to determine what questions to ask, knowledge of which features vary among species and which are similar forms the basis for enquiry into why the variation occurs and suggests possible explanatory factors. Collecting data, even from published sources (the 'fieldwork' of much comparative biology), is also not the easy task it might seem at first sight. The variables selected must be meaningful and homologous, and sometimes categorisation or measurement is not straightforward, particularly for behavioural characteristics. Sifting good data from poor often also requires good biological understanding, or appreciation of the observational or measurement techniques used. Many examples in this volume serve to illustrate these points.

The integration of knowledge from other areas of biology, such as physiology and energetics, can also play an important role, in suggesting the causal direction of relationships, as John Stuart Mill explains. For example, Barton (Chapter 7) uses detailed anatomical and physiological information on the form and functioning of the primate visual cortex to support his proposal that visual specialisation was an important factor in primate brain expansion. The relative metabolic costs to mothers of different reproductive stages are an important element in several chapters, including those by Lee and Bean (Chapters 5 and 13).

Overall, the importance of good biological knowledge and understanding is that they are essential to the development of biologically reasonable interpretations, explanations and argument towards causation.

Ultimately, explanations at different levels, say ecological and physiological, must be capable of integration, or at least must not be incompatible. It is also possible that several different causal factors may be involved in producing an effect, including different levels of explanation. Explaining the variation in relative brain size among mammals, including primates, is a case in point. Some of the suggested explanations, such as the hypothesis that greater social complexity needs a larger brain, or that foraging for certain diets requires more memory than others, are possible explanations of *why* variation in brain size has evolved. However, whichever of these is correct, or whatever the combination of their relative selective forces, the energetic question of *how* larger brain size can be developed ontogenetically and sustained in the adult also requires an answer. These are two different questions, and some of the differences in findings in chapters such as those by Ross and Jones, Lee, and Barton (Chapters 4, 5 and 7) must be seen in this context.

Analytical techniques

This volume demonstrates the use of a range of analytical tools, statistical and others, as applied to the comparative method. Some analyses are largely conducted by argument and non-statistical means; others use simple measures such as whether a particular feature is commonly associated with another, in order to establish possible cause-and-effect relationships. However, most use statistical techniques including bivariate regression and correlation, and a range of multivariate methods such as multiple regression, stepwise multiple regression, partial correlation, principal components analysis and clustering techniques. The choice of method is important because it can have an effect on the results. This is partly because of the demands or requirements of statistical techniques themselves, and the need to understand fully how they behave, and why they produce the patterns of significant and insignificant results that they do. For example, how sensitive is a particular technique to error variance in the data (e.g. see Purvis and Webster, Chapter 3)? If the sample were slightly larger, or some of the data slightly different, how much difference could it make to the results obtained? Are some data points particularly influential on the results, and hence is it particularly important that these data are confirmed? Choice of method is also important because of the different models on which the various techniques are based. For example, if two variables *together* influence a feature, perhaps in some species one variable is more influential in producing species-specific variation whereas in other species the other variable is more important. In such a case, overall variation in the 'feature' may not show up as significantly related to either variable independently across the range of a particular sample. Models that might be used to test hypotheses, such as that underlying multiple regression, commonly only test to see whether the variables are *independently* correlated with the feature concerned. However, in this case, the hypothesis would be supported even if the two variables are only significantly correlated with the feature when their *joint* effects are considered, which requires a different model. The maternal energy hypothesis for brain size, which is referred to by both Barton (Chapter 7) and Lee (Chapter 5) is a case in point. It proposes that brain size depends on maternal investment during gestation, as measured by a combination of the mother's basal metabolic rate and gestation length. Increase in either variable would increase maternal investment, and so their combined effects must be investigated to test the hypothesis fully.

Several chapters, explicitly or implicitly, involve the use of cladistic methodology to determine the phylogenetic patterns of character variation. Robson-Brown (Chapter 2) discusses this in some detail and in particular provides an overview of what is conceptually required for such analyses. Using the example of sleep patterns, she demonstrates how cladistics can help to identify the relative importance of phylogenetic constraints and present ecological conditions on various sleep parameters. For example, she shows that total sleep times and the amount of paradoxical sleep do not easily change evolutionarily with changes in the security of species' sleeping conditions.

This is related to a longstanding concern in comparative biology, which is the potential problem of phylogenetic inertia; in other words, the possibility that species retain some features because of their ancestry, rather than these features being adaptations to present conditions. If this has occurred, species should not be treated as independent points in analyses seeking causal correlations or associations between factors.

A number of chapters use the method of phylogenetic contrasts, and Purvis and Webster (Chapter 3) provide a clear explanation of the reasons why such a method may be needed, and its basic functioning, as developed in the most commonly used form, Comparative Analysis by Independent Contrasts (CAIC). Mace and Holden (Chapter 15) discuss an alternative based on a maximum likelihood approach. Phylogenetic contrasts by CAIC has become *de rigueur* in many areas of comparative biology, but, like any other analytical technique, it needs to be used with care, with an appreciation of how it behaves, or why it produces the results it does in any particular case. Plotting out results and examining which points are the important contributors to significant (or non-significant) findings can be very instructive. It can, for example, enable the identification of outliers that inspection of the raw data may identify as resulting from poor or mistaken data, a weakly supported point on the phylogenetic tree used, an obvious grade effect, etc.

Purvis and Webster (Chapter 3) provide a welcome discussion of the criticisms made of CAIC. They accept some of the problems raised, and they reject others, including the potential problem of grade shifts. In interspecies analyses, which largely do not take the potential problems of phylogenetic inertia into account in a formalised manner, but which rely on biological knowledge and sensitive investigation of the data sets involved to ensure that results are properly founded, grade shifts certainly can be, and are, recognised (*contra* Purvis and Webster). On the other hand, grade shifts are not necessarily easy to identify using CAIC. For example, a grade shift may not be marked and hence not produce a clearly outlying contrast point. In any case, the size of the contrast measurements between species points on either side of a grade shift is a meaningless combination of the effects of the grade shift and the relationship within any one grade between the two variables concerned. This combination may produce pairs of contrast values that stand out as outlying points on a contrast plot, or it may not, and hence a grade effect would be missed. Also, a grade shift may involve multiple points in a phylogeny. Whilst there is a single phylogenetic link between strepsirhines and haplorhines, there are multiple links on the primate phylogeny between insectivores and frugivores, for example. Therefore, if species in the different dietary categories form grades for the variables under consideration, multiple points on the contrast plot should be removed from analyses, but they may well be difficult, or impossible, to distinguish.

The phylogenetic tree used in analyses by phylogenetic contrasts is fundamental to the resultant findings. Although CAIC provides techniques for dealing with incomplete phylogenies (see Chapter 3), complete phylogenies or completed parts of the phylogenies used may not be correct. The effects of at least known 'weak links' should be tested, especially if the contrast points for these links are particularly influential on the overall results. The phylogenetic tree chosen for analyses can easily be a significant factor in the results produced. Mace and Holden (Chapter 15), whilst not specifically referring to CAIC, describe the need to test whether results of phylogenetic methods are dependent on a particular phylogeny or 'model of history'.

The use of a regression model in phylogenetic contrasts by CAIC, as Purvis and Webster (Chapter 3) mention, is also potentially problematic. When other line-fitting techniques, which recognise the potential for error in both variables, were first used for comparative species analyses, some significantly different results were produced, affecting the acceptance or dismissal of hypotheses when correlation levels were not especially high. This problem at least needs to be borne in mind by the users of CAIC until the problems associated with incorporating more suitable line-fitting techniques are solved.

Purvis and Webster (Chapter 3) discuss at some length one particular problem of phylogenetic contrasts by CAIC, that it is especially sensitive to error variance in the data. This, they explain, is expected to be greater at lower phylogenetic levels, or younger nodes on a phylogenetic tree. One means of trying to avoid this potential problem is to limit analyses to older nodes or contrasts at higher phylogenetic levels, as both Ross and Jones, and Barton (Chapters 4 and 7) do. However, this inevitably reduces sample sizes and wastes potentially useful data. Also, in most cases, the raw data for analyses only come from extant species, and all other data points on the phylogenetic tree are estimated from these. The older a node, the more estimation is involved in establishing values for the variables concerned, and the more likely the node itself is affected by inaccurate phylogenetic reconstruction. In trying to overcome a problem inherent in a particular analytical method, it does not seem wise to place too great reliance on what may be unreliable data estimations, or at least on estimations whose accuracy is unknown. CAIC calculates variable values at ancestral nodes by simple averaging of the values of descendants. However, this may be far from accurate. For example, fossil evidence shows us that the ancestor of chimps and humans probably had a relative brain size close to that of chimps, and much smaller than that of humans, but, as Foley (Chapter 14) points out, CAIC would reconstruct its value very inaccurately as the average of the two extant species. Evolutionary rates of change can be very different in different lineages, even closely related ones, and models that assume otherwise introduce an error factor of often unknown magnitude.

As a means of trying to overcome the potential problem of phylogenetic inertia in comparative analyses in biology, the method of phylogenetic contrasts has a clear logical basis – it is a form of the Method of Residues. However, important problems remain with its implementation. At the very least, great care needs to be taken to examine the possible effects of such problems on any analysis undertaken, although in some cases, as outlined above, these are unknown and unknowable. As Purvis and Webster state 'phylogenetic comparative methods [are] not . . . black boxes; an understanding of the methods permits informed choice of which is most suitable for the available data, and an assessment of whether the data and phylogenetic estimates are indeed adequate for good tests of hypotheses' (p. 65). Similar provisos, with the specific suggestion that plots of results should always be inspected carefully, stand as good advice whatever the analytical technique used.

Conclusion

The chapters of this book provide clear demonstrations of the power of the comparative method in socioecology. The centrality of biological knowledge and understanding to its meaningful application is also borne out in many examples. The analyses presented commonly rely on multiple sources of data and background information collected by detailed fieldwork and other studies by many researchers, who all deserve acknowledgement. Without their often long-term efforts, comparative analyses could not be performed. However, the intellectual dependency is not just unidirectional. Without comparative analyses, single-species studies would merely produce uninterpretable descriptions lacking any sense of context, any means of determining important and interesting features, and any basis on which to decide what questions are worth investigating. The comparative method is the fundamental tool of evolutionary biology. It can produce well-founded interpretations and explanations of the natural world. However, it is not easy to negotiate the complications and obscurity, in Stuart Mill's terms, of the 'natural experiments' of evolution that provide most of the raw data. As Martin (1983) said, we should not talk of a spectrum from 'hard' to 'soft' science, but rather from 'hard' to 'difficult' science.

Acknowledgements

The author is grateful to Phyllis Lee for the invitation to write this chapter. She was first introduced to the comparative method by R.D. Martin, and working with him and other colleagues and students, particularly Caroline Ross and Kate Jones, has provided her with rich and varied experience of its explanatory power. The author has had very helpful discussions with Josep R. Llobera, who first introduced her to John Stuart Mill, particularly using outgroup comparisons with the social sciences. She would like to thank all these people.

References

Martin, R.D. (1983). Unpublished Inaugural Lecture, University College London.
Stuart Mill, J. (1872, 1967). *A System of Logic: Ratiocinative and Inductive. Being a Connected View of the Principles of Evidence and the Methods of Scientific Investigation.* London: Longman.

2 *Cladistics as a tool in comparative analysis*

KATE ROBSON-BROWN

Introduction

In recent years, our understanding of primate evolutionary biology and adaptive strategies has benefited greatly from advances in comparative biology (see, for example, Clutton-Brock and Harvey, 1984; Ridley, 1986; Dunbar, 1992). Nevertheless, it is now widely recognised that because closely related species may share many features inherited from a common ancestor, as statistical units these species are not independent (Ridley, 1983; Harvey and Pagel, 1991).

To take account of this problem, a range of methods has been developed that places evolutionary history, in the form of phylogeny, at the heart of comparative analysis. These methods differ in their application of evolutionary assumptions and statistical procedure. Cladistic methods involve the generation of bifurcating trees that describe the relationships between species or other taxa. These trees may then be used as a framework for the comparative analysis of other, independent features of those species. Most procedures for generating trees require data that are discrete, rather than continuous, and many cladistic debates ultimately focus on the details of these character definitions. Cladistic methods that involve parsimony or compatibility criteria proceed by tracing historical relationships among taxa based upon patterns of shared, derived characters. In the same way, mapping other characters onto the tree framework permits the assessment of whether these characters are shared and derived, revealing the evolutionary history of those characters.

This approach may be adopted as a basis for testing adaptational hypotheses within a comparative framework, or for assessing the temporal sequence of changes in descriptive characters. It also provides a means of studying the evolutionary association of characters within a clade, patterns of convergent and divergent adaptation, the phylogenetic constraints on character diversification, and is a useful tool for generating macro-evolutionary predictions about character evolution that can be tested

23

experimentally (Brooks and McLennan, 1991; McLennan, 1991; Greene, 1994).

The power of cladistic analysis has been recognised in many fields of biology, contributing to studies of biogeography (Eldredge and Cracraft, 1980; Lauder, 1981, 1982, 1990; Huey and Bennett, 1987), molecular evolution (Ritland and Clegg, 1987; Harrison, 1991, rates of evolution (Felstein, 1985; Donoghue, 1989), behaviour and behavioural ecology (Dobson, 1985; Greene, 1986; Coddington, 1988; Gittleman, 1988; Carpenter, 1989); ecology (Ridley, 1983; Losos, 1990; Wanntorp *et al.*, 1990; Futuyama and McCafferty, 1991), and ontogeny (Fink, 1982; Larson, 1984). Underlying these different applications is a common interest to recognise the importance of the comparative context, and to build an understanding of the dynamics of particular biological systems upon it. This does not require an exclusively cladistic approach. Comparative analysis is not the preserve of cladistic methodology, and the application of cladistics is best managed as part of a statistical analysis tailored to the particular needs of the data and the requirements of the problem being addressed.

For many biologists, the importance of phylogenetic context to comparative analysis is not in dispute where the data in hand refer to molecular or morphological characteristics, although the relationship between these sources of information is currently under close scrutiny.[1] However, interesting questions and doubts surface when the data describe behavioural or socioecological characteristics (Clutton-Brock and Harvey, 1984; Di Fiore and Rendall, 1994; Robson Brown, 1995), where there is uncertainty about the meaning and independence of character units (Di Fiore and Rendall, 1994) or the applicability of evolutionary models (Dobson, 1985). It is not the intention here to provide an introduction to phylogenetic, cladistic systematics or a critique of computer applications (see examples in Hennig, 1950, 1966; Felenstein, 1979, 1983; Maddison, Donoghue and Maddison, 1984; Ridley, 1986; de Queiroz, 1988; Sober, 1988; Harvey and Pagel, 1991; Swofford, 1993; Smith, 1996). Instead, the aim is to address some issues surrounding the applicability of this analytical tool to the context of primate socioecology – in particular, the relevance of the phylogenetic context, and the definitions of behavioural homology.

The relevance of phylogeny to socioecology

Socioecology constitutes the interrelationships between the social characteristics of organisms and their environment. Outside the field of primatology, arguments in favour of applying comparative methods based on cladistics to this context tend to refer back to an emphasis on evolution as the unifying concept of biology (Funk and Brooks, 1990). This view notes that patterns of variation and change over time in organismic diversity provide the fundamental data of evolutionary biology, whatever the special features of different subdisciplines might be. Within the subdiscipline of primate socioecology, concern is focused on ecological, behavioural, and life history information, and the patterns of association between them. Research into such associations can encourage the construction of models of evolution that describe the interaction of these sets in hierarchical terms (Salthe, 1985). Models of evolving hierarchical systems have described, on the one hand, ecological hierarchies, or economic systems, manifested in patterns of energy flow in ecosystems. On the other hand, the genealogical hierarchies describe systems of information flow manifested by genealogical relationships, and therefore the phylogenetic hierarchy. Adopting such a view might encourage the investigation of the genealogical context of behavioural and ecological diversity or, alternatively, the ecological and behavioural context of genealogy. Comparative analysis is the medium through which this type of investigation may proceed.

The suggestion that behavioural strategies mediate between organismal and environmental hierarchies, even to the extent of *constructing* niches (Odling-Smee, 1995; Laland, Odling-Smee and Feldman, 1996), encourages the view that cladistic analysis may act not only as a robust framework for comparative analysis, but also as an important means of generating information about the relationship between these sources of information, and their underlying systems. Yet, socioecology refers to ecological, behavioural and life history traits rather than genes and molecular data, and there is some concern about whether such traits are amenable to evolutionary explanation derived from cladistic analysis. The question remains whether cladistic methods should be applied only where the relationship between phylogenetic context and character variation is perceived to be unproblematic and direct, and where models of evolutionary change are already perceived to correlate with those prescribed by cladistic procedure. Alternatively, can cladistic methods be applied as part of a more broad and exploratory comparative analysis, intended as a framework for hypothesis testing (Brooks and McLennan, 1991)?

Some confidence in the latter view may be drawn from the fact that complex traits such as behaviour have long been seen as amenable to general evolutionary explanation. Darwin himself demonstrated a profound interest in the evolutionary context of behaviour of all forms, human, animal and plant (Burkhardt, 1983a, 1983b). This interest was manifest not in a secondary application of the theory of natural selection to behaviour, but in the incorporation of behavioural information at the most fundamental level of his theorising. He presented the habits and instincts of organisms as characters for selection and as agents of organic change (Darwin, 1859; Richards, 1981). Darwin's lead did not, however, immediately inspire the development of research programmes devoted to the evolutionary interpretation of behaviour. This lack may be explained by methodological difficulties, the reaction occasioned by the anecdotal writings of some naturalists, or Darwin's own reluctance to explain the generation of phenotypic variation (Burkhardt, 1983b). Nevertheless, one of the most important legacies of Darwin's work is an appreciation that behaviour should not be an afterthought to evolutionary theory (Nitecki and Kitchell, 1986; Plotkin, 1988).

Subsequently, Whitman (1898) and Elton (1930) demonstrated that an animal's 'choice' of the environment was important in evolution, and in the mid-1930s Lorentz, in collaboration with Tinbergen and Heinroth, began to stress the need for a 'phylogenetic view' within animal psychology (Heinroth, 1971). Over the next decades, their efforts to establish rigorous controls for the observation of behaviour and their demonstration of explanatory models largely shaped the field of ethology as we understand it today (Heinroth, 1911; Lorenz, 1935, 1937, 1950, 1965; Tinbergen, 1951, 1959a). Lorenz explicitly proposed that an active phenotype may play a causal evolutionary role. This encouraged the investigation of the 'purposefulness' of behaviour, its relation to the whole pattern of an animal's activities, and its evolutionary history. Tinbergen developed a methodology for incorporating concepts of environmental, contextual variation within this field of behavioural theory, particularly with reference to species-specific signalling systems (Tinbergen, 1959b, 1967). Here, particular cases of ecological adaptiveness were analysed within a comparative framework and assessed in terms of '*survival value*'.

Tinbergen proposed the well-known 'four questions' that ethology should address: function, mechanism, ontogeny and evolutionary history. Behavioural ecology developed partly in response to a demand for research on the function of behaviour, and to prove that an understanding of behaviour such as mating or foraging leads to insights into population and community ecology. Often, the emphasis tends towards explaining equilibria within behavioural systems rather than their transformation. This amounts to an explicit movement away from phylogenetic issues. Other methods that involve the simulation of complex systems or the construction of interpretative schemes for the anslysis of such systems do incorporate a concept of time depth, but explicit reference to the relationship between phylogenetic and behavioural processes is still avoided. This tends to limit what can be described of the relationship between behavioural characteristics of organisms and their environment because the development of adaptive characteristics has three stages: the origin, diversification and maintenance of traits. Microevolutionary studies concentrate on the maintenance of traits in current environments in which processes shaping interactions between organism and environment can be measured. A 'historical' approach would complement these studies by using phylogeny as a template for reconstructing the historical patterns of trait origin and diversification, and for testing of both equilibrium and transformational hypotheses.

Socioecology and homology

There is a sound theoretical basis for the cladistic analysis of socioecological traits, and in addition a growing number of successful studies on organisms as phylogenetically distant as sweat bees and prairie dogs, which should encourage the application of this methodology to primate data. This suggests that the qualities of socioecological information do not necessarily render the application of cladistic methodology inappropriate. The phylogenetic context of socioecological traits is clearly important to the understanding of adaptive strategies and their transformation. Phylogenetic systematic methods may therefore be as revealing for the comparative analysis of socioecological traits as those derived from any other part of the phenotype.

Accepting that phylogenetic systematic methods are applicable to socioecology, however, does not dispel all reservations concerning procedure and statistical methodology. Phylogenetic inference is in no sense neutral, and ideally the evolutionary models employed should be sensitive to the requirements of the data sets to which they are applied. It is unclear, however, how this matching of information can be achieved. Much controversy arises over the special requirements of different types of information, such as morphological or molecular data (Goodman, Miyamoto and Czelusniak, 1987; Bishop and Friday, 1987). There is also no universally accepted procedure for the placement of boundaries between character states when continuous characters such as lengths or ratios are used. There is certainly a need for experimental phylogenetic analysis based on many different types of data to address these issues.

Cladistic methods rely on the assumption that it is possible to identify characters which are 'the same', or homologous, in different species. One major reservation that stands in the way of the use of cladistic methods in socioecological contexts is the problem of defining homology for behavioural features that describe social, cognitive, or foraging strategies, and ecological features that describe the environment in qualitative rather than quantitative terms. This problem has to be addressed because the concept of homology is crucial to the understanding of phylogenetics and cladistics in general. Without a means of homology recognition, we are unable to confirm evolutionary patterns of behaviour, the ecological context, or to contribute to debate about how these relate to general evolutionary processes. Equally, when inferring phylogeny, differing interpretations arise partly from the differences in the evolutionary model that underlies the anslysis. Such models use different definitions of the terms 'homology' and 'novelty', which clearly affects character selection and description.

Within any data set suitable for systematic analysis, there is likely to be both character congruence and character conflict. Character congruence may be synapomorphous (derived) or symplesiomorphous (primitive), and these categories refer to homologous traits. Some cladists have argued that homology *is* synapomorphy, therefore bringing together the major concept of comparative anatomy, namely homology, with the historical perspective of Hennig's (1966) special similarity (Farris, 1977; Patterson, 1982). Character conflict may arise when traits are homoplasic. In effect, such character conflict may be generated in two distinct ways. In taxa that are not necessarily closely related but that may be subject to similar pressures, similar traits serving a similar function may reveal themselves on close inspection to be analogous rather than homologous. The evolutionary process that might produce such an effect is termed convergence. Internal constraints in these conditions are different, so similar adaptations that might arise tend to be recognisable as products of separate evolutionary history. Secondly, where there is the formation of similar traits in similar taxa (and the 'same' character), for example in sibling taxa, the process is known as parallel evolution (Patterson, 1982, 1988). Adaptations that arise as a result of this process may be extremely similar in structure and function.

These definitions are useful ones, but this should not give the impression that they represent the consensus view. Early definitions were not immediately relevant to socioecological traits; Owen (1843) defined a homologue as 'the same organ in different animals under every variety of form and function', and an analogue as 'a part or organ in different animals which has the same function' but derived from a non-homologous base (Owen, 1843, p. 379). These definitions have been elaborated on by Haas and Simpson (1946), who describe homology as a 'similarity between parts, organs, or structures of different organisms, attributable to common ancestry'. More useful to socioecology is Boyden's (1947) suggestion that homology should be based on characters' correspondence with representatives in a common ancestor of the organisms being compared, and from which they were descended in evolution (de Beer, 1971; Galizia, 1993; Greene, 1994). This may lead to a very general definition, such as proposed by Van Valen (1982) as 'correspondence caused by continuity of information'. If it is assumed that a branching tree-like structure is a true representation of the world, these relations defined by the homologies yield a hierarchical system, which is the actual tree pattern. So, 'historical' procedure effectively defines the relationship between similarity and homology with reference to evolutionary mechanisms. As a result, the practical assessment of homology has fallen into two main schools: working 'operational' definitions that assess character correspondence based on the matching of landmarks; and 'taxic' definitions that make assessments on the basis of monophyletic groups (Smith, 1990). Clearly, although there may be difficulties in formulating robust concepts of homology, efforts have been made to minimise these problems in the practical assessment of characters.

The problem of behavioural, social or cultural homology has attracted considerable comment in the literature (e.g. Pribram, 1958; Atz, 1970; Hodos, 1976; Lauder, 1986, 1994; Greene, 1994) and many criteria for homology have been proposed. Almost all these studies suggest that homologous behaviours may be recognised by the application of criteria that are largely independent of phylogenetic patterns of character distribution. These criteria fall into three categories.

First, behaviours can be considered homologous if the neural and neuromuscular control systems of the behaviours are similar (e.g. Baerends, 1958; Hodos, 1976). This suggestion relies on the assumption that the central nervous system and the motor patterns it produces are in some way more conservative than the behaviours themselves. It also assumes a tight coupling of neural structure and behaviour. Neither of these assumptions is directly supported by neuropsychological evidence, and the success or failure of this neutral criterion depends entirely on the phylogenetic distribution of the neural components of a behaviour in relation to the phylogenetic distribution of the behaviour itself. The status of such a relation may be very variable. Behaviours might even be homologous where the neural substrates are not (Lauder, 1986).

The second criterion suggests that two behaviours are homologous if the gross morphological features used in the behavious are homologous (e.g. Haas and Simpson, 1946; Atz, 1970). Again, there are so few data bearing on the problem of congruence between morphological and behavioural novelties that excessive reliance on morphology as arbiter of homology may well lead to erroneous conclusions. The particular drawback to this criterion is that the circumstances in which the functions of behaviour are homologous are unclear. Hodos (1976) avoids this difficulty by employing confused definitions of 'homology' and 'analogy'. He states that 'the insertion of food into the mouth by a man and a monkey would be both homologous (because the hands of monkeys and humans are derived from hands of their common ancestors) and analogous (because the behaviours serve the same purpose)' (Hodos, 1976, p. 160). As Lauder (1986) has pointed out, one might question the utility of a criterion that leads to the determination of a behaviour as both homologous and convergent.

A final criterion suggest that two behaviours are likely to be analogous if the biological role (or function) of those behaviours is similar, or if the two behaviours could have been subjected to similar selection pressures (e.g. Mayr, 1958). This definition is difficult to apply because it refers to selection in a broad way, stressing its significance but also highlighting the lack of available evidence concerning the nature of variation in behaviour, its genetic basis and its relation to fitness.

One conclusion of the homology debate is that, according to the existing criteria, socioecological characters should be considered amenable to homology assessment. The cladistic analysis of socioecological traits should not be rejected because of the homology debate alone. However, criticism of these homology criteria has occasionally been strong, with Atz (1970) and Brown (1975) claiming that social and general behavioural homologies cannot be evaluated, and Klopfer (1969, 1975) even suggesting that these characters do not evolve at all. These criticisms have to some extent held back cladistic analysis in these fields, with reports being rather thinly scattered through the literature of the 1970s and 1980s (Dewsbury, 1975; Arnold, 1977; Greene, 1983; Wenzel, 1992), but with the application of modern phylogenetic systematics some reservations have been overcome (Barlow, 1977; Drummond, 1981; Lauder, 1981, 1982, 1986; Gould and Vrba, 1982; de Queiroz and Wimberger, 1993). The primary test of homology is congruence with other characters, and there is no reason to assume that one group of characters will not be useful within cladistic analysis (Donoghue, 1992; Wenzel, 1992). It should not be assumed that experiential effects on social or ecological characters is so great that phylogenetically congruent similarities might not indicate genealogical relationships. Nor should this assumption be based on the view that other types of characters, such as morphometric characters, are free of these effects. Both these assumptions require further scrutiny (Hollister, 1918; Lauder, 1986, 1994; Wenzel, 1992; Green, 1994; Nelson, 1994; Rieppel, 1994).

There is great potential for the use of cladistics in the comparative analysis of primate socioecology. Cladistic analysis may provide a valuable means of *generating* new information about the evolution of different types of primate traits. Nevertheless, there is a need for research programmes that explicitly test the question of historical congruence among characters referring to behaviours, functions and morphological patterns in an attempt to identify general features of transformations in various aspects of the phenotype. The current literature of comparative biology as a whole is optimistic, suggesting that, for many taxa, social or soioecological characters are not necessarily more variable, nor more difficult to describe, measure and compare across taxa than morphological characters (Greene, 1994). To demonstrate this, one example of the application of cladistics to primate comparative analysis is presented in the following section.

Example: the phylogenetic tree as a template for mapping primate sleep traits

Sleep is a deceptively complex phenomenon, which has physiological, behavioural, psychological and environmental components. It can be identified by sustained physical quiescence in a species-specific posture and site accompanied by reduced responsiveness to external stimuli and characteristic electroencephalogram (EEG) patterns, but where a wakeful condition may be quickly achieved (Zepelin, 1994). The timing and structure of daily sleep are species-specific and refer to the activity pattern of that species, such as diurnal, nocturnal or arrhythmic. The transition from a waking state into sleep in mammals is associated on a cortical level with a change from irregular, low-voltage, fast EEG waves to high-voltage, slow waves. This is called non-REM (NREM) or slow-wave sleep (SWS). Occasionally, the cortical waves change to a desynchronised, regular, low-voltage, fast pattern known as rapid eye movement sleep (REM), or paradoxical sleep (PS). In primates, non-REM sleep has been subdivided into stages 1–4, with stages 3 and 4 referring to slow-wave sleep. These states may be considered homologous across primate species because the structure of sleep and the neural substrate that underlies it are thought to be relatively conservative and following a general mammalian pattern (Zepelin, 1994).

Theories about the function of mammalian sleep tend to focus on the immobilisation or adaptive non-responding of individuals in a sleeping state. For example, restorative theory suggests that sleep, and in particular paradoxical sleep, relieves a deficit in body tissues (Berger, 1975, 1984), brain tissues (Jouvet, 1975; Horne, 1983) or psychological systems (Crick and Mitchison, 1983), associated with a wakeful state. Immobilisation theory suggests that sleep and endothermy are linked and that the primary function of sleep is energy conservation (Zepelin and Rechtschaffen, 1974; Berger, 1975, 1984; Zepelin, 1994; Berger and Phillips, 1995). Adaptive non-responding theory considers sleep to be a social and behavioural mechanism that promotes survival by ensuring temporal integration of a species in its ecological niche, preventing activity at times when it would be wasteful of energy or risk predation (Bert and Pegram, 1969; Allison and Van Twyver, 1970, 1972; Bert, Pegram and Balzamo, 1972; Meddis, 1977, 1983; Siegel, 1995).

The phylogenetic context of mammalian sleep traits is well documented, and raises many interesting questions concerning the relationship between ecology, social behaviour, life history and physiology. For example, studies on mammals have shown that metabolic rate and body size account for much interspecies variability in sleep quotas (Zepelin, 1994; Berger and Phillips, 1995), and that precociality is positively correlated with paradoxical sleep levels (Elgar, Pagel and Harvey, 1988). This example does not attempt to recreate these studies, but simply to demonstrate a cladistic interpretation of some of this information for primates.

Data

Data describing the sleep traits of 12 primate taxa were drawn from the literature: *Eulemur mongoz, Euoticus senegalensis, Nycticebus coucang, Perodicticus potto, Saimiri sciureus, Aotus trivigatus, Macaca* sp., *Papio* sp., *Cercopithecus aethiops, Erythrocebus patas, Pan troglodytes* and *Homo sapiens*. Data were collected from the literature on total daily sleep time, paradoxical sleep, sleep cycle length, and sleep conditions.

Table 2.1 presents the scored traits describing total daily sleep time, paradoxical sleep, sleep cycle length and sleep conditions. Estimates of sleep patterns based on behavioural observations of captive or wild adult individuals, and laboratory EEG recordings on adults, were included. Where more than one estimate was available for a species, the mean values were taken. Sleep condition scores refer to Allison and Van Twyver (1970, 1972) scoring scale, but all other traits were described by continuous data. To transform these continuous data into discrete scores suitable for cladis-

Table 2.1. *Primate traits*

Taxon	Total sleep time (h)	Paradoxical sleep time (h)	Sleep cycle (min)	Sleep conditions	Paradoxical sleep (%)
Nycticebus coucang	3	x	x	0	x
Perodictus potto	3	x	x	0	x
Galago senegalensis	2	1	0	0	11.2
Petterus mongoz	2	0	x	0	9.4
Aotus trivigatus	4	2	x	0	10.6
Salmiri sclureus	2	2	x	0	14.5
Cercopithecus aethiops	2	1	x	0	6.8
Erythrocebus patas	2	1	x	0	10.1
Macaca	2	1	1	0	11.9
Papio	1	1	1	2	10.6
Pantroglodytes	1	2	2	1	14.4
Homo sapiens	0	2	2	1	23.8

Adams and Barrett (1974); Allison and Cicchetti (1976); Allison and Van Twyver (1970, 1972); Balzamo (1973); Balzamo and Bert (1975); Balzamo *et al.* (1972, 1978); Berger (1984); Berger and Phillips (1995); Bert and Pegram (1969); Bert (1973); Bert *et al.* (1967, 1972, 1975); Campbell and Tobler (1984); Crowley *et al.* (1972); Elgar *et al.*(1993); Elgar *et al.* (1988, 1990); Horne (1983); Jouvet (1975); McNew *et al.* (1971); Meddis (1977, 1983); Perachio (1971); Reite *et al.* (1965); Siegel (1995); Tobler (1990, 1995); Zepelin (1994); Zepelin and Rechtschaffen (1974).

tic analysis, the 'gap weighting' method was employed (Chappill, 1989; Thiele, 1993). This method preserves the ordinal properties of the data and divides data sets up using rank order of states and the magnitude of gaps between states.

The phylogeny used as a basis for this comparative analysis is adapted from Purvis' composite estimate (1995). Excel 5.0a and MacClade 3.01 (Maddison and Maddison, 1992) were used for data management and statistics. The character states were treated as ordered.

Comments

Several interesting results arise from this graphic example. The phylogenetic context of total sleep time and paradoxical sleep is notable. Total sleep time appears to be longer in Strepsirhines than in Haplorhines, with two interesting derived states being the owl monkey, with a very extended sleep time, and humans, with a reduced sleep time. More paradoxical sleep is also characteristic of the Haplorhines, with Old World monkeys showing lower levels than New World monkeys and the apes. Haplorhines are often described as more altricial than strepsirhines, and it is known that during periods of mammalian paradoxical sleep, thermoregulation and breathing regularity are compromised, just as is observed in fetal states. It may be, therefore, that increases in paradoxical sleep levels are associated with shifts in the maturational timetable. In short, a cladistic treatment of this information suggests that a significant shift in the character of sleep occurred at or soon after the Strepsirhine–Haplorhine split, although at this stage the mechanisms and causes of the event are unknown and merit further investigation.

Species differences in mammalian sleep characteristics are often assumed to relate directly to ecological conditions. It is suggested, for example, that cetacean unihemispheric sleep is an adaptation to the marine environment, and that, in general, predator species with secure sleeping conditions have longer periods of uninterrupted sleep and paradoxical sleep than prey species with insecure conditions (Allison and Van Twyver, 1970; Hediger, 1980). Figure 2.1 displays the findings for primates mapped against phylogeny. This shows clearly that although total sleep time and amount of paradoxical sleep are positively correlated with respect to phylogeny, sleeping condition is not associated with these characters. Baboons, although they suffer from insecure sleeping conditions, do not experience reduced paradoxical sleep: their levels of paradoxical sleep are the same as those of their sister taxon in this study, the macaques. Phylogenetic constraint, rather than ecological constraint, may be the significant factor here.

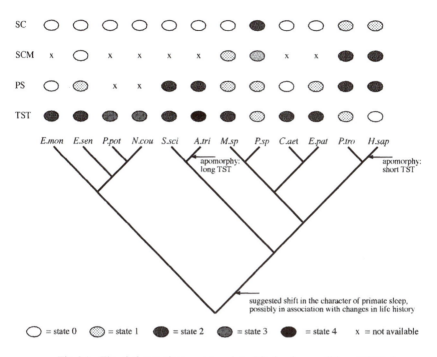

Fig. 2.1 The phylogenetic tree as template. SC, sleeping conditions; SCM, sleep cycle minimum; PS, paradoxical sleep length; TST, total sleep time; *E. mon = Eulemur mongoz; E. sen = Euoticus senegalensis; P. pot = Periodictus potto; N. cou = Nycticebus coucang; S. sci = Saimiri sciurius; A. tri = Aotus trivigatus; M.sp = Macaca; P.sp = Papio; C. aet = Ceropithecus aethiops; E. pat = Erythrocebus patas; P. tro = Pan troglodytes; H. sap = Homo sapiens.*

This example demonstrates that a 'historical' approach to comparative analysis that uses phylogeny as a template for reconstructing the patterns of character change can be very rewarding. In other fields of behavioural and socioecology, the opportunities presented by a 'historical' views have already been rediscovered (Lauder, 1986; Brooks and McLennan, 1991; Harvey and Pagel, 1991; McLennan, 1993; Gittleman and Decker, 1994; Greene, 1994). For example, it has been shown that phylogenetic trees may be used as templates for testing whether or not ecological and behavioural similarities among members of a clade (in this case gasterosteid fishes) result from convergence or inheritance from a common ancestor (McLennan, Brooks and McPhail, 1988).

As the present example has suggested, if there are phylogenetic constraints on ecological or behavioural diversification, it may be that taxa express the same traits regardless of the environment in which they are located. Those traits may have arisen in 'ancestral' environments since abandoned. Cladistic analysis may show a decoupling of the products of genealogical inheritance and environmental constraints, and some have suggested that ecological and behavioural traits showing phylogenetic constraint demonstrate that change in state for those traits can be viewed as conservative elements in evolution (McLennan *et al.*, 1988; Funk and Brooks, 1990; Packer, 1991). In addition, a cladistic approach permits the investigation of how closely phylogenesis in a clade is associated with changes in ecology and behaviour, and therefore how closely ecological evolution and behavioural evolution match phylogenetic patterns (Carothers, 1984). It poses the question as to whether one can obtain a robust cladogram by analysing these traits alone, as Di Fiore and Rendall (1994) suggest for primates, and whether such cladograms agree with phylogenies derived from morphological or biochemical data (Page and Charleston, 1997), as Dobson (1985) demonstrated for prairie dogs. This, in turn, may lead to investigations as to how often adaptive changes in structure and in function actually appear on the same branch of the phylogenetic tree (Chevrud, Dow and Leutenegger, 1985; Silén-Tullberg and Møller, 1993), and whether there are ecological and behavioural traits that characterise species-rich clades. In these contexts, cladistic analysis provides a valuable contribution to the battery of methods available for comparative analysis, and there is every reason to believe that primate socioecology is well placed to contribute to this growing field.

Note

1 Molecules and Morphology in Systematics. International Conference, 24–28 March 1997, Paris. Proceedings to be published as a special issue of *Molecular Phylogenetics and Evolution* 1998.

References

Adams, P. and Barrett, E. (1974). Nocturnal sleep in squirrel monkeys. *Electroencephalography and Clinical Neurophysiology* **36**, 201–4.
Allison, T. and Cicchetti, D.V. (1976). Sleep in mammals: ecological and constitutional correlates. *Science* **194**, 732–4.
Allison, T. and Van Twyver, H. (1970). The evolution of sleep. *Natural History* **79**, 56–65.
Allison, T. and Van Twyver, H. (1972). Electrophysiological studies of the echidna, *Tachyglossus aculeatus*. I: Waking and sleeping. *Archive of Italian Biology* **110**, 145–84.

Arnold, S. (1977). The evolution of courtship behavior in New World salamanders with some comments on Old World salamanders. In *The Reproductive Biology of Amphibians*, ed. D. Taylor and S. Guttman, pp. 141–83. New York: Plenum Press.

Atz, J. (1970). The application of the idea of homology to animal behavior. In *Development and Evolution of Behavior*, ed. L. Aronson, E. Tobach, D. Lehrman and J. Rosenblatt, pp. 53–74. San Francisco: W.H. Freeman.

Baerends, G. (1958). Comparative methods and the concept of homology in the study of behavior. *Archives Neerlandaises de Zoologie Supplement* **13**, 401–17.

Balzamo, E. (1973). Etude des états de vigilance chez *Papio cynocephalus* adulte. *Comptes Rendus des Seances de la Société de Biologie et de ses Filiales (Paris)* **167**, 1168–72.

Balzamo, E. and Bert, J. (1975). Sleep in *Papio anubis*: its organisation and lateral geniculate spikes. *Sleep Research* **4**, 138.

Balzamo, E, Bradley, R.J., Bradley, D.M., Pegram, G.V. and Rhodes, J.M. (1972). Sleep ontogeny in the chimpanzee from birth to two months. *Electroencephalography and Clinical Neurophysiology* **33**, 47–60.

Balzamo, E., Vuillon-Cacciuttolo, G., Petter, J.-J. and Bert, J.J. (1978). Etats de vigilance chez deux *Lemuridae*: rythmes EEG et organisation obtenues par télémesure. *Sleeping Waking* **2**, 237–45.

Barlow, G. (1977). Modal action patterns. In *How Animals Communicate*, ed. T. Sebeok, pp. 98–134). Bloomington, Indiana: Indiana University Press.

Berger, R.J. (1975). Bioenergetic functions of sleep and activity rhythms and their possible relevance to ageing. *Federal Proceedings* **34**, 97–102.

Berger, R.J. (1984). Slow wave sleep, shallow torpor and hibernation: homologous states of diminished metabolism and body temperature. *Biology and Psychology* **19**, 306–26.

Berger, R.J. and Phillips, N.H. (1995). Energy conservation and sleep. *Behavioural Brain Research* **69**, 65–73.

Bert, J. (1973). Similitudes et differences du sommeil chez deux babouins, *Papio hamadryas* et *Papio papio*. *Electroencephalography and Clinical Neurophysiology* **35**, 209–12.

Bert, J., Balzamo, E., Chase, M. and Pegram, V. (1975). The sleep of the baboon, *Papio papio*, under natural conditions and in the laboratory. *Electroencephalography and Clinical Neurophysiology* **39**, 657–62.

Bert, J., Collomb, H. and Martino, A. (1967). L'électroencephalogramme du sommeil d'un prosimien. Sa place dans l'organisation du sommeil chez les primates. *Electroencephalography and Clinical Neurophysiology* **23**, 342–50.

Bert, J. and Pegram, V. (1969). L'électroencephalogramme du sommeil chez les cercopithecinae, *Erythrocebus patas* et *Cercopithecus aethiops sabaeus*. *Folia Primatologica* **11**, 151–9.

Bert, J., Pegram, V. and Balzamo, E. (1972). Comparaison du sommeil de deux macaques (*Macaca radiata* et *Macaca mulatta*). *Folia Primatologica* **17**, 202–8.

Bishop, M. and Friday, A. (1987). Tetrapod relationships: the molecular evidence. In *Molecules and Morphology in Evolution: Conflict or Compromise?* ed. C. Patterson, pp. 123–39. Cambridge: Cambridge University Press.

Boyden, A. (1947). Homology and analogy. *American Midland Naturalist* **37**, 648–69.

Brooks, D. and McLennan, D. (1991). *Phylogeny, Ecology and Behavior: a Research Program in Comparative Biology.* Chicago: University of Chicago Press.

Brown, J. (1975). *The Evolution of Behavior.* New York: W.W. Norton.

Burkhardt, R. (1983a). The development of evolutionary ethology. In *Evolution from Molecules to Men*, ed. D. Bendall, pp. 429–44. Cambridge: Cambridge University Press.

Burkhardt, R. (1983b). Darwin on animal behaviour and evolution. In *The Darwinian Heritage*, ed. D. Kohn, pp. 200–36. Princeton, NJ: Princeton University Press.

Campbell, S.S. and Tobler, I. (1984). Animal sleep: a review across phylogeny. *Neuroscience and Biobehavioral Reviews* **8**, 269–300.

Carothers, J. (1984). Sexual selection and sexual dimorphism in some herbivorous ancient dispersals. *American Naturalist* **124**, 244–54.

Carpenter, J. (1989). Testing scenarios: wasp social behavior. *Cladistics* **4**, 291–6.

Chappill, J.A. (1989). Quantitative characters in phylogenetic analysis. *Cladistics* **5**, 217–34.

Chevrud, J.M., Dow, M.M. and Leutenegger, W. (1985). The quantitative assessment of phylogenetic constraints in comparative analyses: sexual dimorphism in body weight among primates. *Evolution* **39**, 1335–51.

Clutton-Brock, T.H. and Harvey, P.H. (1984). Comparative approaches to investigating adaptation. In *Behavioural Ecology: an Evolutionary Approach*, 2nd edition, ed. J. Krebs and N. Davies, pp. 7–29. Sunderland, MA: Sinauer Associates.

Coddington, J. (1988). Cladistic tests of adaptational hypotheses. *Cladistics* **4**, 3–22.

Crick, F. and Mitchison, G. (1983). The function of dream sleep. *Nature* **304**, 111–13.

Crowley, T., Kripke, D., Halberg, F., Pegram, G. and Schildkraut, J. (1972). Circadian rhythms of *Macaca mulatta*: sleep, EEG, body and eye movement and temparature. *Primates* **13**, 149–68.

Darwin, C. (1859). *On the Origin of Species by Natural Selection.* London: John Murray.

de Beer, G. (1971). *Homology, an Unsolved Problem.* Oxford: Oxford University Press.

de Queiroz, A. (1988). Systematics and the Darwinian revolution. *Philosophy of Science* **55** 238–59.

de Queiroz, A. and Wimberger, P. (1993). The usefulness of behavior for phylogeny estimation: levels of homoplasy in behavioural and morphological characters. *Evolution* **47**, 46–60.

Dewsbury, A. (1975). Diversity and adaptation in rodent copulatory behavior. *Science* **190**, 947–54.

Di Fiore, A. and Rendall, D. (1994). Evolution of social organisation: a reappraisal for primates by using phylogenetic methods. *Proceedings of the National Academy of Science USA* **91**, 9941–5.

Dobson, F. (1985). The use of phylogeny in behavior and ecology. *Evolution* **39**, 1384–8.

Donoghue, M. (1989). Phylogenies and the analysis of evolutionary sequences, with examples from seed plants. *Evolution* **43**, 1137–56.

Donoghue, M. (1992). Homology. In *Keywords in Evolutionary Biology*, ed. E.

Keller and E. Lloyd, pp. 170–9. Cambridge, MA: Harvard University Press.

Drummond, H. (1981). The nature and description of behavior patterns. In *Perspectives in Ethology*, ed. P. Bateson and P. Klopfer, pp. 1–33. New York: Plenum Press.

Dunbar, R. (1992). The evolutionary implications of social behaviour. In *The Role of Behaviour in Evolution*, ed. H. Plotkin, pp. 165–189. Cambridge, MA: MIT Press.

Edgar, D.M., Dement, W.C. and Fuller, C.A. (1993). Effect of SCN lesions on sleep in squirrel monkeys: evidence for opponent processing in sleep–wake regulation. *Journal of Neurosciences* **13**, 1065–79.

Eldredge, N. and Cracraft, J. (1980). *Phylogenetic Patterns and the Evolutionary Process*. New York: Columbia University Press.

Elgar, M.A., Pagel, M.D. and Harvey, P.H. (1988). Sleep in mammals. *Animal Behaviour* **36**, 1407–19.

Elgar, M.A., Pagel, M.D. and Harvey, P.H. (1990). Sources of variation in mammalian sleep. *Animal Behaviour* **40**, 991–5.

Elton, C. (1930). *Animal Ecology and Evolution*. Oxford: Oxford University Press.

Farris, J. (1977). Phylogenetic analysis under Dollo's law. *Systematic Zoology* **26**, 77–88.

Felsenstein, J. (1979). Alternative methods of phylogenetic inference and the interrelationships. *Systematic Zoology* **28**, 49–62.

Felsenstein, J. (1983). Parsimony in systematics: biological and statistical issues. *Annual Review of Ecology and Systematics* **14**, 313–33.

Felsenstein, J. (1985). Confidence limits on phylogenies with a molecular clock. *Systematic Zoology* **34**, 152–61.

Fink, W. (1982). The conceptual relationship between ontogeny and phylogeny. *Paleobiology* **8**, 254–64.

Funk, V.A. and Brooks, D. (1990). Phylogenetic systematics as the basis of comparative biology. *Smithsonian Contributions to Botany* **73**, 1–45.

Futuyama, D. and McCafferty, S. (1991). Phylogeny and the evolution of host plant associations in the leaf beetle genus *Ophraella*. *Evolution* **42**, 217–26.

Galizia, G. (1993). Evolutionary models and phylogenetic inference: a morphometric study of insectivora. Unpublished PhD dissertation, University of Cambridge.

Gittleman, J.L. (1988). The comparative approach in ethology: aims and limitations. *Perspectives in Ethology* **8**, 55–83.

Gittleman, J.L. and Decker, D.M. (1994). The phylogeny of behaviour. In *Behaviour and Evolution*, ed. P. Slater and T. Halliday, pp. 80–105. Cambridge: Cambridge University Press.

Goodman, M, Miyamoto, M. and Czelusniak, J. (1987). Pattern and process in vertebrate phylogeny revealed by coevolution of molecules and morphologies. In *Molecules and morphology in evolution: Conflict or compromise?* ed. C. Patterson, pp. 141–76. Cambridge: Cambridge University Press.

Gould, S. and Vrba, E. (1982). Exaptation – a missing term in the science of form. *Paleobiology* **8**, 4–15.

Greene, H. (1983). Dietary correlates of the origin and radiation of snakes. *Experientia* **35**, 747–8.

Greene, H. (1986).Diet and arboreality in the Emerald Monitor, *Varanus prasinus*, with comments on the study of adaptation. *Fieldiana Zoologica, New Series* **31**, 1–12.

Greene, H. (1994). Homology and behavioural repertoires. In *Homology: the Hierarchical Basis of Comparative Biology*, ed. B Hall, pp. 369–91. San Diego: Academic Press.

Haas, O. and Simpson, G. (1946). Analysis of some phylogenetic terms, with attempts at redefinition. *Proceedings of the American Philosophical Society* **90**, 319–49.

Harrison, R. (1991). Molecular changes at speciation. *Annual Review of Ecology and Systematics* **22**, 281–8.

Harvey, P. and Pagel, M. (1991). *The Comparative Method in Evolutionary Biology.* Oxford: Oxford University Press.

Hediger, H. (1980). The biology of natural sleep in animals. *Experientia* **36**, 13–16.

Heinroth, O. (1911). Beiträge zur Biologie, namentlich Ethologie und Psychologie der Anatiden. *Verhandlungen des V, Internationalen Ornithologischen-Kongresses Berlin* 589–702.

Heinroth, O. (1971). *Oskar Heinroth. Vater der Verhsltensforschung, 1871–1945.* Stuttgart: Wissenschaftliche Verlagsgesellschaft.

Hennig, W. (1950). *Grundzüg einer Theorie der phylogenetischen Systematik.* Berlin: Deutscher Zentralverlag.

Hennig, W. (1966). *Phylogenetic Systematics.* Urbana, IL: University of Illinois Press.

Hodos, W. (1976). The concept of homology and the evolution of behavior. In *Evolution, Brain and Behavior: Persistent Problems*, ed. R. Masterson, W. Hodos and H. Jerison, pp. 153–68. Hillsdale, NJ: Lawrence Erlbaum Associates.

Hollister, N. (1918). East African Mammals in the United States National Museum: Part I Insectivora, Chiroptera and Carnivora. *Bulletin of United States National Museum* **99**, 1–194.

Horne, J.A. (1983). Mammalian sleep function with particular reference to man. In *Sleep Functions in Humans and Animals: an Evolutionary Perspective*, ed. A. Mayes, pp. 262–312. New York: Van Nostrand Rheinhold.

Huey, R.B. and Bennett, A.F. (1987). Phylogenetic studies of coadaptation: preferred temperatures versus optimal performance temperatures of lizards. *Evolution* **41**, 1098–115.

Humphries, C.J. and Parenti, L.R. (1986). *Cladistic Biogeography.* Oxford: Clarendon Press.

Jouvet, M. (1975). The function of dreaming: a neurophysiologist's point of view. In *Handbook of Psychobiology*, ed. M.S. Gazzaniga and C. Blakemore, pp. 499–527. New York: Academic Press.

Klopfer, P. (1969). Review of R.F. Ewer, *Ethology of Mammals. Science* **165**, 887.

Klopfer, P. (1975). Review of J Alcock, *Animal Behavior: An Evolutionary Approach. American Scientist* **63**, 578–80.

Laland, K.N., Odling-Smee, F.J. and Feldman, M.W. (1996). The evolutionary consequences of niche construction – a theoretical investigation using 2-locus theory. *Journal of Evolutionary Biology* **9**, 239–316.

Larson, A. (1984). Neontological inferences of evolutionary pattern and process in the salamander family *Plethodontidae. Evolutionary Biology* **17**, 119–217.

Lauder, G. (1981). Form and function: structural analysis in evolutionary morphology. *Paleobiology* **7**, 430–42.

Lauder, G. (1982). Historical biology and the problem of design. *Journal of Theoretical Biology* **97**, 57–67.

Lauder, G. (1986). Homology, analogy and the evolution of behaviour. In *Evolution of Animal Behaviour*, ed. M. Nitecki and J. Kitchell, pp. 9–40. Oxford: Oxford University Press.

Lauder, G. (1990). Functional morphology and systematics: studying functional patterns in an historical context. *Annual Review of Ecology and Systematics* **21**, 317–40.

Lauder, G. (1994). Homology, form and function. In *Homology: the Hierarchical Basis of Comparative Biology*, ed. B. Hall, pp. 152–97. San Diego: Academic Press.

Lorenz, K. (1935). Der Kumpan in der Umwelt des Vogels. *Journal für Ornithologie* **83**, 137–213; 289–413.

Lorenz, K. (1937). Biologische Fragestellung in der Tierpsychologie. *Zeitschrift für Tierpsychologie* **1**, 24–32.

Lorenz, K. (1950). The comparative method in studying innate behavior patterns. *Symposium of the Society of Experimental Biology* **4**, 221–68.

Lorenz, K. (1965). *Evolution and Modification of Behavior*. Chicago: University of Chicago Press.

Losos, J. (1990). A phylogenetic analysis of character displacement in Caribbean *Anolis* lizards. *Evolution* **44**, 558–69.

Maddison, W.P., Donoghue, M. and Maddison, D.R. (1984). Outgroup analysis and parsimony. *Systematic Zoology* **33**, 83–103.

Maddison, W. and Maddison, D. (1992). *MacClade Version 3.01*. Sunderland, MA: Sinauer Associates.

Mayr, E. (1958). Behavior and systematics. In *Behaviour and Evolution*, ed. A. Roe and G. Simpson, pp. 341–62). New Haven.

McLennan, D. (1991). Integrating phylogeny and experimental ethology: from pattern to process. *Evolution* **45**, 1773–89.

McLennan, D. (1993). Phylogenetic relationships in the *Gasterosteidae*: an updated tree based on behavioral characters with a discussion of homoplasy. *Copeia* **2**, 218–326.

McLennan, D., Brooks, D. and McPhail, J.D. (1988). The benefits of communication between comparative ethology and phylogenetic systematics: a case study using gasterosteid fishes. *Canadian Journal of Zoology* **66**, 2177–90.

McNew, J.J., Howe, R.C. and Adey, W.R. (1971). The sleep cycle and subcortical-cortical EEG relations in the unrestrained chimpanzee. *Electroencephalography and Clinical Neurophysiology* **30**, 489–503.

Meddis, R. (1977). *The Sleep Instinct*. London: Routledge and Kegan Paul.

Meddis, R. (1983). The evolution of sleep. In *Sleep Functions in Humans and Animals: an Evolutionary Perspective*, ed. A. Mayes, pp. 57–106. New York: Van Nostrand Rheinhold.

Nelson, G. (1994). Homology and systematics. In *Homology: the Hierarchical Basis of Comparative Biology*, ed. B. Hall, pp. 102–51. San Diego: Academic Press.

Nitecki, M.H. and Kitchell, J.A. (1986). *Evolution of Animal Behaviour*. Oxford: Oxford University Press.

Odling-Smee, F.J. (1995). Niche construction, genetic evolution and cultural change. *Behavioural Processes* **35**, 195–205.

Owen, R. (1843). *Lectures on the Comparative Anatomy and Physiology of the Invertebrate Animals, Delivered at the Royal College of Surgeons*. London: Green and Longmans.

Packer, L. (1991). The evolution of social behaviour and nest architecture in sweat bees of the subgenus *Evylaeus* (*Hymenoptera*: *Halictidae*): a phylogenetic approach. *Behavioral Ecological Sociobiology* **29**, 153–60.

Page, R.D.M. and Charleston, M.A. (1997). From gene tree to organismal phylogeny: reconciled trees and the gene tree/species tree problem. *Molecular Phylogenetics and Evolution* **7**, 231–40.

Patterson, C. (1982). How does phylogeny differ from ontogeny? In *Development and Evolution*, ed. B. Goodman, H. Holder and C. Wylie, pp. 1–31. Cambridge: Cambridge University Press.

Patterson, C. (1988). The impact of evolutionary theories on systematics. In *Prospects in Systematics*, ed. D. Hawksworth, pp. 59–91. Oxford: Clarendon Press.

Perachio, A. (1971). Sleep in the nocturnal primate *Aotus trivigatus*. *Proceedings of the Third International Congress of Primatology* **2**, 54–60.

Plotkin, H.C. (1988). *The Role of Behavior in Evolution*. Cambridge, MA: MIT Press.

Pribram, K. (1958). Comparative neurology and the evolution of behavior. In *Behavior and Evolution*, ed. A. Roe and G. Simpson, pp. 140–164. New Haven: Yale University Press.

Purvis, A. (1995). A composite estimate of primate phylogeny. *Philosophical Transactions of the Royal Society of London, B* **348**, 405–21.

Reite, M., Rhodes, J., Kavan, E. and Adey, W. (1965). Normal sleep patterns in macaque monkeys. *Archives of Neurology* **12**, 133–44.

Richards, R. (1981). Instinct and intelligence in British natural theology: some contributions to Darwin's theory of the evolution of behaviour. *Journal of the History of Biology* **14**, 190–230.

Ridley, M. (1983). *The Explanation of Organic Diversity: the Comparative Method and Adaptations for Mating*. Oxford: Oxford University Press.

Ridley, M. (1986). The number of males in a primate troop. *Animal Behaviour* **34**, 1848–58.

Rieppel, O. (1994). Homology. topology and typology: the history of modern debates. In *Homology: the Hierarchical Basis of Comparative Biology*, ed. B. Hall, pp. 64–101. San Diego: Academic Press.

Ritland, K. and Clegg, M. (1987). Evolutionary analysis of plant DNA sequences. *The American Naturalist* **130**, S74–100.

Robson Brown, K.A. (1995). A Phylogenetic Systematic Analysis of Hominid Behaviour. Unpublished PhD dissertation, University of Cambridge.

Rosen, D. (1976). A vicariance model of Caribbean biogeography. *Systematic Zoology* **24**, 431–64.

Salthe, S. (1985). *Evolving Hierarchical Systems*. New York: Columbia University Press.

Siegel, J.M. (1995). Phylogeny and the function of REM sleep. *Behavioural Brain Research* **69**, 29–34.

Sillén-Tullberg, B. and Møller, A. (1993). The relationship between concealed ovulation and mating systems in anthropoid primates: a phylogenetic analysis. *The American Naturalist* **141**, 1–25.

Smith, A.B. (1996). *Systematics and the Fossil Record*. Oxford: Blackwell Scientific Publications.

Smith, G. (1990).Homology in morphometrics and phylogenetics. In *Proceedings of the Michigan Morphometrics Workshop*, ed. F. Rohlf and F. Bookstein, pp. 325–38. Ann Harbor: The University of Michigan Museum of Zoology.

Sober, E. (1988). *Reconstructing the Past*. Cambridge, MA: MIT Press.

Swofford, D.L. (1993). A computer program for phylogenetic inference using maximum parsimony. *Journal of General Physiology* **102**, A9.

Thiele, K. (1993). The Holy Grail of the perfect character: the cladistic treatment of morphometric data. *Cladistics* **9**, 275–304.

Tinbergen, N. (1951). *The Study of Instinct*. Oxford: Clarendon Press.

Tinbergen, N. (1959a). Behaviour, systematics, and natural selection. *Ibis* **101**, 318–30.

Tinbergen, N. (1959b). Comparative studies of behaviour in gulls (Laridae): a progress report. *Behaviour* **15**, 1–70.

Tinbergen, N. (1967). Adaptive features of the black-headed gull *Larus ridibundus L. Proceedings of the XIV International Ornithological Congress* **43**, 59.

Tobler, I. (1990). Napping and polyphasic sleep in mammals. In *Napping: Biological, Psychological and Medical Aspects*, ed. D. Dinges and R. Broughton, pp. 9–30. London: Raven Press.

Tobler, I. (1995). Is sleep fundamentally different between mammalian species? *Behavioural Brain Research* **69**, 35–41.

Van Valen, L. (1982). Homology and causes. *Journal of Morphology* **173**, 305–12.

Wanntorp, H., Brooks, D., Nilssen, T. *et al.* (1990). Phylogenetic approaches in ecology. *Oikos* **57**, 119–32.

Wenzel, J.W. (1992). Behavioral homology and phylogeny. *Annual Review of Ecology and Systematics* **23**, 361–81.

Whitman, C. (1898). Animal behavior. In *Biological Lectures, Wood's Hole*, ed. C. Whitman, pp. 285–338. Boston: Ginn and Co.

Zepelin, H. (1994). Mammalian sleep. In *Principles and Practice of Sleep Medicine*, 2nd edn, ed. M.H. Kryger, T. Roth and W.C. Dement, pp. 30–64. Philadelphia: W.B. Saunders.

Zepelin, H. and Rechtschaffen, A. (1974). Mammalian sleep, longevity, and energy metabolism. *Brain Behavioural Evolution* **10**, 425–70.

3 Phylogenetically independent comparisons and primate phylogeny

ANDREW PURVIS AND ANDREA J. WEBSTER

Introduction

Comparisons among species provide a wealth of information about evolution that could not otherwise be obtained. Experiments cannot generally be conducted over evolutionary time, but interspecies variation can be viewed as the results of natural experiments (Harvey and Pagel, 1991; Harvey and Purvis, 1991). Unfortunately, nature tends to use lousy experimental designs (Harvey and Pagel, 1991; Rees, 1995; Nee, Read and Harvey, 1996), complicating the analysis of multispecies data. The difficulties stem from the obvious fact that closely related species tend to be similar, whereas standard statistical tests of association assume independence of data points (Felsenstein, 1985). This chapter explores the analogy between comparative and experimental data, to highlight the pseudoreplication inherent in analyses that treat species or higher taxa as independent points. To be statistically valid, comparative tests of evolutionary hypotheses have to consider the relationships among the species, i.e. their phylogeny. How much difference phylogenetic methods can make is illustrated, using an example concerning the evolution of primate bacula. An outline is given of the logic of the most popular current phylogenetic comparative approach – independent comparisons or independent contrasts (IC) (Felsenstein, 1985) – and one implementation of the IC approach, Comparative Analysis by Independent Contrasts (CAIC), is described (Purvis and Rambaut, 1995).

Because estimates of phylogeny underpin comparative tests, we consider how they can be constructed and what features they should ideally possess. An estimate (Purvis, 1995a) of the relationships among all extant primate species is discussed.

The chapter also addresses some recent misconceptions about IC analyses, and emphasises that comparisons that do not consider phylogeny are likely to mislead and cannot be justified. However, we recognise that IC still requires further development in several areas, one of which – the need

44

to deal with error variance in species values – is illustrated in some detail.

This chapter is restricted to analyses in which at least some variables are continuous. For those interested in analyses in which traits vary discretely, Ridley and Grafen (1996) provide a review of the current state of play.

Pitfalls of multispecies data

This section considers a comparative hypothesis for which phylogenetic and non-phylogenetic tests give very different results. Use is made of the analogy between comparative analyses and experiments (Harvey and Pagel, 1991; Read and Nee, 1995; Rees, 1995; Nee *et al.*, 1996) to show why the non-phylogenetic result cannot be taken at face value.

Dixson and Purvis (in press) investigated the evolution of the baculum (penile bone) in primates. One hypothesis is that some lineages have evolved relatively long bacula for their body sizes because such bacula confer an advantage in sperm competition. Bacula are predicted to be longer relative to body size in lineages in which females are polyandrous than in lineages in which females mate with only a single male. Figure 3.1a plots log baculum length against log adult male body weight for many primate species (data from Dixson and Purvis, in press). It appears that males in polyandrous species do indeed have large bacula for their body size. A t-test on the residuals, comparing the relative baculum lengths of polyandrous and monandrous species, is highly significant ($t_{59} = 6.09$, $p < 0.0001$). Is this evidence for adaptation? Not necessarily, because of non-independence of data points. Figure 3.1b shows the same data, but now the symbols denote the taxa to which species belong. Callitrichids have relatively short bacula and tend to be monandrous, whereas strepsirhines have relatively long bacula and typically are polyandrous. A simple analogy with experimental design shows that the species within each of these taxa are not true replicates, but pseudoreplicates.

Figure 3.2a shows a very bad design for an experiment to assess whether a fertiliser improves crop yield. The fertiliser is applied to six plots (F +), six further plots are left as controls (F −), and the yields are later assessed. Suppose the F + plots give a significantly greater mean yield. Is that because of the fertiliser? We cannot tell, because the plots do not differ only in whether or not we fertilised them. The F + plots are all in one place, so their greater yield could equally well be due to better soil, or better drainage, or better shelter, or fewer pests, or more light – or any other reason. Our treatments are totally confounded by every other difference between the locations. The analogy to comparative data sets is important.

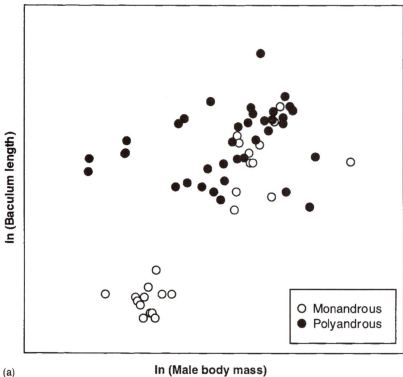

(a)   **ln (Male body mass)**

Fig. 3.1   (a) Baculum length and male body weight across primate species. Data have been logarithmically transformed. Symbols denote mating systems. (b) The same data, but symbols now denote taxonomic membership. Note how species in the same taxon tend to cluster together.

Differences in Y between callitrichids and strepsirhines might be due to the X-variable we have measured (mating system), but could equally well be due to any other difference between the two clades. Adding more callitrichids and more strepsirhines will not increase the level of replication – for this question, species within each clade represent pseudoreplicates.

In Figure 3.2b, the six pairs of plots are true replicates. Within each, the plots receive different treatments but should be similar in other respects such as soil quality. Differences among locations in such variables no longer matter because we are making comparisons *within* locations rather than between them. Again, the analogy is important for comparative analyses. The experimentalists' locations are our clades. To make truly

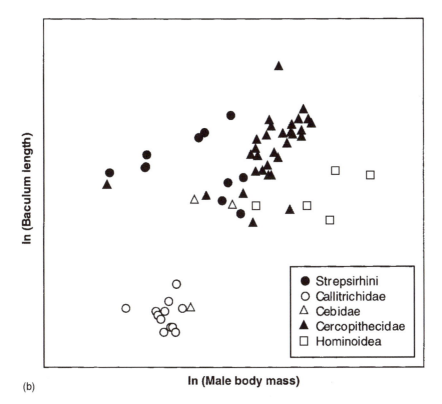

(b)   **In (Male body mass)**

replicated comparisons, we must analyse variation within clades rather than between them: we need to compare pairs of lineages that differ in the *X*-variable (here, mating system), and base such comparisons on phylogeny to ensure their independence. If the same association between baculum length and mating system is found repeatedly in different pairs of lineages, then we can have confidence that the association is meaningful, rather than the result of unmeasured confounding or covarying traits. Note that the power of the test will depend upon the number of matched pairs, not the number of species: sampling effort (whether in field, laboratory or library) should be directed towards maximising the number of informative independent comparisons. Phylogeny should therefore inform the collection of data as well as their analysis. Although this point is obvious, it is often neglected. An excellent example of an efficient design can be found in Clayton and Cotgreave (1994).

What happens in the baculum example when independent matched pairs are compared? Dixson and Purvis (in press) were able to make ten such

a)

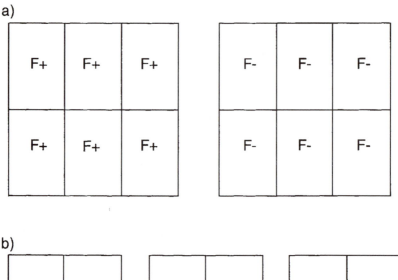

b)

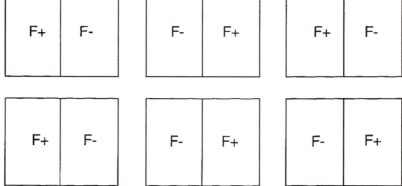

Fig. 3.2   Two designs for an experiment to assess the effect of a fertiliser upon crop yield. Fertiliser is added to plots labelled F + ; F − plots are controls. See text for explanation.

comparisons between a polyandrous lineage and a related monandrous one. Relatively large bacula were associated with polyandry in five of these comparisons, but with monandry in the other five (sign test: $p = 1$). The analysis of proper replicates shows no effect of mating system on baculum length, in contrast to the highly significant p-value obtained when species are treated as independent data points. However, further matched-pairs analyses showed that relative baculum length is associated with the duration and number of intromissions during copulation, suggesting a

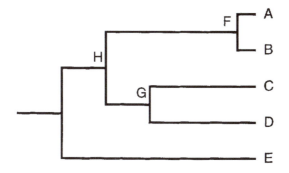

Fig. 3.3   A phylogeny of five species (A–E). See text.

different explanation for interspecific variation in baculum length (Dixson and Purvis, in press).

Independent comparisons

Phylogenetically independent comparisons were originally proposed by Felsenstein (1985), and have since been extended, developed or adapted in several ways (e.g. Burt, 1989; Grafen, 1989; Pagel, 1992; Martins, 1994, 1997; McPeek, 1995; Purvis and Rambaut, 1995; Díaz-Uriarte and Garland, 1996). As originally proposed, the method is intended for use with continuous variables. Put very simply, it computes differences in Y and X between sister lineages, and tests whether the differences in Y are correlated with those in X.

Figure 3.3 shows a phylogeny of five species, A to E. Because A is closely related to B, we would expect them to have similar phenotypes, both for the traits we are studying and for other characteristics that might affect our Y-variable. Felsenstein (1985) pointed out that the differences between A and B have evolved since they diverged (point F), and will be independent of the differences between C and D (which evolved along other branches of the phylogeny, since point G). So two ICs can easily be made. This logic leads very quickly to a method that is used widely (e.g. Møller and Birkhead, 1992; Grubb and Metcalfe, 1996), namely, looking at the association between Y and X among the species within each of several, say, genera. A sign test is used to assess the null hypothesis of no correlation, which predicts that positive and negative associations will be found equally (Felsenstein, 1985).

Such comparisons cannot be used in parametric tests because, although independent, they do not have the same expected variance: A and B are more closely related than C and D, so we would expect differences between them to be smaller. Felsenstein (1985) pointed out how parametric testing, and more comparisons, are possible if a model of evolution is adopted that specifies how the absolute difference between two species increases over time. For mathematical simplicity, Felsenstein (1985) considered a Brownian motion (random walk) model of character evolution. Under this model, the variance of Y_A-Y_B is proportional to the total length of the path through phylogeny linking species A and B. This relationship allows the comparisons to be scaled to have common expected variance, as required by parametric tests, if branch lengths are available. Additionally, if the model is reasonable, further sister-clade comparisons can be made: F can be compared with G, and H with E. Values of X and Y for F, G and H are computed as weighted means of the descendant species, the weights being derived from the branch lengths (see Felsenstein, 1985, for algorithms and reasons for them). In this way, a data set of n non-independent species values is transformed into $n - 1$ ICs for each variable, which can be used in standard parametric tests such as regression and correlation. Note that the regressions and correlations must be forced through the origin (Grafen, 1992; Garland, Harvey and Ives, 1992).

CAIC

CAIC (Purvis and Rambaut, 1995) is a comparative analysis package for Apple Macintoshes. It implements two algorithms, known for no good reason as CRUNCH and BRUNCH, that are descended from Felsenstein (1985). Whereas Felsenstein's (1985) method assumes complete information about phylogeny, CRUNCH and BRUNCH can be used when phylogenetic information is incomplete. Additionally, BRUNCH does not require that change in Y and X can be modelled by Brownian motion.

The CRUNCH algorithm comes from Pagel (1992). When the phylogeny is completely bifurcating, the algorithm is the same as Felsenstein's (1985). But estimates of phylogeny often contain polytomies – multifurcations – which usually reflect our ignorance of the precise order of splitting (see Purvis and Garland, 1993, for discussion). Pagel (1992) recognised (as did Grafen, 1989, and Burt, 1989) that, to ensure independence, only one comparison could be made at each node in the phylogeny, irrespective of how many branches descend from it, and proposed the following approach (for algorithms, see Pagel, 1992; Purvis and Rambaut, 1995). The taxa immediately descended from the polytomy are split into

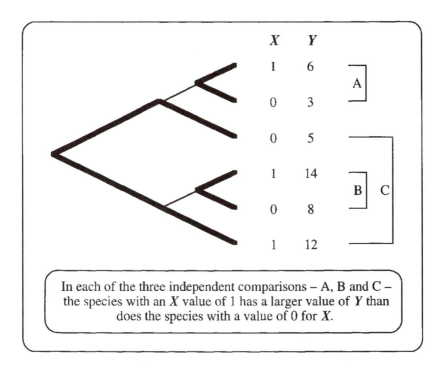

X	Y
1	6
0	3
0	5
1	14
0	8
1	12

In each of the three independent comparisons – A, B and C – the species with an *X* value of 1 has a larger value of *Y* than does the species with a value of 0 for *X*.

Fig. 3.4   A phylogeny of six species illustrating how BRUNCH partitions variation among species into independent comparisons. From the *CAIC User's Guide*.

two groups on their values of a variable selected by the user (usually the *X*-variable): those above the mean are put into one group, and those below into another. The two groups are then compared.

The BRUNCH algorithm (Purvis and Rambaut, 1995), borrows from Pagel (1992) and Burt (1989). It is mainly intended for continuous *Y* and discrete *X*; however, it can also be useful when both variables are continuous but cannot be modelled by Brownian motion. Figure 3.4 shows roughly how it works. As many ICs as possible are made between lineages differing in their value of the *X* variable. Comparisons are independent if lines through the phylogeny linking the species to be compared neither touch nor cross (Burt, 1989). A sign test is used to test whether *Y* and *X* are associated. If *Y* fits Brownian motion, a t-test can be used on the scaled contrasts in *Y* (under the null hypothesis, their mean should be zero). Not all comparisons are between sister taxa (see Fig. 3.4), but they are still matched pairs.

CAIC takes as its input a text file of data. The first column contains the species names, and species must be in alphabetical order. Subsequent columns, of which there can be up to 128, hold data for continuous or discrete variables. Missing values are coded as ' − 9'. The data file is then collated against a pre-prepared phylogeny of the group being studied. All the species in the file must be present in the phylogeny, but the reverse is not true. CAIC next asks the user to select the columns (up to 40 at a time) for which ICs will be computed, and the algorithm to be used. ICs are then calculated from the species that have no missing values for *any* of the columns selected; it is therefore sensible to select only those columns that bear directly on the hypothesis under test.

As well as producing ICs, CAIC generates other information that can be used to test many of its assumptions. First, it specifies the node at which the comparison was made. Comparative tests assume that any relationship between Y and X is the same throughout the phylogeny; this assumption can be tested by ANCOVAs (Garland *et al.*, 1993; Purvis and Rambaut 1995) or contingency tables. CAIC also specifies the age of the node at which each IC was made, the comparison's expected variance before it was scaled, and the estimate of each variable at the node. If the Brownian model is reasonable, there should be no pattern when any of these is plotted against the scaled ICs (Garland *et al.*, 1992; Purvis and Rambaut, 1995). If a pattern is found that is not removed by transformation of either data or branch lengths, the user might prefer to use BRUNCH or a method that permits other models of evolution (e.g. Martins, 1995).

CAIC is freely available by anonymous ftp from
http://www.bio.ic.ac.uk/evolve/docs/
WWW users can find CAIC's home page at
http://www.bio.ic.ac.uk/evolve/software

Other independent comparison programs

Other programs perform different variants of IC. Felsenstein's CON-TRAST program (distributed with PHYLIP) implements his 1985 paper. Grafen's (1989) phylogenetic regression permits both discrete and continuous X-variables to be tested and controlled for. Martins' (1995) COM-PARE has a wider array of evolutionary models. Garland *et al.*'s (1993) programs allow transformation of the phylogeny branch lengths to make the Brownian motion assumption as reasonable as possible. Of these programs, only phylogenetic regression (Grafen, 1989) is valid when the phylogeny is incompletely resolved.

Estimates of phylogeny

Comparative tests must be underpinned by estimates of phylogeny. What characterises a good estimate? First and foremost, it should not make incorrect statements of relationship, i.e. every node in the estimated phylogeny should correspond to a genuine clade. If this requirement is not met, the wrong taxa will be compared. Second, the estimate of phylogeny should be as well resolved as possible, because only one IC can be computed at each node. Third, estimated phylogenies should ideally be robust, with much evidence supporting each statement of relationship. Lastly, more powerful tests are facilitated by estimates of branch length, perhaps in units of time or molecular change. Biologists embarking upon a comparative study are often disappointed to find that no phylogeny has yet been published on their group that meets these requirements. What should they then do?

One option is to infer a phylogeny from molecular sequence data taken from all the species in the study. This seems like a huge task but is becoming more feasible, for small data sets at least, as DNA sequencing becomes more routine and sequence data accumulate on databases such as Genbank and EMBL. Gittleman *et al.* (1996) give one example of this approach, and Hillis, Moritz and Mable (1996) provide an up-to-date guide on how to collect the sequences and infer a tree from them.

A generally less time-consuming option is to piece together an estimate from published trees (hereafter, source trees), each containing a subset of the species in the data set. But how should this patchwork be put together? A common approach is to survey the literature and select the best available source tree for each major taxon and discard the rest; 'best' here might mean most recent, most well resolved, most robust, based on the most data, or constructed using the best algorithms (see Quicke, 1993; Swofford *et al.*, 1996). However, the choice of a 'best' phylogeny is subjective and, unless the criteria are stated explicitly, unrepeatable.

Repeatability, at least, can be ensured by using an algorithm to combine all available source trees. Various 'consensus' algorithms are available (see Quicke, 1993) but, because each source tree effectively has a veto, the consensus tree tends to become less and less well resolved as more source trees are added (Purvis, 1995b).

Baum (1992) and Ragan (1992) independently proposed a technique, Matrix Representation with Parsimony (MRP), that circumvents this problem. In MRP, each node in each source tree is recoded as a binary character. Species descended from this node score a 1; those not descended from the node score a 0; and those not featured in the source tree are scored as missing. This new matrix is analysed using parsimony to give a composite tree, which may be well resolved even when source trees conflict. In Baum and Ragan's original proposal, large trees can have undue influence on the composite tree. Modifications have been proposed to amend this flaw (Purvis, 1995b; Ronquist, 1996), but none has so far succeeded (Bininda-Emonds and Bryant, 1998). The best way to combine source trees is currently a topic of active debate (see Sanderson, Purvis and Henze, 1998, for a recent review), but algorithms do at least permit repeatability. Additionally, there is some evidence (see next section) that the composite phylogeny can be reasonably robust to the choice of algorithm.

Sometimes, no estimates of phylogeny will have been published. Taxonomies can be used as phylogenetic hypotheses of last resort. Although they are not likely to be as good as proper phylogenies (taxonomies may not even be intended to reflect phylogeny – see, for example, Ridley (1986) – and are likely to be poorly resolved), tests using them are likely to be much better than non-phylogenetic analyses.

A composite primate phylogeny

Purvis (1995a) generated the first composite phylogeny of all extant primate species by combining 112 source trees using his modification of MRP (Purvis, 1995b). The composite tree is largely bifurcating, having 160 nodes out of a possible 202, but some component clades – particularly Galagidae and Colobinae – are much less well resolved. The timings of 90 of the nodes were estimated from molecular, fossil or karyotypic data and, for the purposes of comparative analysis, timings of the remaining nodes were guessed by interpolation (for details, see Purvis, 1995a). Figure 3.5 shows the result.

Although the algorithm used to construct the tree is flawed (Ronquist, 1996), does this matter? We applied Baum (1992) and Ragan's (1992) original algorithm to the 112 source trees, keeping all other aspects of the analysis the same as in Purvis (1995a). The resulting 'new' composite tree (Appendix 3.1) was very similar to the 'old' one in Purvis (1995a): of 160 nodes in the old tree, only 12 conflicted with the new; all of these had been only weakly supported in the original paper (as judged by the bootstrap scores). Seven polytomies in the old tree were resolved, while two nodes in the old tree were collapsed to polytomies in the new. Table 3.1 lists these differences. Our impression is that details of the method will matter more when the conflict among source trees is greater; when the signal of the 'true'

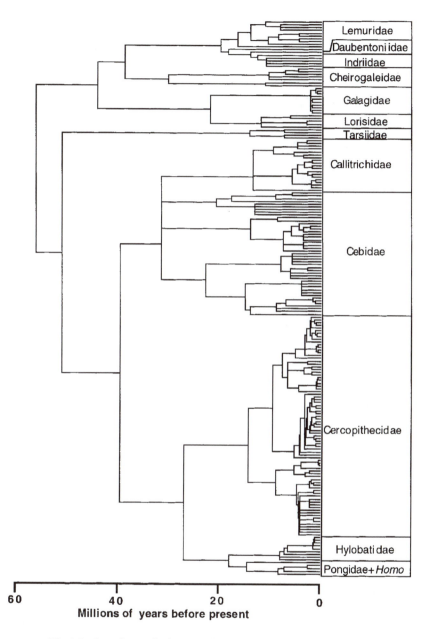

Fig. 3.5   An estimate of primate phylogeny. After Purvis (1995a).

Table 3.1. *Comparison between Purvis' (1995a) estimate of primate phylogeny and that produced by the re-analysis described in the text*

Purvis 1995	Re-analysis
Hapalemur unresolved	*H. griseus* and *H. simus* sister species
Hapalemur sister to *Lemur catta*; *Varecia* sister group to these plus *Petterus* (*Eulemur*)	*Varecia* sister to *L. catta*; *Hapalemur* sister to these plus *Petterus* (*Eulemur*)
Within *Petterus* (*Eulemur*), *P. coronatus* and *P. mongoz* are sister species, as are *P. fulvus* and *P. macaco*. *P. rubriventer* is outgroup to the other four	Within *Petterus* (*Eulemur*), *P. coronatus* and *P. rubriventer* are sister species. *P. mongoz* is now sister to the *P. fulvus*/*P. macaco* clade
Indri indri in four-way polytomy with the members of *Propithecus*	*I. indri* sister to *P. verreauxi*, making *Propithecus* paraphyletic
Cheirogaleus major and *C. medius* in a polytomy with the *Microcebus*/*Mirza* clade	*Cheirogaleus major* and *C. medius* are sister species
Three-way polytomy between *Saguinus inustus*, *S. nigricollis* and the *S. fuscicollis*/*S. tripartitus* clade	*S. nigricollis* is sister to the *S. fuscicollis*/*S. tripartitus* clade
Callicebus sister to *Aotus*	*Callicebus* sister to all other platyrrhines
Three-way polytomy between *Callicebus personatus*, *C. hoffmannsi* and the *C. cinerascens*/*C. moloch* clade	*Callicebus personatus* is sister to the clade comprising *C. cinerascens* and *C. moloch*
Macaca nemestrina sister to a clade of nine species including *M. arctoides*. Sister group to all these is a four-species clade including *M. nigra*. Sister to all of these is *M. silenus*	Four-way polytomy of *M. nemestrina*, *M. silenus*, the nine-species clade and the four-species clade
Macaca assamensis and *M. thibetana* in a polytomy with an *M. radiata*/*M. sinica* clade	*M. assamensis* and *M. thibetana* sister species
Macaca fuscata and *M. mulatta* sister species	Three-way polytomy of *M. fuscata*, *M. mulatta* and *M. cyclopis*
Cercopithecus lhoesti and *C. preussi* comprise the sister clade of a 17-species group of *Cercopithecus*. This clade forms a polytomy with *C. aethiops* and *C. solatus*. The next sister group is *Erythrocebus patas*, and *Miopithecus talapoin* is the next	The *C. lhoesti*/*C. preussi* clade is linked to *C. solatus*, then *C. aethiops*, then *E. patas* (making *Cercopithecus* paraphyletic). This clade is in a polytomy with the 17-species group and *M. talapoin*
Polytomy of *Hylobates agilis*, *H. lar* and *H. muelleri*	*H. agilis* and *H. lar* sister species

phylogeny is strong, even suboptimal methods (and there is not yet any optimal method) will be able to detect it.

The list of species followed in the composite tree is from Corbet and Hill (1991; their taxon names are used in this chapter for consistency, but see Groves, 1993). There is, however, disagreement over the species richness of many groups. For instance, Corbet and Hill (1991) list 64 species of New World monkey, but a more recent taxonomy (Rylands, Mittermeier and Luna, 1995) lists 98. Changes in species numbers are due sometimes to new discoveries, sometimes to the perceived requirement of species monophyly, and sometimes to differences between 'lumpers' and 'splitters', who tend to decrease or increase numbers of species, respectively.

The composite phylogeny is freely available in the format required by CAIC via the World Wide Web; the url is http://www.bio.ic.ac.uk/tes/peopleap.htm. It has already been used in several comparative studies of adaptation (e.g. Barton, 1996; Fa and Purvis, 1997) and macroevolution (Purvis, Nee and Harvey, 1995; Gittleman and Purvis, 1998). The same site also holds several other CAIC-format phylogenies of various groups, and the composite tree produced by the re-analysis described above.

Misconceptions about phylogenetic comparative analyses

Several misconceptions have arisen about independent comparisons analyses in recent years. We take this opportunity to address them.

IC discards character variation that is 'due to phylogeny' (e.g. Lord, Westoby and Leishman, 1995). Some comparative methods (e.g. Stearns, 1983; Cheverud, Dow and Leutenegger, 1985; Gittleman and Kot, 1990) discard variation in this way. They split interspecific variation among species into two components, one representing phylogenetic differences among higher taxa and the other representing adaptive differences among close relatives. Only the latter is used in the test. IC does not discard variation like this.

Use of within-genus means instead of species values removes the problem of non-independence. Averaging species within genera will reduce the level of pseudoreplication, but closely related genera may still be unusually similar (Harvey and Pagel, 1991). To return to the data set of Dixson and Purvis (in press), the association between relative baculum length and mating sytem is highly significant ($p = 0.015$), even when species are averaged within genera.

Grade shifts make IC analyses unsuitable (Martin, 1996). On the contrary, grade shifts (when the slope of Y on X is the same in two major taxa but the intercept differs) make non-phylogenetic analyses perform particularly badly: if species of both grades are pooled, the slope estimate produced by non-phylogenetic analysis is meaningless. When comparisons are based on phylogeny, however, grade shifts are easy to test for and to pinpoint: a hypothesis that a particular clade is grade-shifted predicts that the comparison between that clade and its sister group will be an outlier (i.e. will have an unusually large standardised residual). Similarly, outlying contrasts can indicate the possible existence of previously unsuspected grades.

Phylogenetic analyses are necessary only when the p-value of a non-phylogenetic analysis is marginal. The baculum analysis is enough to demonstrate that this view is ill-founded; many simulations (e.g. Grafen, 1989; Purvis, Gittleman and Luh, 1994) make the same point.

Phylogenetic analyses are less powerful than non-phylogenetic analyses. Simulations (e.g. Grafen, 1989; Purvis *et al.*, 1994) show that the reverse is probably true when the elevated Type I error rates of non-phylogenetic analyses are taken into account.

Phylogeny need not be considered if stabilising selection is the reason why related species are similar. This view is understandable but incorrect: phylogeny must be considered no matter why close relatives are similar (Burt, 1989; Grafen, 1989; Harvey and Pagel, 1991). That is because the values of *Y* may depend not only on the *X*-variable(s) being considered, but also on any number of other factors that themselves reflect phylogeny (i.e. close relatives have similar values). Only comparisons based on phylogeny will be able to control for the effects of such unmeasured confounding variables.

Phylogenetic analyses are unnecessary when studying ecological, rather than evolutionary, questions. This is correct only if close relatives are not unusually similar in the traits being considered. Otherwise, phylogeny is important even if the traits change over ecological time (Harvey, 1996). For example, geographic range size (Letcher and Harvey, 1994) and extinction risk (Bennett and Owens, 1997) both show phylogenetic pattern, presumably because at least some factors shaping them take similar values in close relatives.

IC analyses must be horribly invalid because characters do not evolve by Brownian motion (Wenzel and Carpenter, 1994). However, the model is testable and, if it is violated significantly, remedial action can often be taken or a BRUNCH-style analysis performed. Recent simulations (Díaz-Uriarte and Garland, 1996) asked how badly IC is misled when data evolved under other models are analysed under Brownian motion. The results are encouraging: when the recommended assumption-testing was done and necessary steps taken, the Type I error rates never exceeded twice the nominal *p*-value at $\alpha = 0.05$, for any of the 14 models considered. Although validity is compromised slightly, IC remains much more valid than non-phylogenetic analyses.

Outstanding problems

Although IC has proved to be extremely useful, not all problems presented by comparative data have been solved. We would like to draw attention to one remaining issue in particular, that of data error. IC basically assumes that species values are known parametrically but, in reality, they are estimates and so subject to random error. This section illustrates some of the consequences of this mismatch between the model and reality. The conclusion will be that IC analyses, especially parametric tests like Felsenstein's (1985) original proposal, make greater demands of their data than do non-phylogenetic tests. This, of course, is an argument for obtaining better data, not for using worse tests.

We have used data sets differing in quality to 'test' some null hypotheses widely accepted to be false. Five estimates of each of five life history variables (adult body mass, gestation length, interbirth interval, neonatal mass, and age at sexual maturity) were compiled for up to 21 primate species. Most data came from Hayssen, van Tienhoven and van Tienhoven (1993). In sexually dimorphic species, female values of life history traits were used. Where ranges were given, the midpoint was used. All values were log-transformed. From this compilation, five data sets were then generated. To make them differ in quality, the sample size on which they were based was varied. The best data set contains the averages of all five values available: we call this the $n = 5$ data set. The other data sets, which we call $n = 4$, $n = 3$, $n = 2$ and $n = 1$, contain averages of four, three, two or one randomly selected values, respectively. As n decreases, so the sampling error associated with each species value will tend to increase. Note that the number of species does not change with n (although it does vary among life history variables).

Gestation length, interbirth interval, neonatal mass and age at sexual maturity are known from comparative studies of much larger data sets to correlate positively with body mass (Harvey, Read and Promislow, 1989; Purvis and Harvey, 1995). The five data sets were analysed both non-phylogenetically and using CRUNCH with Purvis's (1995a) phylogeny. In non-phylogenetic tests, all associations were significant at the $p < 0.001$ level, with the fit of the points improving slightly as n increased (Table 3.2a, upper part).

With IC, data quality had a much more marked effect (Table 3.2, lower part). Figures 3.6a to 3.6e show how the significance of the association between gestation length and adult mass depends on sample size. The trend is the same for all four Y-variables: the $n = 1$ data set produces the weakest correlation, and the strength of the association tends to increase as n rises.

Why is IC so sensitive to measurement and sampling error? IC assumes that differences among taxa have arisen through evolution, rather than through sampling error. When distant relatives are being compared, the error matters little because it is small relative to the evolved difference (Fig. 3.7a). But when close relatives are compared, the sampling error can easily

Table 3.2. *The relationships between each of four life history variables and body size for the five data sets, n = 1 to n = 5*

	Gestation length (16 df)			Interbirth interval (8 df)			Neonatal mass (12 df)			Age at sexual maturity (14 df)		
	b	t	p	b	t	p	b	t	p	b	t	p
Analyses treating species as independent points												
$n = 1$	0.114	5.11	< 0.001	0.288	3.97	< 0.001	0.821	12.94	< 0.001	0.321	4.07	< 0.001
$n = 2$	0.084	4.45	< 0.001	0.341	5.60	< 0.001	0.793	13.48	< 0.001	0.322	4.90	< 0.001
$n = 3$	0.085	5.52	< 0.001	0.359	6.43	< 0.001	0.778	14.15	< 0.001	0.329	5.61	< 0.001
$n = 4$	0.088	5.92	< 0.001	0.361	7.19	< 0.001	0.800	15.37	< 0.001	0.326	5.79	< 0.001
$n = 5$	0.087	6.24	< 0.001	0.352	8.07	< 0.001	0.806	14.78	< 0.001	0.322	6.33	< 0.001
Analyses of phylogenetically independent contrasts												
$n = 1$	0.018	0.48	0.64	0.181	1.13	0.29	0.728	5.99	< 0.001	0.249	2.17	0.048
$n = 2$	0.017	0.55	0.59	0.278	2.52	0.036	0.677	6.23	< 0.001	0.205	2.36	0.033
$n = 3$	0.028	1.24	0.23	0.287	3.30	0.011	0.680	8.93	< 0.001	0.224	2.48	0.026
$n = 4$	0.034	1.74	0.10	0.279	3.76	0.006	0.711	10.66	< 0.001	0.195	2.54	0.023
$n = 5$	0.042	2.42	0.028	0.280	3.96	0.004	0.714	9.63	< 0.001	0.223	3.29	0.005

See text for explanation; see also Figure 3.6.
b = slope as estimated by ordinary least-squares (through the origin in the lower half of the table); t = t-statistic for the slope; p = probability.

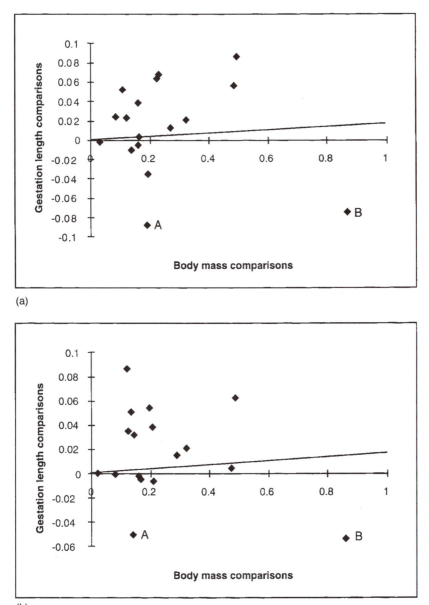

(a)

(b)

Fig. 3.6 Five plots of independent contrasts in gestation length and adult body weight, from the five data sets $n = 1$ (3.6a) to $n = 5$ (3.6e). Note how later figures have steeper slopes and fewer points below the X axis. Point A is a comparison between two very closely related *Macaca* species; B is a comparison between two galagids. See text for explanation. See also Table 3.2.

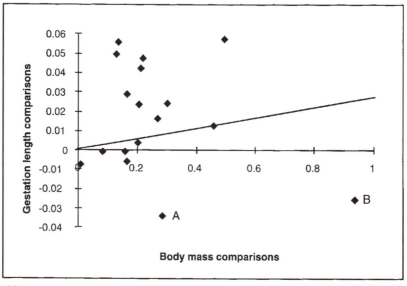

(c)

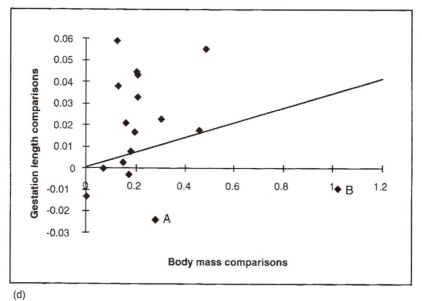

(d)

Fig 3.6 (*continued*)

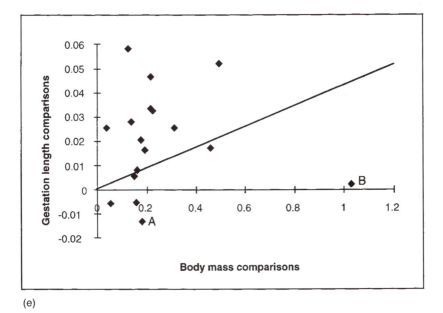

(e)

overwhelm small evolved differences between species (Fig. 3.7b). Because of the hierarchical nature of phylogenies, it is likely that more comparisons will be between close relatives than between distant ones. To make matters worse, the differences are all converted into rates of change (by dividing by branch length) in an attempt to give them a common variance as required by parametric statistics. Differences between very close relatives, which may largely have arisen through error rather than through evolution, may be hugely exaggerated when converted into rates, as they have apparently evolved over a very short time. This problem can be seen in the outliers in Figs. 3.6a–e, which include comparisons between the gorilla and the chimp, two closely related macaque species (labelled 'A' on Figs. 3.6a–e), and within the genus *Saguinus*.

The implications are clear. When closely related taxa are being compared, data quality is crucial if IC tests are to be powerful. Rounding or the use of modes in place of means – both common in comparative data sets (Gittleman, 1989) – is particularly likely to produce very influential comparisons. Comparative biologists should be wary of uncritically accepting values published in secondary and perhaps even primary sources. They should also inspect contrasts plots to see if comparisons among close relatives tend to have high leverage: if they do, a branch length transformation may be in order. Even quite large data sets may not permit powerful IC analyses unless the data are good; users must bear in mind that

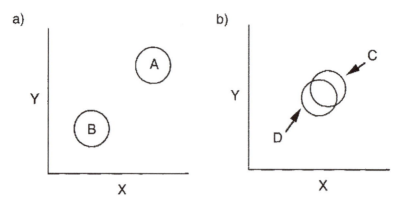

Fig. 3.7   An illustration of why data error matters most for comparisons among close relatives. The axes represent two variables under study, and the circles encompass the combinations of X and Y found in populations of each of two species. In part (a), the species being compared are phenotypically quite different. Although Y and X are estimated with sampling error, there is no doubt that they are positively correlated between species A and B. In (b), the species are more closely related: now it is quite possible that estimates of Y and X based on small samples will lead to the incorrect conclusion that they are negatively correlated between species C and D.

failure to reject the null hypothesis does not mean that the null hypothesis is correct.

Our analysis also highlights another need for caution in IC analyses. The comparison between *Galago senegalensis* (the lesser bush baby) and *Otolemur crassicaudatus* (the thick-tailed bush baby) was an extreme outlier in many of the contrasts analyses (it is the point labelled 'B' on Figs. 3.6a–e). We suspect that the most recent common ancestor of these two species probably lived longer ago than the 1.8 Myr estimated by Purvis (1995a). Errors in the branch lengths will lead to misestimation of rates of change; again, we recommend inspection of the contrast plots and analyses to see if any contrasts have unusually high leverage.

Error variance in X-variables is already known to flatten estimates of least-squares regression slopes in non-phylogenetic and phylogenetic analyses alike (Harvey and Pagel, 1991); in IC analyses, slope estimates from contrasts at older nodes are commonly steeper than estimates from more recent nodes (Purvis and Harvey, 1995; Taggart *et al.*, 1998). Regression models other than least-squares are discussed by Harvey and Pagel (1991) and Garland et al. (1992), but more work remains to be done in this area.

Conclusion

Biologists have been urged to treat interspecies comparisons with the same respect as would be given to experimental results (Maynard Smith and Holliday, 1979). However, because 'natural experiments' are badly designed, only carefully designed comparisons deserve such respect. Comparisons must be based on a phylogeny to ensure independence and must be based on explicit models of character change if parametric tests are to be applied. Many reservations about phylogenetic comparative methods are based on misconceptions. Although the methods are not perfect, the problems with contrasts cannot be used as justifications for ignoring phylogeny. To use another quote, 'phylogenies are fundamental to comparative biology. There is no doing it without taking them into account' (Felsenstein, 1985). We urge users of phylogenetic comparative methods not to view them as 'black boxes': an understanding of the methods permits informed choices of which is most suitable for the available data, and an assessment of whether the data and phylogenetic estimates are indeed adequate for good tests of hypotheses.

Acknowledgements

We are very grateful to Phyllis Lee for her invitation to write this chapter and for her editorial comments. Christophe Thébaud suggested Figure 3.7. This work was supported by the Royal Society and the Natural Environment Research Council (GR3/8515 and studentship).

References

Barton, R.A. (1996). Neocortex size and behavioural ecology in primates. *Proceedings of the Royal Society of London, Series B* **263**, 173–7.

Baum, B.R. (1992). Combining trees as a way of combining data sets for phylogenetic inference, and the desirability of combining gene trees. *Taxon* **41**, 1–10.

Bennett, P.M. and Owens, I.P.F. (1997). Variation in extinction risk among birds: chance or evolutionary predisposition? *Proceedings of the Royal Society of London, Series B* **264**, 401–8.

Bininda-Emonds, O.R.P. and Bryant, H.N. (1998). Properties of matrix representation with parsimony analyses. *Systematic Biology* **47**, 507–18.

Burt, A. (1989). Comparative methods using phylogenetically independent contrasts. *Oxford Surveys in Evolutionary Biology* **6**, 33–53.

Cheverud, J.M., Dow, M.M. and Leutenegger, W. (1985). The quantitative assessment of phylogenetic constraints in comparative analyses: sexual dimorphism in body weight among primates. *Evolution* **39**, 1335–51.

Clayton, D.H. and Cotgreave, P. (1994). Relationship of bill morphology to grooming behaviour in birds. *Animal Behaviour* **47**, 195–201.

Corbet, G.B. and Hill, J.E. (1991). *A World List of Mammalian Species*. Oxford: Oxford University Press.

Díaz-Uriarte, R. and Garland, T. J. (1996). Testing hypotheses of correlated evolution using phylogenetically independent contrasts: sensitivity to deviations from Brownian motion. *Systematic Biology* **45**, 27–47.

Dixson, A.F. and Purvis, A. (in press). Sexual selection and genital morphology in male and female primates. *Folia Primatologica*.

Fa, J.E. and Purvis, A. (1997). Body size, diet and population density in Afrotropical forest mammals: a comparison with neotropical species. *Journal of Animal Ecology* **66**, 98–112.

Felsenstein, J. (1985). Phylogenies and the comparative method. *American Naturalist* **125**, 1–15.

Garland, T., Dickerman, A.W., Janis, C.M. and Jones, J.A. (1993). Phylogenetic analysis of covariance by computer simulation. *Systematic Biology* **42**, 265–92.

Garland, T., Harvey, P.H. and Ives, A.R. (1992). Procedures for the analysis of comparative data using phylogenetically independent contrasts. *Systematic Biology* **41**, 18–32.

Gittleman, J.L. (1989). The comparative approach in ethology: aims and limitations. In *Perspectives in Ethology*, Volume 8, ed. P.P.G. Bateson and P.H. Klopfer, pp. 55–83. New York: Plenum Press.

Gittleman, J.L., Anderson, C.G., Kot, M. and Luh, H.-K. (1996). Comparative tests of evolutionary lability and rates using molecular phylogenies. In *New Uses for New Phylogenies*, ed. P.H. Harvey, A.J. Leigh Brown, J. Maynard Smith and S. Nee, pp. 289–307. Oxford: Oxford University Press.

Gittleman, J.L. and Kot, M. (1990). Adaptation: statistics and a null model for estimating phylogenetic effects. *Systematic Zoology* **39**, 227–41.

Gittleman, J.L. and Purvis, A. (1998). Body size and species-richness in carnivores and primates. *Proceedings of the Royal Society of London, Series B* **265**, 113–19.

Grafen, A. (1989). The phylogenetic regression. *Philosophical Transactions of the Royal Society of London, Series B*, **326**, 119–57.

Grafen, A. (1992). The uniqueness of the phylogenetic regression. *Journal of Theoretical Biology* **156**, 405–23.

Groves, C.P. (1993). Order Primates. In *Mammal Species of the World: a Taxonomic and Geographic Reference*, ed. D.E. Wilson and D.M. Reeder, pp. 243–77. Washington DC: Smithsonian Institute Press.

Grubb, P.J. and Metcalfe, D.J. (1996). Adaptation and inertia in the Australian tropical lowland rain-forest flora: contradictory trends in intergeneric and intrageneric comparisons of seed size in relation to light demand. *Functional Ecology* **10**, 512–20.

Harvey, P.H. (1996). Phylogenies for ecologists. *Journal of Animal Ecology* **65**, 255–63.

Harvey, P.H. and Pagel, M.D. (1991). *The Comparative Method in Evolutionary Biology*. Oxford: Oxford University Press.

Harvey, P.H. and Purvis, A. (1991). Comparative methods for explaining adaptations. *Nature* **351**, 619–24.

Harvey, P.H., Read, A.F. and Promislow, D.E.L. (1989). Life history variation in placental mammals: unifying the data with theory. *Oxford Surveys in Evolutionary Biology* **6**, 13–31.

Hayssen, V., van Tienhoven, A. and van Tienhoven, A. (1993). *Asdell's Patterns of Mammalian Reproduction*. Ithaca: Comstock.

Hillis, D.M., Moritz, C. and Mable, B.K. (1996). *Molecular Systematics*, 2nd edn. Sunderland, MA: Sinauer Associates.

Letcher, A.J. and Harvey, P.H. (1994). Variation in geographical range size among mammals of the Palearctic. *American Naturalist* **144**, 30–42.

Lord, J., Westoby, M. and Leishman, M. (1995). Seed size and phylogeny in six temperate floras: constraints, niche conservatism, and adaptation. *American Naturalist* **146**, 349–64.

Martin, R.D. (1996). Scaling of the mammalian brain: the maternal energy hypothesis. *News in Physiological Sciences* **11**, 149–56.

Martins, E.P. (1994). Estimating the rate of phenotypic evolution from comparative data. *American Naturalist* **144**, 193–209.

Martins, E.P. (1995). *COMPARE: Statistical Analysis of Comparative Data, Version 1.0*. Eugene: Department of Biology, University of Oregon.

Martins, E.P. (1997). Phylogenies and the comparative method: a general approach to incorporating phylogenetic information into the analysis of interspecific data. *American Naturalist* **149**, 646–67.

Maynard Smith, J. and Holliday, R. (1979). *The Evolution of Adaptation by Natural Selection*. London: The Royal Society.

McPeek, M.A. (1995). Testing hypotheses about evolutionary change on single branches of a phylogeny using evolutionary contrasts. *American Naturalist* **145**, 686–703.

Møller, A.P. and Birkhead, T.R. (1992). A pairwise comparative method as illustrated by copulation frequency in birds. *American Naturalist* **139**, 644–56.

Nee, S., Read, A.F. and Harvey, P.H. (1996). Why phylogenies are necessary for comparative analysis. In *Phylogenies and the Comparative Method in Animal Behavior*, ed. E.P. Martins, pp. 399–411. New York: Oxford University Press.

Pagel, M.D. (1992). A method for the analysis of comparative data. *Journal of Theoretical Biology* **156**, 431–42.

Purvis, A. (1995a). A composite estimate of primate phylogeny. *Philosophical Transactions of the Royal Society of London, Series B* **348**, 405–21.

Purvis, A. (1995b). A modification to Baum and Ragan's method for combining phylogenetic trees. *Systematic Biology* **44**, 251–5.

Purvis, A. and Garland, T. (1993). Polytomies in comparative analyses of continuous characters. *Systematic Biology* **42**, 569–75.

Purvis, A., Gittleman, J.L. and Luh, H.K. (1994). Truth or consequences – effects of phylogenetic accuracy on 2 comparative methods. *Journal of Theoretical Biology* **167**, 293–300.

Purvis, A. and Harvey, P.H. (1995). Mammal life history: a comparative test of Charnov's model. *Journal of Zoology (London)* **237**, 259–83.

Purvis, A., Nee, S. and Harvey, P.H. (1995). Macroevolutionary inferences from primate phylogeny. *Proceedings of the Royal Society of London, Series B* **260**, 329–33.

Purvis, A. and Rambaut, A. (1995). Comparative analysis by independent contrasts (CAIC): an Apple Macintosh application for analysing comparative data. *Computer Applications in Bioscience* **11**, 247–51.

Quicke, D.L.J. (1993). *Principles and Techniques of Contemporary Taxonomy*. London: Chapman and Hall.

Ragan, M.A. (1992). Phylogenetic inference based on matrix representation of trees. *Molecular Phylogenetics and Evolution* **1**, 53–8.

Read, A.F. and Nee, S. (1995). Inference from binary comparative data. *Journal of Theoretical Biology* **173**, 99–108.

Rees, M. (1995). EC–PC comparative analyses? *Journal of Ecology* **83**, 891–3.

Ridley, M. (1986). *Evolution and Classification: the Reformation of Cladism*. Harlow: Longman.

Ridley, M. and Grafen, A. (1996). How to study discrete comparative methods. In *Phylogenies and the Comparative Method in Animal Behavior*, ed. E.P. Martins, pp. 76–103. New York: Oxford University Press.

Ronquist, F. (1996). Matrix representation of trees, redundancy, and weighting. *Systematic Biology* **45**, 247–53.

Rylands, A.B., Mittermeier, R.A. and Luna, E.R. (1995). A species list for the New World Primates (Platyrrhini): distribution by country, endemism, and conservation status according to the Mace–Lande system. *Neotropical Primates* **3** (Supplement).

Sanderson, M.J., Purvis, A. and Henze, C. (1998). Phylogenetic supertrees: assembling the tree of life. *Trends in Ecology and Evolution* **13**, 105–9.

Stearns, S.C. (1983). The influence of size and phylogeny on patterns of covariation among life history traits in the mammals. *Oikos* **41**, 173–87.

Swofford, D.L., Olsen, G.J., Waddell, P.J. and Hillis, D.M. (1996). Phylogenetic inference. In *Molecular Systematics*, 2nd edn, ed. D.M. Hillis, C. Moritz and B.K. Mable, pp. 407–514. Sunderland, MA: Sinauer Associates.

Taggart, D.A., Breed, W.G., Temple-Smith, P.D., Purvis, A. and Shimmin, G. (1998). Testis mass, sperm number and sperm length: their relationship to reproductive strategies in marsupials and monotremes. In *Sperm Competition and the Evolution of Animal Mating Systems*, ed. T.R. Birkhead and A.P. Møller, pp. 623–66. New York: Academic Press.

Wenzel, J.W. and Carpenter, J.M. (1994). Comparing methods: adaptive traits and tests of adaptation. In *Phylogenetics and Ecology*, ed. P. Eggleton and R.I. Vane-Wright, pp. 79–101. New York: Academic Press.

Appendix 3.1
The relationships among extant primates according to the re-analysis described in this chapter

(((((((Microcebus murinus,Microcebus rufus),Mirza coquereli),(Cheirogaleus major,Cheirogaleus medius)),(Allocebus trichotis,Phaner furcifer)),((((Lemur catta,Varecia variegata),((Petterus coronatus,Petterus rubriventer),(Petterus mongoz,(Petterus fulvus,Petterus macaco)))),(Hapalemur aureus,(Hapalemur griseus,Hapalemur simus))),((((Propithecus diadema,Propithecus tattersalli,(Propithecus verreauxi,Indri indri)),Avahi laniger),Daubentonia madagascariensis),Lepilemur mustelinus))),((((Nycticebus coucang,Nycticebus pygmaeus),(Loris tardigradus,Arctocebus calabarensis)),Perodicticus potto),(Galago granti,(Galago alleni,(Galagoides zanzibaricus,Galagoides demidoff)),(Galago senegalensis,Galago moholi),(Otolemur crassicaudatus,Otolemur garnettii),(Euoticus elegantulus,Euoticus inustus)))),(((Tarsius bancanus,Tarsius syrichta),(Tarsius pumilus,Tarsius spectrum)),((((((((((Callithrix argentata,Callithrix humeralifer),Callithrix jacchus),Cebuella pygmaea),(Leontopithecus chrysomelas,(Leontopithecus chrysopygus,Leontopithecus rosalia))),(((((Saguinus fuscicollis,Saguinus tripartitus),Saguinus nigricollis),Saguinus inustus),(((Saguinus bicolor,Saguinus midas),(Saguinus oedipus,Saguinus leucopus)),((Saguinus mystax,Saguinus labiatus),Saguinus imperator)))),Callimico goeldii),((Saimiri oerstedii,Saimiri boliviensis,Saimiri sciureus,Saimiri ustus,Saimiri vanzolinii),((((Cebus capucinus,Cebus albifrons),Cebus olivaceus),Cebus apella))),(Aotus azarae,Aotus trivirgatus)),(((Pithecia aequatorialis,Pithecia irrorata,Pithecia monachus,Pithecia pithecia,Pithecia albicans),((Cacajao calvus,Cacajao rubicundus),Cacajao melanocephalus,Chiropotes albinasus,Chiropotes satanas)),(((Lagothrix lagothricha,Lagothrix flavicauda,((((Ateles geoffroyi,Ateles fusciceps),Ateles belzebuth),Ateles paniscus)),Brachyteles arachnoides),(Alouatta villosa,Alouatta seniculus,Alouatta palliata,Alouatta fusca,Alouatta caraya,Alouatta belzebul)))),((((Callicebus olallae,Callicebus donacophilus),Callicebus oenanthe),((((((Callicebus moloch,Callicebus cinerascens),Callicebus personatus),Callicebus hoffmannsi),Callicebus brunneus),(Callicebus dubius,Callicebus calligatus,Callicebus cupreus)),Callicebus torquatus,Callicebus modestus)),((((((((Macaca maurus,Macaca tonkeana),(Macaca ochreata,Macaca nigra)),((Macaca mulatta,Macaca fuscata,Macaca cyclopis),Macaca fascicularis)),((Macaca sinica,Macaca radiata),Macaca arctoides,(Macaca assamensis,Macaca thibetana)),Macaca nemestrina,Macaca silenus),Macaca sylvanus),((Theropithecus gelada,(Papio hamadryas,(Papio ursinus,Papio cynocephalus,(Papio papio,Papio anubis)))),((Cercocebus aterrimus,Cercocebus albigena),((Cercocebus torquatus,Cercocebus galeritus),(Mandrillus sphinx,Mandrillus leucophaeus))))),(((((((Cercopithecus cephus,Cercopithecus ascanius,Cercopithecus erythrotis),(Cercopithecus erythrogaster,Cercopithecus petaurista)),(Cercopithecus nictitans,Cercopithecus mitis)),((((Cercopithecus mona,Cercopithecus campbelli),((Cercopithecus wolfi,Cercopithecus denti),Cercopithecus pogonias)),Cercopithecus neglectus),Cercopithecus hamlyni)),((Cercopithecus salongo,Cercopithecus dryas),Cercopithecus diana)),(Erythrocebus patas,(Cercopithecus aethiops,(Cercopithecus solatus,(Cercopithecus lhoesti,Cercopithecus preussi)))),Miopithecus talapoin),Allenopithecus nigroviridis)),((((((Presbytis comata,Presbytis melalophos,Presbytis rubicunda,Presbytis frontata),Presbytis cristata,Presbytis geei,(Presbytis vetulus,Presbytis johnii),Presbytis obscura,Presbytis phayrei,Presbytis pileata,Presbytis potenziani,Presbytis francoisi,Presbytis aurata),Presbytis entellus),(Nasalis larvatus,Simias concolor)),(((Pygathrix roxellana,Pygathrix brelichi),Pygathrix avunculus),Pygathrix nemaeus)),((((Colobus guereza,Colobus polykomos),Colobus angolensis),Colobus satanas),(Procolobus verus,(Colobus kirkii,Colobus badius))))),((((((((Hylobates lar,Hylobates agilis),Hylobates muelleri),Hylobates moloch),Hylobates pileatus),Hylobates klossii),Hylobates hoolock),Hylobates concolor,Hylobates syndactylus),(Pongo pygmaeus,(Gorilla gorilla,((Pan troglodytes,Pan paniscus),Homo sapiens)))))))));

Part 2
Comparative life history and biology

Editor's introduction

In 1959, Le Gros Clark characterised the Order Primates by their general-ised limb structure, the retention of give digits, and a number of derived morphological and physiological traits. In particular, these traits were the elaboration of the visual system and reduction of olfaction, the expansion of the brain, especially the cerebral cortex, invasive placentation, reduced litter size and the prolongation of postnatal life. Whereas a number of these traits is also shared with other living non-primate taxa, the suite of traits associated with reproductive rates and cognitive capacities among pri-mates continues to intrigue.

Possibly the greatest expansion in research interest over the 20 years since Clutton-Brock and Harvey attempted their comparative study of primate socioecology in 1977 has been in the areas of life histories and reproduction. This section presents some recent work. But, there is a more fundamental question of interest: do these specialisations in brains, life histories and reproductive biology arise from ecological adaptations? Do they enable ecological adaptations? And do they constrain primate social systems? By exploring such questions, life history studies can provide a background to the comparative socioecological work that follows in Part 3. Thus, in order to understand the function of social systems and their evolutionary patterns, we also need a perspective on the conditions (physiological and morphological) and constraints on the organisms.

As Blurton Jones et al. (Chapter 6) note, primate life histories were first compared by Adolph Schultz in 1934. Even with the scant data then available on gestation length, duration of lactation, age at first reproduc-tion and life span, his proposed pattern illustrated the primate phenomenon of a 'time strategy' rather than an 'energy strategy'. These observations have been replicated now for 60 years, controlling for body size evolution, for phylogeny and ecology, but interest in causality and consequence in life history evolution remains a major focus of research.

What are we missing in order to understand primate life histories? Probably the most significant gap is in understanding mortality patterns. If mortality determines the pace of a life history, then research should focus on when and why primates die. But we are continually faced with the problem that most of such rates are population specific, and our method-ologies for comparative population analysis within species are, as yet, problematic (but see Mace and Holden, Chapter 15, for humans).

In this part's chapters, patterns of primate reproduction and growth and their relations to ecology and energetics are explored. Primate brain evol-ution is placed into its socioecological context, and the distinctive human life history is reviewed. While by no means a complete portrait of current life history research, background life history adaptations are introduced. Two chapters on sex in primates – one relating evolved characteristics of reproductive physiology to social systems (Chapter 8) and another on intrasexual competition and its consequences (Chapter 9) – follow. Re-search on a variety of birds and mammals suggests that ecology can structure reproductive opportunities, either through proximate constraints on female energetics or through facultative variation in the extent and nature of intrasexual association, and hence competition and investment. Because sexual reproduction requires the (temporary) co-operation of at least two individuals, sex, mating systems and sociality are theoretically linked. Thus, ecological influences on reproductive biology and behaviour are of considerable theoretical importance.

But, it should be noted at the outset, that mating systems are not, at least for primates, robust descriptions of social systems. Rather, the mating system appears to operate within the constraints of sociality, and in par-ticular can be more influenced by interactions with non-social group members than by within-group dynamics (van Schaik *et al.*, Chapter 8). If any conclusion can be reached here, it is that sex does not structure societies; societies structure associations, which in turn have consequences for competition, patterns of investment in mates and offspring,and for reproductive physiology. The complexity of the 'primate pattern' of sex and sociality becomes increasingly apparent, and explanations of causality are generally lacking. These are questions that need to be tackled in the future.

4 Socioecology and the evolution of primate reproductive rates

CAROLINE ROSS AND KATE E. JONES

Introduction

A thought experiment: take two animals, one male and one female. Provide their offspring with unlimited food and allow them to breed. If animals die only of old age, how many animals will you have after 2 years, 5 years, 10 years, 20 years or 50 years? Figure 4.1 shows the results of this experiment using the intrinsic rate of population increase (r_m, see Table 4.1, p. 80, for a definition) for six primate species. It is clear from this that some animals can increase their population size more rapidly than others. The bushbaby (*Galago moholi*) population contains over 2000 females after only 12 years and 3600 million (3.6×10^{14}) individuals after 50 years, whereas the gorilla population contains only 29 females after 50 years.

The intrinsic rate of population increase depends on three variables: age at first reproduction (AR), birth rate (b) and age at last reproduction (L). Birth rate has an important influence on the rate of population increase, which is illustrated by the examples given of *Macaca silenus* and *Macaca sylvanus* the females of which start breeding at about five years of age. However, in *Macaca sylvanus*, birth rate is higher, leading to a higher intrinsic rate of population increase. Animals that start breeding at an early age also show a higher intrinsic rate of population increase than do those that start breeding later. This has a knock-on effect over the generations, as a female that breeds early produces female infants that also breed early, thus increasing the total number of females that are producing infants. For example, both *Macaca sylvanus* and *Erythrocebus patas* breed annually, but *E. patas* reaches maturity nearly two years before *M. sylvanus*. Over time, this difference in AR has a large effect on the population size, with the *E. patas* population taking 33 years to grow to 2000 females and the *M. sylvanus* population taking 41 years (Fig. 4.1). In contrast, variation in the age at last reproduction makes little difference to the intrinsic rate of population increase. Titus (*Callicebus moloch*) and patas monkeys (*E. patas*) both start breeding at three years of age and produce an infant annually, but the patas monkeys live far longer than the titis (21 as

73

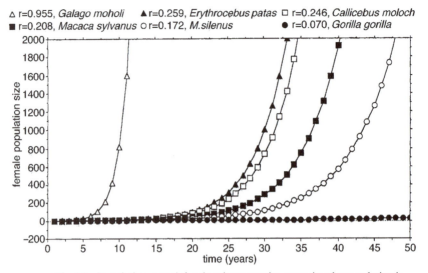

Fig. 4.1   Population growth for six primate species, assuming the population is started with a single adult female and the sex ratio is 1:1. The rate of population growth is calculated by assuming that there is no mortality until the age at last reproduction is reached. Curves were calculated using the following life history variables: *Callicebus moloch* ($\alpha = 3.0$ years, $b = 1.0$ offspring/year, $L = 12.0$ years $r_m = 0.27$), *Erythrocebus patas* (3.0, 1.0, 21.6, 0.26), *Galago moholi* (1.0, 3.2, 16.5, 0.96), *Gorilla gorilla* (10.0, 0.26, 50.0, 0.07), *Macaca silenus* (4.9, 0.72, 38.0, 0.17), *Macaca sylvanus* (4.8, 1.0, 22.0, 0.21).

compared to 12 years maximum recorded longevity). Despite this, Figure 4.1 shows that the population growth of the patas is only slightly more rapid than that of the titis, with the latter taking only two years more to reach a female population of 2000 (35 years compared to 33 years).

Life history theory seeks to explain this variation, asking not only, 'Why do gorillas breed slowly and bushbabies breed rapidly?' but also, 'Why do the reproductive rates of primates differ from those of other animals?'. This chapter examines these questions, both by reviewing previous studies and by presenting the results of some new analyses. It starts with a brief overview of life history theory and then explores the primate data in the light of these ideas.

Reproductive rates and life history theory

Design constraints, body size and phylogeny

Variation in reproductive rates is correlated with body size. Generally speaking, large animals take longer to reach maturity and breed more slowly than do small animals. There are usually allometric relationships between life history variables and body weight (e.g. Peters, 1983; Calder, 1984), many of which have been described in primates (e.g. Kirkwood, 1985; Harvey, Martin and Clutton-Brock, 1987; Ross, 1988; Lee, Majluf and Gordon, 1991). For many parameters, the association between body weight and a life history parameter may be very strong; e.g. body weight variation in primates explains about 80% of the variation in female age at first reproduction. Explanations for these high correlations between life history parameters and body size fall into two categories: those that link body size with life history parameters because of reasons of 'design constraint', and those that suggest an 'adaptive link'.

The 'design constraint' model suggests that body size acts as a constraint on life history evolution, so that selection for body size 'carries along' life history characters with it. Life-history parameters are therefore constrained within certain limits by the size of the organism (e.g. Western, 1979; Western and Ssemakula, 1982). Although design constraints will be imposed by an animal's size, other aspects of its physiology and anatomy will also constrain the evolution of reproductive rates within certain limits. For example, eutherian mammalian reproduction does not allow the evolution of egg laying and birds are unlikely to evolve asexual reproduction. Constraints such as these are likely to be shared by closely related organisms that have a similar anatomy and physiology and these related species are therefore likely to have similar life histories.

Alternatively, the correlation between body size and life history parameters may be adaptive as the body size of an organism may influence the way in which it experiences its environment (Pianka, 1970). For example, a large-bodied animal will not be as threatened by environmental fluctuations as will a smaller animal, all other things being equal. In this way, body size itself can influence life history characters.

Despite the strong correlations between body size and reproductive rates, neither the design constraint model nor an adaptive link between body size and reproductive rates can fully explain life history variation. As noted by Pagel and Harvey (1993), this is for two basic reasons. Firstly, it does not explain why body size should vary: if being big must lead to slow reproduction, why should animals waste time and resources growing large? Secondly, it does not explain the variation in life histories that is not correlated with body weight. This second point is important as several studies have shown that life history traits co-vary predictably, so that animals fall along a continuum from those with high mortality, short lifespans, high birth rates and fast development to those with low mortality, long lifespans, low birth rates and slow development, even if the influence of body size is removed (e.g. Stearns, 1983; Promislow and Harvey, 1990). Numerous models have been proposed that try to explain these observations. These models assume that natural selection will act to optimise life histories, so as to maximise the lifetime reproductive success of individuals (i.e. they assume that life histories are not only varying due to 'design constraints') and are similar in that they recognise the importance of trade-offs in life history evolution.

Trade-offs and primate reproduction

If breeding had no costs to the animal, it would be selected to breed as early as possible, as rapidly as possible and for as long as possible so as to maximise its contribution to future generations. Both common sense and a large body of evidence (Boyce, 1988; Stearns, 1992) tell us that producing offspring is not free of cost, and resources devoted to an infant are resources that cannot be used elsewhere. Selection for large infant size will result in fewer infants; the limited resources animals have for reproduction means that they must trade numbers of infants against infant size. Similarly, trade-offs may be made between allocating resources to the production of young or into parental survival. If investing less than the maximum possible in one reproductive attempt increases the survival of the parent, a lowering of reproductive effort per reproductive attempt may result in greater numbers of offspring being produced in an individual's lifetime.

As primates produce only one or two young per litter, there is little scope for trading infant numbers against infant quality. The evolutionary 'decision' to trade the disadvantages of producing few young against the advantages of producing large, 'high-quality' young was probably made early in the line of primate ancestry. Primates do, however, produce twins, some species typically do so and even those that typically produce singletons may produce twins occasionally. Within species, there is a considerable amount of evidence to suggest that an increase in litter size results in smaller young being produced (e.g. humans: Wilson, 1979). There is also evidence of trade-offs when the comparative evidence is considered, with a multiple regression showing that both body weight (W) and litter size contribute to the variation in primate neonatal weight (N) ($n = 82$ species, 80 contrasts, $r^2 = 0.58$; W: coefficient $= 0.613$, $p < 0.0001$; N: coefficient $= -0.266$, $p = 0.0005$, see below for methods).

Several intraspecific studies of primates have tried to find evidence for trade-offs between other reproductive parameters, particularly maternal survival versus fecundity and early reproduction versus fecundity. The results of these studies are variable, with only limited evidence that such trade-offs do occur. Some evidence that fecundity may trade-off against maternal survival comes from Altmann, Hausfater and Altmann's (1988) study of savannah baboons (*Papio cynocephalus*) in which females caring for infants have higher mortality rates than those without infants. A cost of early reproduction was shown in rhesus macaques (*Macaca mulatta*) on Cayo Santiago, where young primaparous females (< 4 years) suffer greater mortality of their infants than older primaparous females (> 4 years) (Sade, 1990). Younger females do tend to have a longer interbirth interval to their second infant than do older females, suggesting that there is a trade-off between early reproduction and fecundity (although the result is not statistically significant). However, Sade (1990) found no difference in maternal survival between young and old primaparous rhesus macaques.

One reason for the differences found between studies is that simple trade-offs between two variables are often likely to be complicated by variation in a third variable (e.g. animals' rank or age), so that trade-offs may not always be observed easily. Despite this, the available evidence does seem to show that primates have the flexibility to trade-off reproductive variables against one another. This is important as it indicates that intraspecific variation in life histories is likely when individuals of a species are experiencing different environments that select for different balances of possible trade-offs. What then, are the selective pressures whose variation brings about diversity in life history strategies?

r and K selection theory and age-specific mortality

The theory of *r* and *K* selection was originally proposed by MacArthur and Wilson (1967) to explain the processes of island colonisation and was extended to include other situations in which mortality is primarily density independent (Pianka, 1970). Where a population is expanding into unoccupied habitats, or into an empty niche caused by a 'population crash', an individual with a high intrinsic rate of population increase will rapidly fill the 'space' with its descendants, faster than slower breeding survivors can. Hence, such conditions will select for a high intrinsic rate of population increase and '*r* selection' occurs. Alternatively, when a population is at the carrying capacity (*K*) of its habitat, MacArthur and Wilson reasoned that there would be a high incidence of density dependent mortality, i.e. mortality due to competition for limiting resources rather than stochastic events. Hence, selection would favour individuals that use the available resources most efficiently, thus maximising the carrying capacity. A *K*-selected population will be expected to have a higher competitive ability and, thus, a lower reproductive effort than will an *r*-selected species.

K-selected populations will also be expected to produce a smaller number of young per litter so that their parental investment per offspring can be maximised, conferring increased competitive ability on the offspring. A *K*-selected species will therefore be characterised by having a later age at first reproduction, longer development time, lower birth rate and fewer, larger young than a more *r*-selected species.

r and *K* selection theory was widely used in early studies of life history variation to explain the observed fast–slow continuum, with slow-growing, slow-reproducing and long-lived animals being assumed to be *K* selected and fast-growing, fast-reproducing and short-lived animals *r* selected. However, the theory has now been largely replaced by more recent models that place emphasis on the effects that mortality has on individuals of different ages, rather than on the causes of mortality (i.e. density dependent or density independent). This change has mainly arisen because a large number of studies has shown that the *r/K* model does not explain the pattern of life histories seen in many taxa and that artificial selection experiments do not produce the results predicted by *r/K* theory (Stearns, 1977, 1992). This lack of correspondence between prediction and observation suggest that there are deficiencies in the *r/K* model.

Another major problem is that the terms 'density-dependent' and 'density-independent' mortality try to describe the reasons *why* mortality occurs but they do not describe *who* mortality affects or when it acts. In many natural populations, mortality is not evenly distributed amongst animals of all ages. In mammals, the usual pattern is that the very young and very old are more likely to die than are adults in their prime. Models that include information on the way in which mortality patterns are distributed have been shown to explain life history variation far more comprehensively than has the *r/K* model. Despite the replacement of *r/K* theory by other models, there is still general recognition that selection may be expected to act differently when mortality (particularly juvenile mortality) is primarily density dependent rather than density independent (Boyce, 1988; Stearns, 1992). In particular, the prediction that density-dependent selection is likely to lead to a decreased reproductive effort and an increased need for competitive infants may still be incorporated into other models (Purvis and Harvey, 1995). One model that does not incorporate density-independent mortality has, nevertheless, received some support from empirical comparative evidence.

Charnov's model

Charnov (1993) presents a model of mammalian life history evolution. The model explains life history allometry by assuming that the age at maturity is determined by adult mortality rates, which are in turn determined by the environment. When mortality rates are high, animals are expected to mature rapidly, so as to minimise their chances of dying, and thus maximise their lifetime reproductive success. Charnov assumes that growth continues up to the age of maturity and then stops, with resources then being diverted to reproduction. Thus, the model predicts that animals that mature late will have a larger body size than those that mature early, and that this will influence fecundity. In a stable population, fecundity is balanced by mortality and hence fecundity will be expected to be positively correlated with juvenile mortality rates.

Charnov's model makes several predictions about the type of relationships one would expect to find between body weight, mortality rates and reproductive rates. In some cases, studies of mammals appear to support this model (Charnov, 1993; Purvis and Harvey, 1995), although Kozlowski and Weiner (1997) suggest that the model may only be appropriate in certain limited situations. One problem with Charnov's model is that it predicts that relative size at independence (the ratio of weaning weight to adult weight, δ) is a constant, something that is not always supported by the data (Purvis and Harvey, 1995). However, Charnov's predicted value of $\delta = 0.33$ is found for primates (Lee *et al.*, 1991; Charnov, 1993) and his model may therefore be appropriate for use in this group.

Charnov and Berrigan (1993) and Charnov (1993) pointed out that primates are unusual amongst mammals in having an allometric exponent for α (α might be female juvenile period length or age at first reproduction in Charnov (1993); it is unclear which is used in the analysis) against body weight that is greater than 0.25 (calculated from species values). Charnov (1993) has linked this difference in the scaling of α to primates having an unusual form of the basic growth function that relates body weight (W) to growth rate (DW/dT): $DW/dT = AW^{0.75}$. The value of the constant A is approximately 1 for most mammals, but for primates Charnov calculates the value as 0.42. This suggests that primates differ from other mammals in having very low production rates, so that they take longer to grow both themselves and their infants than do other mammals, something that may also account for both the late maturation and the slow breeding rates of primates. Some of Charnov's predictions are tested below for a larger primate data set than originally used, and some possible causes and consequences of the low growth rate of primates are discussed.

Table 4.1. *Description of variables used in analyses*

Parameter	Symbol	Definition
Adult female body weight	W	Mean adult female body weight (g). This is the body weight of the population concerned when the mortality data are being discussed (Table 4.3). In other cases it is the species mean (Table 4.2).
Age at first reproduction	AR	Mean female age at first reproduction (years). This is for the population concerned when the mortality data are being discussed (Table 4.3). In other cases it is the species mean (Table 4.2).
Juvenile period	α	Mean female age at first reproduction (years) minus weaning age (years). Charnov (1993) refers to this variable as 'age at maturity' and may use it interchangeably with AR when discussing primate data. The value of α is for the population concerned when the mortality data are being discussed (Table 4.3). In other cases it is the species mean (Table 4.2).
Female birth rate	b	Mean female birth rate (offspring/year). It is calculated assuming 0.5 primary sex ratio. This is for the population concerned when the mortality data are being discussed (Table 4.3). In other cases it is the species mean (Table 4.2).
Interbirth interval	IBI	Mean time (years) between births when first infant survives to weaning age.
Maximum longevity	L	Maximum recorded longevity (years) for the species.
Intrinsic rate of natural increase	r_m	A species maximum rate of population growth that is possible when resources are not limiting. It is calculated by iteratively solving Coles (1954) equation (where $r : r_m$): $$1 = \frac{e^{-r}}{2} + \frac{be^{-r\alpha}}{2} - \frac{be^{-r(L+1)}}{2}$$
Average instantaneous adult mortality rate	M	Mean mortality rate (per year) after age at first reproduction, estimated from life tables following methods of Purvis and Harvey (1995).
Crude adult mortality rate	$M2$	Crude mortality rate (per year) after age at first reproduction,(number of adults alive at time 1 − number of adults alive at time 2)/time interval.
Pre-reproductive mortality rate	Z	Crude mortality rate (per year) from birth to age at first reproduction (or as near to AR as possible), estimated as M2.
Average infant mortality rate	IM	Mean mortality rate (per year) from birth to end of infancy (weaning age). Estimated as M2.
Survival to reproductive age	$S(AR)$	Proportion of live births surviving to AR.
Young per litter	Y	Mean number of infants born per litter.
Brain weight	Brn	Mean adult brain weight (g).
Percentage folivory	% Fol	Mean amount of leaves included in the diet of wild animals (may include averages from more than one study site).

Primate reproductive rates and life history theory

The remainder of this chapter seeks to use primate life history data to answer some questions raised by this very brief review of life history theory:

1. Do age at first reproduction and birth rate vary with mortality rates as predicted?
2. Do environmental variables, which might be expected to correlate with mortality patterns, predict reproductive rates?
3. Do life history variables vary with body size in the way predicted by Charnov's (1993) model?
4. Why has a late age at maturity evolved in primates?
5. How can primates use behavioural strategies to increase their reproductive rates?

Data used

The variables discussed in this chapter are listed in Table 4.1.

Reproductive rate data

Data on reproductive rates are summarised in Table 4.2. These data were collected from a variety of literature sources and include data from both wild and captive populations. Analyses were carried out using both the complete data set and data from wild animals only. Because the results from both sets of analyses were not substantially different, only those from the complete data set are given here. More detailed discussion of these data can be found in papers by Ross (1988, 1992a).

Mortality rate data

Table 4.3 summarises data on mortality rates of 32 wild primate populations, representing 25 species. Data are given for three parameters used in many life history models: pre-reproductive mortality rates (Z), adult mortality rates (M), and survival to reproductive age, S(AR). Data on infant mortality rates (IM) have also been used as these are more widely available than Z or M and might be expected to vary in a similar way to Z. These data rarely meet the criteria necessary to test life history models (Caughley, 1977); many are not from stable populations, estimates are often made from observations of a few individuals and are frequently subject to

Table 4.2. *Reproductive rate and dietary data*

Species	W	IBI	Y	Birth rate	AR	L	r_m	Wean age	% Fol
Alouatta caraya	4882	—	1	—	3.71	—	—	—	—
Alouatta palliata	5824	1.88	1.1	0.59	3.58	20	0.178	1.73	55.3
Alouatta seniculus	5807	1.39	1	.72	4.58	25	.178	1.02	53
Aotus trivirgatus	724	0.75	1	1.33	2.38	20	0.346	0.21	25
Arctocebus calabarensis	253.7	0.5	1	2	1.12	13	0.654	—	0
Ateles fusciceps	9163	2.25	1	0.44	4.86	24	0.133	1.33	—
Ateles geoffroyi	7669	2.66	1	0.38	5.62	27.3	0.113	2.25	20
Ateles paniscus	8554	4	1	0.25	5	33	0.09	—	7.9
Avahi laniger	875	1	1	1	—	—	—	—	—
Brachyteles arachnoides	8070	2.82	1	0.35	7.5	—	0.12	1.75	77
Callicebus moloch	1004	1	1	1	3	12	0.27	0.16	31.3
Callimico goeldii	582	0.47	1	2.14	1.32	17.9	0.629	—	—
Callithrix argentata	353	0.62	2	3.22	1.67	—	0.697	—	—
Callithrix jacchus	287	0.52	2.1	4.06	1.5	11.7	0.846	0.25	0
Cebuella pygmaea	79	0.5	2.1	4.2	1.88	11.67	0.74	0.25	—
Cebus albifrons	2067	1.5	1	0.67	4.02	44	0.178	0.75	3.6
Cebus apella	2201	1.79	1	0.56	5.5	44	0.14	—	1.6
Cebus capucinus	2578	1.6	1	0.63	4	46.9	0.172	—	15
Cebus olivaceus	2500	2.17	1	0.46	6	—	0.131	2.00	—
Cercocebus albigena	6209	2.12	1	0.47	4.08	32.7	0.143	1.00	10
Cercocebus atys	6225	1.08	1	0.92	3.14	18	0.245	—	—
Cercocebus galeritus	5473	—	1	—	6.5	19	—	—	13.5
Cercocebus torquatus	7420	1.08	1	0.92	4.67	27	0.201	—	—
Cercopithecus aethiops	3469	1.33	1	0.75	5	24	0.175	1.00	12
Cercopithecus ascanius	2943	4.33	1	0.23	5	22.5	0.091	—	13
Cercopithecus cephus	2805	—	1	—	5	22	—	1.00	6.1
Cercopithecus diana	4533	1	1	1	5.42	34.8	0.193	1.00	6.2
Cercopithecus lhoesti	4700	1.33	1	0.75	—	—	—	—	—

Cercopithecus mitis	4280	3.92	1	0.26	5.92	20	0.095	1.91	18.3
Cercopithecus neglectus	4081	1.62	1	0.62	4.67	22	0.163	1.00	9.4
Cercopithecus pogonias	3021	—	1	—	5	20	—	—	1.2
Cheirogaleus medius	173	1	2	2	1.19	9	0.636	—	—
Colobus badius	7421	2.12	1	0.47	4.08	—	0.145	2.16	78.3
Colobus guereza	8102	1	1	1	4.75	22.25	0.209	1.07	82
Colobus polykomos	7662	1.04	1	0.96	8.5	30.5	0.15	—	—
Daubentonia madagascarensis	2800	—	1	—	—	23.25	—	—	0
Erythrocebus patas	6317	1	1	1	3	21.58	0.261	1.00	0
Galago alleni	262	—	1.3	—	1.04	8	—	0.58	—
Galago crassicaudatus	1120.3	1	1.6	1.6	2.21	15	0.401	0.37	0
Galago demidovii	62.8	1	1.2	1.2	0.97	13	0.476	—	0
Galago garnetti	738.5	0.57	1.76	1.58	17	—	0.505	0	—
Galago moholi	179	0.5	1.6	3.2	1	16.5	0.955	0.27	0
Galago senegalensis	195	0.51	1	1.97	2	14	0.503	—	—
Galago zanzibaricus	132.3	0.5	1.2	2.4	1	—	0.788	—	0
Gorilla gorilla	82475	3.83	1	0.26	10.04	50	0.07	2.75	86
Hapalemur griseus	916	3.5	1.5	—	2.38	12	—	—	—
Homo sapiens	55000	2.69	1	0.29	14	100	0.062	2.00	31
Hylobates lar	5464	3	1	0.37	9.31	31.5	0.092	—	43.8
Hylobates syndactylus	10568	2.53	1	0.33	9	35	0.087	—	57
Indri indri	6250	1.5	1	0.39	—	—	—	1.00	—
Lagothrix lagotricha	5585	1.5	1	0.67	5	25.92	0.164	0.86	—
Lemur catta	2290	1.5	1.2	0.8	2.01	27.1	0.267	0.29	43.6
Lemur fulvus	2428	1.5	1	0.67	2.16	30.08	0.228	0.37	49
Lemur macaco	2428	1	1	1	2.18	27.08	0.3	—	—
Lemur mongoz	1890	—	1	—	2.52	25.33	—	—	—
Leontopithecus rosalia	559	0.5	2	4	2.38	14.17	0.617	0.25	1.6
Lepilemur mustelinus	602	—	1	—	1.88	—	—	0.21	91
Loris tardigradus	255.5	0.5	1	2	1.5	12	0.563	0.46	0
Macaca arctoides	8523	1.48	1	0.68	3.84	30	0.184	—	—
Macaca fascicularis	3574	1.07	1	0.94	3.86	37.08	0.222	1.15	16
Macaca fuscata	9100	1.5	1	0.67	5.54	33	0.155	1.00	41

Table 4.2. (*continued*)

Species	W	IBI	Y	Birth rate	AR	L	r_m	Wean age	% Fol
Macaca mulatta	5445	1	1	1	4.5	27	0.213	1.00	—
Macaca nemestrina	5571	1.11	1	0.9	3.92	26.29	0.216	1.00	19
Macaca niger	4600	1.48	1	0.68	5.44	18	0.165	—	—
Macaca radiata	3700	1	1	1	4	30	0.226	1.00	—
Macaca silenus	5000	1.38	1	0.72	4.9	38	0.171	1.00	—
Macaca sinica	3590	1.5	1	0.67	5	30	0.163	85	0
Macaca sylvanus	8283	1	1	1	4.8	22	0.208	—	40
Mandrillus leucophaeus	8450	1.23	1	0.81	5	28.6	0.181	—	—
Mandrillus sphinx	11350	1.46	1	0.69	4	46.33	0.182	0.96	—
Microcebus coquereli	302	1	1.5	1.5	1	15.25	0.559	0.11	—
Microcebus murinus	72	1	1.9	1.9	1	15.42	0.667	0.11	2
Miopithecus talapoin	1120	1	1	1	4.38	27.67	0.216	0.49	95
Nasalis larvatus	9593	—	1	—	4.5	13.5	—	—	0
Nycticebus coucang	630	1	1	1	—	16	—	0.49	—
Pan paniscus	33 200	4.82	1	0.21	—	26.83	—	—	—
Pan troglodytes	40 300	5.5	1	0.18	13	53	0.053	4.00	20.2
Papio cynocephalus	11 532	1.75	1	0.57	5.5	40	0.142	1.00	7.8
Papio hamadryas	10 404	2	1	0.5	6.1	35.6	0.126	1.00	7
Papio papio	16 166	1.16	1	0.86	—	40	—	—	—
Papio ursinus	14 773	—	1	—	3.67	45	—	—	12.2
Perodicticus potto	935	0.97	1.1	1.14	2.03	22.33	0.338	0.41	0
Pithecia pithecia	1604	1.58	1	0.63	2.08	13.75	0.231	—	0
Pongo pygmaeus	37 078	6.5	1	0.15	9.68	57.33	0.052	1.12	22
Presbytis entellus	10 280	1.68	1	0.59	3.42	25	0.179	1.25	48
Presbytis senex	5797	1.67	1	0.6	—	—	—	0.62	60
Propithecus diadema	3630	2	1	0.48	4	—	—	—	41
Propithecus verreauxi	3183	1	1	1	3.5	18.17	0.244	0.49	41
Pygathrix nemaeus	8180	1.36	1	0.74	—	10.25	—	—	—

Saguinus fuscicollis	350	1	2	2	2.33	—	0.443	0.25	3.2
Saguinus geoffroyi	483	0.66	2	3.04	—	—	—	—	4
Saguinus labiatus	520	0.82	2	2.44	—	—	—	0.19	—
Saguinus midas	558	0.55	2	3.64	2	13.25	0.662	—	0
Saguinus oedipus	425	0.58	1.9	3.26	1.89	13.5	0.649	—	—
Saimiri sciureus	699	1.17	1	0.86	2.5	21	0.257	0.14	0
Tarsius bancanus	109	0.65	1	1.54	2.52	12	0.361	0.22	0
Tarsius spectrum	220	0.42	1	2.4	1.42	12	0.649	0.19	0
Tarsius syrichta	120	—	1	—	—	13.5	—	0.23	—
Theropithecus gelada	11 427	2.14	1	0.47	4	19.25	0.15	1.23	91.8
Varecia variegata	2700	1	1.8	1.8	1.95	13	0.46	0.25	—

Abbreviations as given in Table 6.1. Data on brain size can be found in Barton (Chapter 7); group sizes were taken from Smuts *et al.* (1987).

Table 4.3. *Mortality rate data*

Species/site/references[1]	IM (age I)	M/M2	Z (age J)	S(AR) (age J)	Birth rate	AR	α	W	Male weight
Alouatta palliata Canas, Costa Rica Glander (1980)	0.38 (1.0)	0.03	—	—	0.53	3.73	2.7	5.00	6.00
Alouatta palliata[2] Barro Colorado Island, Panama Froehlich et al. (1981)	0.40 (1.0)	0.15	0.16	0.35	0.37	4.00	3.00	6.60	7.08
Alouatta seniculus Crockett & Pope (1993); Robinson (1988)	0.21 (1.0)	0.05	0.16 (2.0)	0.35	0.70	5.17	4.17	4.50	6.25
Ateles paniscus chamek Manu, Peru McFarland Symington (1988)	0.33 (1.0)	0.03	—	—	0.35	—	—	8.44	9.11
Brachyteles arachnoides Minas gerais, Brazil Strier (1991, 1993); Strier et al. (1993)	0.05 (1.5)	—	—	—	0.35	7.50	5.75	8.07	9.61
Cebuella pygmaea Maruti River Basin, Peru Soini (1982)	0.67 (0.5)	—	—	—	—	—	—	0.12	0.11
Cebus olivaceus Hato Masaguaral, Venezuela Robinson (1988); O'Brien & Robinson (1993)	0.18 (1.0)	0.05	0.06	0.62	0.46	6.00	4.00	2.52	3.29
Cercopithecus aethiops Amboseli, Kenya Cheney et al. (1981, 1988)	0.58 (1.0)	0.16	0.16 (4.0)	0.27 (6.0)	0.70	5.06	4.06	2.98	4.26
Cercopithecus mitis Cape Vidal, South Africa Macleod, personal communication	0.23 (1.0)	0.04	—	—	0.50	—	—	3.50	6.80
Colobus guereza Bole Valley, Ethiopia	0.38 (0.3)	0.04	—	—	—	—	—	9.20	13.5

Dunbar & Dunbar (1974) *Erythrocebus patas* Laikipia, Kenya Chism et al. (1984)	0.78 (1.0)	—	—	—	1.00	3.00	2.41	6.50	12.4
Gorilla gorilla beringei Karisoke, Rwanda Watts (1989, 1991); Harcourt et al. (1981); Watts & Pusey (1993)	0.34 (3.0)	0.04	0.05 (8.0)	0.59 (8.0)	0.26	10.00	6.50	97.50	162.50
Lemur catta Beza Mahafaly, Madagascar Sussman (1991); Sauther (1991); Sauther & Sussman (1993)	0.52 (1.0)	0.21	0.24	0.29	1.00	3.00	2.75	2.21	2.21
Lemur catta Berenty, Madagascar Rasamimanana & Rafidinarivo (1993)	—	—	0.64 (2.0)	—	0.69	—	—	2.21	2.21
Macaca fascicularis Keambe, Sumatra van Schaik & van Noordwijk (1985); van Noordwijk et al. (1993)	0.22 (1.0)	0.04	0.09 (4.0)	0.66 (4.0)	—	—	—	3.59	5.36
Macaca fuscata[2] Mt Ryozen, Japan Sugiyama & Ohsawa (1982)	0.37 (2.0)	0.03	0.06 (7.0)	0.55 (7.0)	0.34	6.74	5.74	8.92	—
Macaca fuscata Mt Kawarade, Japan Ikeda (1982)	0.29 (1.0)	—	0.25 (4.0)	0.02	0.52	5.00	4.00	7.75	10.29
Macaca fuscata[2] Takagoyama, Japan Hiraiwa (1981)	0.00 (1.0)	—	—	—	0.54	5.44	4.44	7.75	10.29
Macaca mulatta Dunga Gali, Pakistan Melnick (1981, in Richard, 1985)	0.46 (1.0)	—	—	—	0.38	—	—	8.80	11.00
Macaca sinica Pollonaruwa, Sri Lanka Dittus (1975)	0.53 (1.0)	0.06	0.17	0.15	0.69	5.50	4.50	3.20	5.68

Table 4.3. (*continued*)

Species/site/references[1]	IM (age I)	M/M2	Z (age J)	S(AR) (age J)	Birth rate	AR	α	W	Male weight
Macaca sylvanus Akfadou, Algeria Menard & Vallet (1993)	0.39 (1.0)	—	0.13 (3.5)	0.49 (5.0)[3]	0.63	5.30	4.80	11.00	16.00
Macaca sylvanus Tigounatine, Algeria Menard & Vallet (1993)	0.25 (1.0)	—	0.10 (3.5)	0.59 (5.0)[3]	0.56	5.50	4.80	11.00	16.00
Pan troglodytes schweinfurthii Gombe, Tanzania Courtenay & Santow (1989); Watts & Pusey (1993)	0.09 (5.0)	0.07	0.04 (14.0)	0.50	0.18	13.00	8.00	33.70	42.70
Papio cynocephalus[2] Mikumi, Tanzania Rhine (1997); Wasser & Wasser (1995)	0.21 (1.0)	—	0.07 (6.0)	0.58	0.56	6.37	4.95	12.30	21.80
Papio cynocephalus Amboseli, Kenya Altmann et al. (1977, 1985, 1988)	0.25 (1.4)	0.11	0.11 (6.0)	0.33 (6.0)	0.57	6.50	5.08	12.30	21.80
Papio cynocephalus anubis Gilgil, Kenya Smuts & Nicolson (1989)	0.22 (1.0)	0.05	—	—	0.90	—	—	12.71	—
Papio hamadryas Erer, Ethiopia Sigg et al. (1982)	0.09 (1.0)	—	0.06 (6.0)	0.64 (6.0)	0.50	6.10	4.60	11.4	21.00
Presbytis entellus Winkler et al. (1984); Hrdy (1977) (body weight data)	0.40 (1.0)	0.08	0.33 (2.0)	0.33 (2.0)	0.78	3.40	2.30	11.70	—
Propithecus diadema edwardsi Ranomafana, Madagascar Wright (1995)	0.48 (1.0)	0.12	0.19	0.25	0.48	4.00	2.50	6.26	5.94

Propithecus verreuxi verreuxi Berenty, Madagascar Richard (1985)	0.49 (1.0)	—	—	—	0.42	4.00	2.50	2.95	3.25
Propithecus verreuxi verreuxi Beza Mahafaly, Madagascar Richard et al. (1991)	0.39 (1.0)	0.19	0.07	0.60	0.75	5.50	4.00	2.95	3.25
Theropithecus gelada Sankaber, Ethiopia Dunbar (1980)	0.05 (1.0)	—	0.03	0.88	0.47	4.00	2.75	11.70	19.00

[1] References are for demographic and life history variables. Body weight data are taken from the same sources, from the source quoted, or from Smith and Jungers (1997). [2] Data not used in analyses. [3] Assuming 3% mortality per year from two to five years.
Symbols as in Table 4.1 except: age I = age at which infancy was considered to have ended (years); age J = age at which juvenile period is considered to be ended (years) if different from AR; birth rate = mean number of offspring born per year; male weight = mean adult male body weight.

error (e.g. because immigration and emigration confuse the overall picture or because ages of animals are not accurately known). For example, lifetable data that could be used to calculate M were available for only six species and none of these would have met the strict criteria suggested by Caughley (1977) and in other species $M2$ was used in place of M.

Despite these problems, the available data have been used to investigate the relationship between mortality patterns and life histories in primates for two reasons. Firstly, biases in the data are unlikely to be systematic and the errors that are present are more likely to obscure relationships than they are to create false correlations. Secondly, restricting the sample to those populations for which rigorous criteria can be met would be to leave a sample that would be too small to carry out any analyses. Some data presented in the table are excluded from further analyses, either because they are from highly disturbed populations or because better data were available for other populations of that species (Table 4.3).

Another potential problem in using these data is that they do not distinguish between different causes of mortality. Several models of life history evolution (e.g. Charnov, 1993) assume that all mortality is due to extrinsic causes, something that is unlikely to be true in many cases. Although this problem needs to be borne in mind when interpreting the results, it is one that cannot be solved given the data available.

Table 4.3 also includes measures of female age at first reproduction, weaning age and birth rate for the same populations as are used for the analysis of the mortality data. Wherever possible, body weight data used are from the same populations; where these were not available, body weight data for a population that was as geographically close as possible, and of the same subspecies, were preferred over the species average given in Table 4.2.

Group size, diet and brain size

Data on group size and diet were taken from Smuts *et al.* (1987); brain size data are from Harvey *et al.* (1987) and Barton (Chapter 7).

Methods

Analyses were carried out using least squares regression and multiple regression. Least squares regression underestimates the value of the slope of the best-fit line when there is error in the measurement of the X variable (as there will be in this data set) but was preferred to major axis regression as the results could be directly compared to other studies (i.e. Charnov,

1993; Purvis and Harvey, 1995). All data were log transformed before analysis and comparative analyses were then carried out using two methods.

1. Species data. All data were analysed using traditional allometric methods to control for the confounding effects of body weight (Harvey and Mace, 1982). Residual values of the variables under analysis have been calculated using species values. Although there is now a great deal of evidence to suggest that the use of species data points may lead to bias in the results of comparative analyses (Harvey and Pagel, 1991) the results of these analyses have been included in order to allow comparison with other studies that also use species data (e.g. Hennemann, 1983; Wootton, 1987; Charnov, 1993).

2. Comparative Analysis by Independent Contrasts (CAIC). This was carried out to control for phylogenetic bias that results when species are treated as independent data points (Harvey and Pagel, 1991). Analyses used a comparative method based on Felsenstein's (1985) method of independent contrasts. This was carried out as detailed in Purvis and Rambaut (1995; see Chapter 3 for further discussion). The composite primate phylogeny, including branch lengths, produced by Purvis (1995) was used for all these analyses.

Contrasts were analysed using least squares regression and multiple regression through the origin. One potential problem in using this method on reproductive rate data is that a large amount of variance occurs at species level, at least some of which is due to random error. As there are more contrasts at the 'top levels' of the phylogeny, these contrasts can have undue influence on the results (Purvis and Harvey, 1995). Thus, all analyses were repeated after splitting the contrasts in half and using only the older contrasts (Purvis and Harvey, 1995). Because the data set varied from one analysis to another, the ages of the excluded contrasts also varied. However, exclusion of 50% of the contrasts never removed contrasts older than ten million years and always included all those younger than five million years. In most cases, exclusion of the younger contrasts increased the strength of the correlations found considerably and the results of these analyses are reported. In a few cases, in which exclusion of the younger contrasts did not increase correlations, the results from the whole data set are given. There was no case in which exclusion of the younger contrasts changed a significant correlation to one that was insignificant or where it was significant but changed in sign.

Table 4.4 Mortality, body weight and life history parameters

Parameter	Predicted relationship	Species values					Contrast values			
		Regression exponent (± S.E.)	p	intercept	r	n	Regression exponent (± S.E.)	p	r	n
(a) Multiple regression of log α versus log W and mortality										
log W		+ 0.197 ± 0.197	0.0303	− 0.120	0.65	18	+ 0.032 ± 0.133	0.8195	0.73	9[a]
log IM	No prediction	− 0.132 ± 0.082	0.1792				− 0.315 ± 0.134	0.0508		
log W		+ 0.172 ± 0.079	0.0531	− 0.174	0.77	14	+ 0.072 ± 0.084	< 0.4380	0.92	6[a]
log Z	Negative	− 0.235 ± 0.111	0.5710				− 0.416 ± 0.102	< 0.0152		
log W		+ 0.222 ± 0.131	0.1498	− 0.290	0.87	8	+ 0.267 ± 0.106	0.0851	0.84	5[a]
log M	Negative	− 0.188 ± 0.245	0.4785				− 0.004 ± 0.204	0.9986		
(b) Multiple regression of log b versus log W and mortality										
log W		− 0.248 ± 0.087	0.0104	+ 0.797	0.69	21	− 0.347 ± 0.113	0.0153	0.78	10[a]
log IM	No prediction	+ 0.180 ± 0.099	0.0853				+ 0.062 ± 0.112	0.5964		
log W		− 0.245 ± 0.078	0.0103	+ 0.921	0.85	13	− 0.375 ± 0.125	0.0404	0.87	6[a]
log Z	Positive	+ 0.246 ± 0.108	0.0461				+ 0.222 ± 0.144	0.1982		
log W		− 0.262 ± 0.122	0.0639	+ 0.992	0.80	11	− 0.330 ± 0.087	0.0125	0.91	7[a]
log M	Positive	+ 0.230 ± 0.196	0.2735				+ 0.446 ± 0.147	0.0291		

[a] Regression calculated after exclusion of younger contrasts (see methods for details).
For explanations of symbols, see Table 4.1.

Results

Do age at first reproduction and birth rate vary with mortality rates as predicted?

Table 4.4 shows the results of multiple regression analyses that regress $\log \alpha$ (length of the juvenile period) and $\log b$ (female birth rate) on $\log W$ plus one of three measure of mortality. Significant results suggest that, when body weight is controlled, there is a negative correlation between pre-reproductive mortality and length of the juvenile period (found with CAIC analyses only) and a positive correlation between adult mortality and birth rate (found with CAIC analysis only) and pre-reproductive mortality and birth rate (found with species data only), i.e. that primates with high rates of mortality reproduce at an early age and have high birth rates. These results agree with the predictions of Charnov (1993) and with a study using a broad range of mammalian data (Promislow and Harvey, 1990).

Another of Charnov's important predictions – that adult mortality rate should be negatively correlated with length of the juvenile period – is not supported. Although the exponents are negative, they do not even approach significance. One reason for the lack of support for relationships involving adult mortality rate may be that the sample size for this variable is very low, with only eleven species (ten contrasts) with both α and M, and even lower sample sizes when the younger contrasts are excluded. We must wait for more data on adult mortality rates to be collected to see if an increase in sample size will lead to a different result.

Although a large sample of good-quality mortality data is not available, there are data available on environmental variables and these might be expected to correlate with mortality patterns. Thus, the finding of relationships between reproductive rates and environmental variables may offer support for the idea that mortality patterns are the primary determinants of variation in reproductive rates.

Do environmental variables predict reproductive rates?

The comparative method has been widely used to investigate links between reproductive rates and habitat type in primates and other mammals. Some of these studies are summarised in Table 4.5. These studies vary in their approach, from those that compare species within genera to those that include a larger number of taxa, but all suggest that reproductive rates vary with environment, with more variable environments usually being associated with higher reproductive rates. However, at least some of the significant results found in the larger comparative studies (Ross, 1988, 1992a)

Table 4.5. *Comparative studies that have found links between primate reproductive rates and environment*

Species included	Finding	Reference
Galago species	Breeding rate increases with increasing variability in temperature. The relationship with rainfall variability and the probability of drought is less clear	Nash (1983)
Lemurs	Lemurs from the moist dense tropical forest have the highest mean litter sizes; those from the drier deciduous woodlands have the lowest mean litter sizes	Rasmussen (1985)
All primates	Species living in more open unforested habitats have higher r_m, relative to body size, than those in forest habitats; climate unpredictability may also be linked to a high r_m	Ross (1988, 1992a)
Macaques	Forest macaques have longer interbirth intervals, a later age at first reproduction, and thus a lower r_m than do opportunistic species living sympatrically	Ross (1992b)
African monkeys	Open grassland species are faster developing and faster breeding than forest species	Rowell & Richards (1979); Chism *et al.* (1984); Cords & Rowell (1987); Cords (1987)

may be related to relative reproductive rates being very similar in some species-rich taxa (e.g. cercopithecines tend to have high reproductive rates and the apes have low reproductive rates). Repeating the analyses using mean subfamily values does remove a part of this problem and gives a similar result (Ross, 1992a), but does not completely remove the problem of taxonomic bias.

When these analyses were repeated using independent contrasts (CAIC), no significant links were found between habitat type and three measures of reproductive rate (AR, birth rate and r_m). These results were insignificant when using all contrasts and when using only older contrasts. CAIC analyses also fail to show significant results when closely related taxa (African monkeys, macaques, bushbabies and lemurs) are dealt with separately.

There are two possible explanations as to why CAIC analysis fails to give the same results as analysis carried out using species or subfamily means as independent points. (1) The links found between environment and life history in the earlier studies may be due to taxonomic bias. (2) Problems with the CAIC method are obscuring 'real' links and although environment is linked to reproductive rate, the measures used in this study are too crude to be used as correlates of mortality patterns. One possible reason why the latter be the case is that CAIC cannot be used to compare categorical data when more than two categories exist (unless some linearity can be assumed). In the analysis by Ross (1992a), habitat was divided into five categories; these were collapsed down to two categories (forest versus non-forest) in the CAIC analysis. Similarly, the analysis of the macaque data carried out by Ross (1992b) only found differences between the reproductive rates of 'forest' and 'opportunistic' macaque species when sympatric pairs of macaques were compared; such an analysis could not be repeated using CAIC. However, in Figure 4.1 the difference clearly shown between the two macaque species illustrates the potential magnitude of this effect.

Do life history variables vary with body size in the way predicted by Charnov's (1993) model?

As noted above, some of Charnov's predictions relating mortality rates to other life history variables are not supported by this data set. Table 4.6 shows the results of several bivariate analyses that test some other specific predictions of Charnov's model of mammalian life history evolution. These analyses support Charnov's (1993) finding that the allometric relationships for primates are different from those found for other mammals, when species data are used. The slope is steeper for both log b and log r_m with log body weight although, as for other mammals, the slopes are both negative. When contrast values are used, the slope values calculated for log b and log r_m remain steeper than those predicted. However, contrary to Charnov's findings, log α does not appear to scale differently in primates than for mammals generally, the contrast value for the slope for log α being identical to the slope that Purvis and Harvey (1995) found for mammals.

The larger data set used here supports Charnov's contention that A (the growth constant) is low for primates. A can be estimated from the regression equation of log α on log W, as the intercept $= -\log A$. In the species analysis found in this study (using units of kilograms and years), this intercept $= 0.331$, so that $A = 0.47$ (i.e. very similar to that of 0.42 calculated by Charnov using a slightly different and smaller sample). The low value of the growth constant for primates is linked to the slow maturation rates of the group, but it does not explain why a low growth rate might be selected for (this is discussed below).

Charnov also uses the scaling relationships of α and b to predict the adult mortality rate (M), so that for mammals: $M = 0.75AW^{-0.25}$ (for primates, Charnov suggests that the exponent may be -0.33 as this reflects the

Table 4.6. *Bivariate relationships between life history parameters and body weight*

Parameter	Predicted slope (Charnov, 1993)	Mammal contrast Slope ± S.E.	Primate species Slope ± S.E.	Intercept	p	r	n	Primate contrasts Slope ± S.E.	p	r	n
log α	+0.25	+0.236 ± 0.04	+0.288 ± 0.03	+0.358	<0.0001	0.71	55	+0.236 ± 0.06	0.0011	0.60	25[1]
log b	−0.25	−0.241 ± 0.06	−0.310 ± 0.03	+0.930	<0.0001	0.81	88	−0.330 ± 0.07	<0.0001	0.60	42[1]
log r_m	−0.25	—	−0.382 ± 0.02	+0.659	<0.0001	0.89	75	−0.285 ± 0.06	<0.0001	0.65	36[1]
log AR	—	—	+0.342 ± 0.02	−0.642	<0.0001	0.87	86	+0.253 ± 0.05	<0.0001	0.42	42[1]
M	−0.25	−0.241 ± 0.05	−0.147 ± 0.17	−1.106	0.4014	0.22	16	−0.137 ± 0.15	0.4001	0.32	8[1]
Z	−0.25	−0.321 ± 0.05	−0.287 ± 0.18	−0.169	0.1306	0.41	15	+0.222 ± 0.30	0.4664	0.20	14[2]
IM	—	—	−0.238 ± 0.12	+0.324	0.0674	0.37	25	−0.313 ± 0.18	0.1132	0.46	12[1]
S(AR)	Not significant	Not significant	+0.142 ± 0.23	−0.577	0.5484	0.16	16	−0.839 ± 0.61	0.1879	0.35	15[2]

[1] Regression calculated after exclusion of younger contrasts (see methods for details).

[2] Regression calculated without exclusion of younger contrasts, where exclusion of younger nodes did not increase the correlation (see methods for details).

Abbreviations of parameters are shown in Table 4.1. All slopes were calculated using least squares regression. Contrast relationships are of the form log$_{10}$ (parameter) = a log$_{10}$ (W); species relationships are of the form log$_{10}$ (parameter) = a log$_{10}$ + c; where a = slope of the line, c = intercept. Mammal contrast slopes are taken from Purvis and Harvey (1995).

higher exponent value he found for log α on log W). However, this prediction is not supported here as the results of this study did not find a significant relationship between M and W (possibly because of the poor quality of the mortality data used).

Charnov also predicts that the pre-reproductive mortality rate (Z) should scale predictably with body weight. The results of this study do not give significant correlations between log Z on log W, although the relationship between log IM and log W approaches significance ($p = 0.06$ for species, 0.11 for contrasts). Despite the lack of significant correlations, it is notable that the slope values for species values are close to the predicted value, as is the slope value for contrast data for log IM on log W. Given the variable quality of the data used here, this may indicate that a larger data set of good-quality data would support Charnov's predictions.

Why has a late age at maturity evolved in primates?

The rate of population increase in primates is low when compared to other mammals of the same size (Ross, 1988; Charnov, 1993). Age at first reproduction has more influence on the rate of population increase than does birth rate or longevity and one of the most striking primate life history characteristics is a relatively late maturation compared to most other mammals. The evolution of late maturation has received a great deal of attention in previous studies of primates, and numerous theories have been put forward to explain it.

Pagel and Harvey (1993) suggest that selection acts primarily on body size and that selection for large body size will automatically lead to later maturity and reduced low adult mortality rates. Once such selection has occurred, animals will have a long juvenile period during which they cannot reproduce. This explains why primates tend to have late ages at maturity relative to smaller mammals but does not explain why they reproduce late compared to other mammals of a similar size. Pagel and Harvey explain this by postulating that because selection for delayed maturity produces time 'in limbo' (i.e. without reproduction), selection will favour those who make best use of this time, e.g. by learning social skills and gaining hunting expertise. If the evolution of 'useful' juvenile behaviour results in decreased mortality rates (for both adults and juveniles) and increased fecundity, there may be additional selection to increase the juvenile period yet further. Thus, a long juvenile period might evolve by a kind of positive feedback whereby a long juvenile period selects for characteristics that in turn lead to further selection for an even longer

Table 4.7. *Reproductive rate parameters versus body size and brain size*

Parameter	r	p	Multiple regression statistics	
			x	p
α	0.770	< 0.0001	W	0.1853
			Brn	0.0113
AR	0.826	< 0.0001	W	0.4191
			Brn	0.0167
L	0.717	0.0005	W	0.7626
			Brn	0.1308
b	0.687	0.00012	W	0.6175
			Brn	0.5020
r_m	0.750	< 0.0001	W	0.9429
			Brn	0.1535

Multiple regression through the origin carried out using CAIC data ($n = 23$ older contrast values) of log (parameter) versus log W and log Brn.
For abbreviations, see Table 4.1.

juvenile period. The salient question is then: why is a juvenile period particularly useful for primates?

Comparative analyses suggest that age at sexual maturity is more highly correlated with brain size than it is with body size (Harvey *et al.*, 1987; Barton, Chapter 7). The data used in this study also show that brain size and age at sexual maturity are closely linked (Table 4.7). This result appears to be robust, being found both when using the largest possible data set and also with smaller subsets of data. It has been suggested by Economos (1980) that such links might be due to intraspecific variation in body weight being greater than that of brain weight. If this is the case, brain weight might be a more accurate measure of size than body weight, and this could explain why it correlates more highly with life history variables such as age at maturity and longevity. However, when body weight effects were removed, correlations with brain weight were not found for some other reproductive variables investigated (Table 4.7), as might be expected if Economos's theory was correct. This suggests that the link between brain size and age at first reproduction may help to explain the evolution of late maturity.

Why, then, should primates with relatively large brains also have a relatively late age at first reproduction? There are three commonly quoted theories that have tried to explain this link. These models suggest that there are relationships between brain size, age at first reproduction, postnatal growth rate, diet and mortality, but the links they postulate vary.

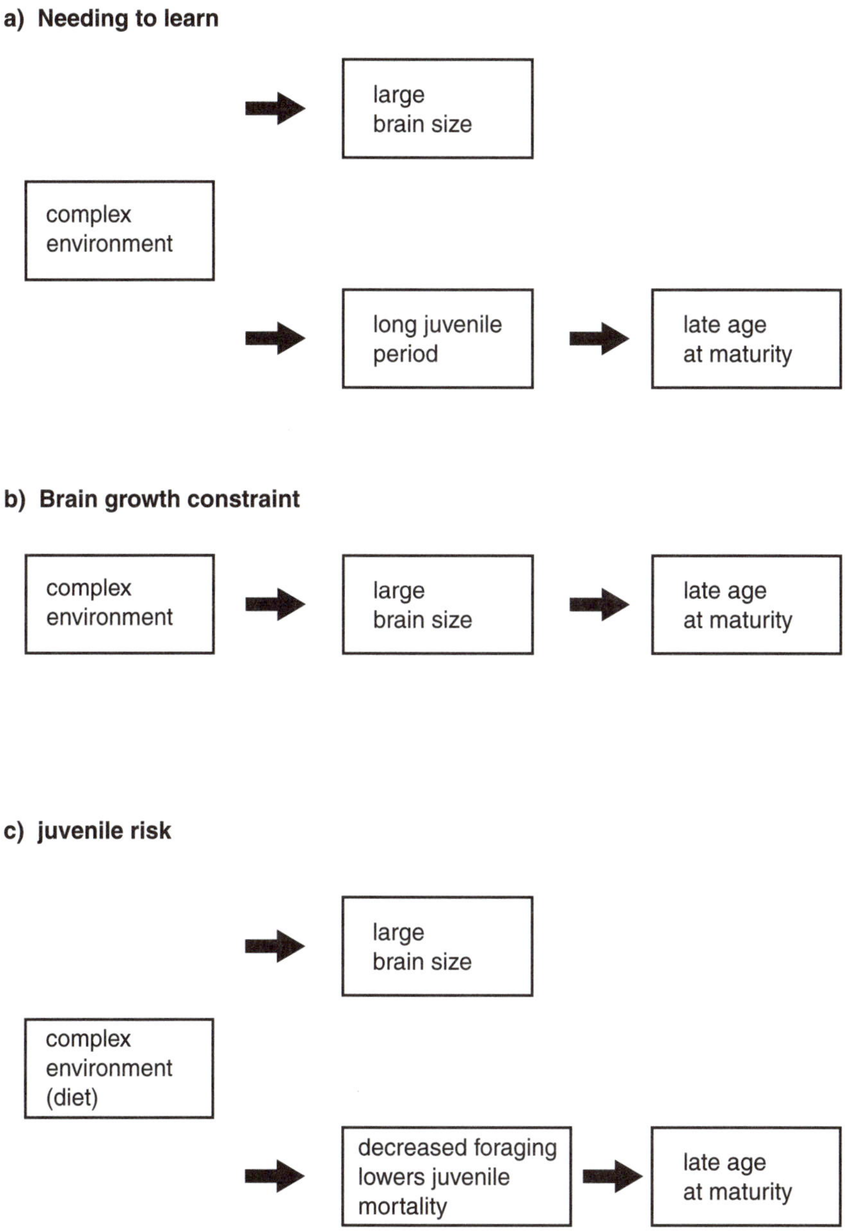

a) Needing to learn

b) Brain growth constraint

c) juvenile risk

Fig. 4.2 Models linking female age at first reproduction to brain size and environmental complexity. See text for details.

1. *Needing to learn.* Animals with large brains are selected to have these large brains because they need to learn (about a complex social and/or physical environment). Thus, they delay maturity until they have learnt enough to become behaviourally mature. This model predicts that animals are selected *both* for large brains and for delayed reproduction, in both cases because they need to be able to cope with complex social or ecological problems before they can successfully breed. Delayed reproduction will be produced by a slow postnatal growth rate, and the majority of learning will take place during the extended infancy/juvenile period (Fig. 4.2a). Environmental factors that might be expected to select for learning ability include diet (some foods are more difficult to obtain than others) and social complexity (animals in large groups need to learn more than those in small groups).

   Joffe (1997) shows that a prolonged juvenile period (relative to overall lifespan) is correlated with a large non-visual neocortex (relative to absolute brain size) in primates and that the absolute length of the juvenile period is correlated with group size. These findings may support the idea that social complexity selects for large brain size but the analyses do not control for the influence of body weight. As both non-visual neocortex and the length of the juvenile period correlate with body weight, the findings may simply reflect the fact that large animals tend to live in larger groups than smaller ones. As body size in primates is known to correlate with a number of ecological correlates, e.g. diet, diurnality, arboreality (Clutton Brock and Harvey, 1977a, 1977b), the link may not reflect a direct relationship between group size and juvenile period.

2. *Brain growth constraint.* Animals with relatively large brains may be unable to grow as fast as those with relatively small brains, as the high energetic costs of having a large brain prevent high rates of production. Thus, delayed reproduction in primates may not be directly adaptive but may be due to selection for a large brain (Fig. 4.2b). However, postnatal brain growth in non-human primates is mostly completed at birth and does not continue after infancy. Thus, brain growth is unlikely to explain a prolonged juvenile period, although maintenance of a relatively large brain may involve a high-energy input and thus slow postnatal growth during both the infant and juvenile periods.

   Charnov and Berrigan (1993) noted that that some non-mammalian taxa with low rates of offspring growth and late maturation do not have large brains. This implies that even if large brain size explains the late maturation in primates, it may not do so in all taxa.

3. *Juvenile risk.* Janson and van Schaik (1993) suggest that primates grow

slowly in order to avoid the need to forage extensively to support a high growth rate. By spending less time on foraging, they avoid high levels of competition with conspecifics and can spend more energy avoiding predators. Thus, the reproductive disadvantage incurred by slow post-natal growth rates is outweighed by the advantage of lowered mortality rates caused by a reduction of predation risk. This model differs from the 'brain growth constraint' model as it assumes that slow growth rates are adaptive and that the link between relative brain size and relative age at first reproduction is caused by a third variable, diet, that influences both (Fig. 4.2c). A folivorous diet is suggested to select for both small brains and a high postnatal growth rate (as leaves are easy to find, increased foraging for them has a smaller influence on mortality risk than does increased foraging for fruit). Janson and van Schaik's model would predict that group size would have an influence on postnatal growth rates, insomuch as group size has an influence on juvenile mortality risk. However, the relationship between group size and mortality risk may vary from one situation to another: large groups may offer protection from predators in some cases, but in others they may lead to higher mortality risk due to increased competition levels.

Testing the models

The predictions of the three models are shown in Table 4.8. The 'brain growth constraint' model differs from the others in that it predicts that holding brain size constant will remove any relationship between environmental variables and delayed reproduction. In contrast, both the 'needing to learn' model and the 'juvenile risk' model predict that diet is a variable that will be expected to be linked to both brain size and delayed reproduction, and that holding diet constant will reduce any correlation between the two. The 'needing to learn' model and the' juvenile risk' model also make the same prediction regarding the type of relationship between diet and brain size. The 'needing to learn' model predicts that food that is difficult to obtain will select for both a large brain size and an increased period of learning. In primates, it is usually assumed that gathering leaves is an easier task than gathering fruit – as the latter is more unpredictable in time and space. This leads to the prediction that there will be a positive correlation between brain size and the degree of folivory. The 'juvenile risk' model also predicts the same patterns – in this case because folivores are expected to be subject to less fluctuation in their food supply and thus to be have a less 'risky' juvenile period. The 'needing to learn' model differs from the others in that it makes a clear prediction that social complexity will correlate with brain size.

Table 4.8. *Multiple regression of $\log_{10} \alpha$ against $\log_{10} W + \log_{10}$ (brain weight), $\log_{10}$ group size and percentage folivory*

α versus	Predicted sign of coefficient for multiple regression:			Results of multiple regression:					
	(i) Needing to learn	(ii) Brain growth constraint	(iii) Juvenile risk	(a) All contrasts (n = 33 contrasts)		(b) Older contrasts (n = 17 contrasts)		(c) Species (n = 34 species)	
				r^2	p (sign of coefficient)	r^2	p (sign of coefficient)	r^2	p (sign of coefficient)
$\log_{10}$ W	No link	+		0.776	0.0474(+)	0.892	0.1738	0.806	0.4709
$\log_{10}$ Brn	+	No link	No link		0.8979		0.0064(+)		0.0032(+)
$\log_{10}$ Grp			No link (see text)		0.2342		0.0242(−)		0.0821(−)
Percentage folivory	—	No link	—		0.0703(−)		0.3383		0.1652

See text for details.

To test these predictions, logged values of female body weight, brain size, group size and the amount of leaves in the diet (percentage folivory) have been used and regressed against the length of the juvenile period (α).

The results (Table 4.8) show some support for the 'brain growth constraint' model, with α showing a significant positive relationship with brain size with both the species and older contrast values, indicating that the evolution of late maturation is linked to the evolution of large brain size. Folivory is only found to be negatively correlated with α when all contrast values are included and, even then, the significance level is above 0.5. This gives little support for the 'needing to learn' and 'juvenile risk' models. Group size does not correlate positively with α, with a negative relationship being found with older contrast values ($p = 0.02$) species values ($p = 0.08$). This indicates that for animals of a given body size, brain size and diet, those that live in larger groups have faster rates of maturation than those living in small groups, the opposite of the result predicted by the 'needing to learn' model. However, the negative relationship ($p = 0.07$) found between folivory and length of the juvenile period may suggest that learning to forage for some foods may need an extended juvenile period.

Thus, these results appear to offer some support for all of the three models, with stronger support for the brain growth constraint model than for the other two.

Strategies to increase reproductive rates

When compared with other mammals of the same size, primates appear to be constrained, by low rates of production, to having slow rates of maturation and low birth rates and thus low reproductive rates. Female primates that are selected to increase their reproductive rates might do so in two ways: by increasing their birth rate or by maturing (and maturing their young) at an earlier age. If having relatively large brains constrains primates to having a low juvenile growth rate, there is little that mothers can do to increase the growth rate of their offspring during the juvenile phase. However, they may be able to increase pre-weaning growth rates if they can allocate large amounts of resources to their young during gestation and/or lactation. Intraspecific evidence suggests that access to resources does, indeed, affect the growth rates of infants (see Chapter 5). Vervet monkeys that have limited access to resources are likely to delay reproduction as compared to others that have access to more resources (Cheney *et al.*, 1988). Factors that relate to resource acquisition such as high rank of the female, or her mother, have also been found to be linked to early maturation in savannah baboons (Altmann *et al.*, 1988) and Japanese macaques (Gouzoules, Gouzoules and Fedigan, 1982). Other studies have shown that better fed animals (e.g. those that are provisioned artificially) have higher birth rates and decreased ages at first reproduction as compared with their non-provisioned conspecifics (Lee and Bowman, 1995).

If mothers cannot increase their reproductive rates by enhancing their access to resources, they may divert resources away from other activities and into growing their infants. If females can 'persuade' others to help them with infant care, they may spend less on carrying and protecting their infants and thus be able to allocate more into infant growth. Interspecific studies do show that species that have high levels of help with infant care are able to grow their young faster and thus increase their birth rate, in comparison to species of the same size lacking such help (Ross and Mac-Larnon, 1995, 1996; Mitani and Watts, 1997). Critically, female primates may use a variety of social and behavioural strategies to overcome the constraints of their physiology and increase their reproductive rate.

Conclusions

Primate reproductive rates are slow when compared to other mammals of the same size. This is primarily due both to the low birth rates and the late maturation of primate species. Several previous studies have suggested that variation in reproductive rates within the primates can be explained by ecology. However, the data presented here do not show strong support for this view and indicate that at least some correlations between ecology and reproductive rate may be artifacts found when species are erroneously treated as independent data points. However, there are limitations imposed on analyses using independent contrasts that mean that a link between ecology and reproductive rates cannot be ruled out entirely. There is some evidence that suggests that, once body weight has been controlled, primates with low rates of pre-reproductive mortality do have later maturation and lower birth rates than do those with higher mortality rates.

Some predictions of Charnov's life history model are supported by this study, or at least not strongly refuted (i.e. when predicted equations are not significant but do show the predicted trends). This suggests that primates are indeed different from many other mammals in having low rates of production, which is linked to their late age at maturity. One of the most important variables that influences reproductive rate – age at first reproduction – appears to be strongly linked to brain size: primates with large brains have a late age at maturity. It is not clear from the analyses presented here whether this is because the maintenance of large brains prevents rapid growth and delays maturity or whether there has been positive selection for delayed maturity, although the former explanation appears to be more strongly supported.

Although primates do appear to be a group that is constrained to have a late maturation, there is evidence that they can use a variety of behavioural and social strategies to increase their reproductive rates. Presence of non-maternal care is linked to increasing growth rates and results in high birth rates. Individuals may also increase their reproductive rates by gaining high status or by using other social means to secure resources. Further understanding of how such behaviour influences the life history of individuals and thus causes intraspecific variation may be an important key to our further understanding of life history evolution.

Acknowledgements

This chapter has been greatly improved by comments from P. Lee and Rob Barton. Andy Purvis provided the authors with the CAIC software and his primate phylogeny. The authors have had useful discussions about this, and related work, with Bob Martin, Ann MacLarnon and Andy Purvis. They thank all of them for their help and time.

References

Altmann, J., Altman, S., Hausfater, G. and McCuskey, S.A. (1977). Life history of yellow baboons: physical development, reproductive parameters and infant mortality. *Primates* **18**, 315–30.

Altmann, J., Hausfater, G. and Altman, S. (1985). Demography of Amboseli baboons, 1963–1983. *American Journal of Primatology* **8**, 113–25.

Altmann, J., Hausfater, G. and Altmann, S.A. (1988). Determinants of reproductive success in savannah baboons. In *Reproductive Success*, ed. T.H. Clutton-Brock, pp. 403–18, Chicago: Chicago University Press.

Boyce, M.S. (1988). Evolution of life histories: theory and patterns from mammals. In *Evolution of Life Histories of Mammals: Theory and Pattern*, ed. M.S. Boyce, pp. 3–30. New Haven; London: Yale University Press.

Calder, W.J. III (1984). *Size, Function and Life History*. Cambridge, MA: Harvard University Press.

Caughley, G. (1977). *Analysis of Vertebrate Populations*. New York: John Wiley.

Charnov E.L. (1993). *Life History Invariants: Some Explorations of Symmetry in Evolutionary Ecology*. Oxford: Oxford University Press.

Charnov, E.L. and Berrigan, D. (1993). Why do female primates have such long lifespans and so few babies? Or life in the slow lane. *Evolutionary Anthropology* **1**, 191–4.

Cheney, D.L., Seyfarth, R.M., Andleman, S. and Lee, P.C. (1981). Behavioural correlates of non-random mortality in free-ranging vervet monkeys. *Behavioral Ecology and Sociobiology* **9**, 153–61.

Cheney, D.L., Seyfarth, R.M., Andleman, S. and Lee, P.C. (1988). Reproductive success in vervet monkeys. In *Reproductive Success*, ed. T.H. Clutton-Brock, pp. 384–402. Chicago: Chicago University Press.

Chism, J., Rowell, T. and Olson, D. (1984). Life history patterns of female patas monkeys. In *Female Primates; Studies by Women Primatologists*, ed. M.F. Small, pp. 175–90. New York: Alan R. Liss.

Clutton-Brock, T.H. and Harvey, P.H. (1977a). Primate ecology and social organisation. *Journal of Zoology, London* **183**, 1–39.

Clutton-Brock, T.H. and Harvey, P.H. (1977b). Species differences in feeding and ranging behaviour of primates. In *Primate Ecology: Studies of Feeding and Ranging Behaviour in Lemurs, Monkeys and Apes*, ed. T.H. Clutton-Brock, pp. 557–84. London: Academic Press.

Cole, L.C. (1954). The population consequences of life history phenonema. *Quarterly Review of Biology* **29**, 103–37.

Cords, M. (1987). Forest guenons and patas monkeys: male–male competition in one-male groups. In *Primate Societies*, eds. B.B. Smuts, D.L. Cheney, R.M. Seyfarth, R.W. Wrangham and T.T. Struhsaker, pp. 98–111. Chicago: University of Chicago Press.

Cords, M. and Rowell, T.E. (1987). Birth intervals of *Cercopithecus* monkeys of the Kakamega Forest, Kenya. *Primates* **28**, 277–81.

Courtenay, J. and Santow, G. (1989). Mortality of wild and captive chimpanzees. *Folia Primatologica* **52**, 167–77.

Crockett, C.M. and Pope, T.R. (1993). Consequences of sex differences in dispersal for juvenile red howler monkeys. In *Juvenile Primates: Life History, Development and Behaviour*, ed. M.E. Pereira and L.A. Fairbanks, pp. 57–74. Oxford: Oxford University Press.

Dittus, W.P.J. (1975). Population dynamics of the toque monkey, *Macaca sinica*. In *Socioecology and Psychology of Primates*, ed. R.H. Tuttle, pp. 125–51. The Hague: Mouton.

Dunbar, R.I.M. and Dunbar, E.P. (1974). Ecology and population dynamics of *Colobus guereza* in Ethiopia. *Folia Primatologica* **21**, 188–208.

Dunbar, R.I.M. (1980). Demographic and life history variables of a population of gelada baboons (*Theropithecus gelada*). *Journal Animal Ecology* **49**, 485–506.

Economos, A.C. (1980). Brain–life span conjecture: a re-evaluation of the evidence. *Gerontology* **26**, 82–9.

Felsenstein, J. (1985). Phylogenies and the comparative method. *American Naturalist* **125**, 1–15.

Froehlich, J.W., Thorington, R.W. Jr and Otis, J.S. (1981). The demography of howler monkeys (*Alouatta palliata*) on Barro Colorado Island, Panama. *International Journal of Primatology* **2**, 207–36.

Glander, K.E. (1980). Reproduction and population growth in free-ranging mantled howling monkeys. *American Journal of Physical Anthropology* **53**, 25–36.

Gouzoules, H., Gouzoules, S. and Fedigan, L.M. (1982). Behavioural dominance and reproductive success in female Japanese monkeys (*Macaca fuscata*). *Animal Behaviour* **30**, 1138–90.

Harcourt, A.H., Fossey, D. and Sabater-Pi, J. (1981). Demography of *Gorilla gorilla*. *Journal of Zoology, London* **195**, 215–33.

Harvey, P.H. and Mace, G.M. (1982). Comparisons between taxa and adaptive trends: a problem of methodology. In *Current Trends in Sociobiology*, ed. King's College Sociobiology Group, pp. 343–62. Cambridge: Cambridge University Press.

Harvey, P.H., Martin, R.D. and Clutton-Brock, T.H. (1987). Life histories in comparative perspective. In *Primate Societies*, eds. B.B. Smuts, D.L. Cheney, R.M. Seyfarth, R.W. Wrangham, and T.T. Struhsaker, pp. 181–96. Chicago: University of Chicago Press.

Harvey, P.H. and Pagel, M. (1991). *The Comparative Method in Evolutionary Biology*. Oxford: Oxford University Press.

Hennemann, W.W. III (1983). Relationship among body mass, metabolic rate and the intrinsic rate of natural increase in mammals. *Oecologia* **56**, 419–24.

Hiraiwa, M. (1981). Maternal and alloparental care in a troop of free-ranging Japanese monkeys. *Primates* **22**, 309–29.

Hrdy, S.B. (1977). *The Langurs of Abu*. Cambridge, MA: Harvard University Press.

Ikeda, H. (1982). Population changes and ranging behaviour of wild Japanese monkeys at Mt. Kawaradake in Kyusha, Japan. *Primates* **23**, 338–47.

Janson, C.H. and van Schaik, C.P. (1993). Ecological risk aversion in juvenile primates: slow and steady wins the race. In *Juvenile Primates: Life History, Development and Behaviour*, ed. M.E. Pereira and L.A. Fairbanks, pp. 57–74. Oxford: Oxford University Press.

Joffe, T.H. (1977). Social pressures have selected for an extended juvenile period in primates. *Journal of Human Evolution* **32**, 593–605.

Kirkwood, J.K. (1985). Patterns of growth in primates. *Journal of Zoology, London* **205**, 123–36.

Kozlowski, J. and Weiner, I. (1997). Interspecific allometries are a by-product of body size optimization. *American Naturalist* **149**, 52–80.

Lee, P.C. and Bowman, J. (1995). Influence of ecology and energetics on primate mothers and infants. In: *Motherhood in Human and Nonhuman Primates: Biological and Social Determinants*, ed. C. Pryce, R.D. Martin and D. Skuse, pp. 47–58. Basel: Karger.

Lee, P.C., Majluf, P. and Gordon I.J. (1991). Growth, weaning and maternal investment from a comparative perspective. *Journal of Zoology, London* **225**, 99–114.

MacArthur, R.H. and Wilson, E.O. (1967). *The Theory of Island Biogeography*. Princeton: Princeton University Press.

McFarland Symington, M. (1988). Demography, ranging patterns and activity budgets of black spider monkeys (*Ateles paniscus chamek*) in the Manu National Park, Peru. *American Journal of Primatology* **15**, 45–67.

Mitani, J.C. and Watts, D. (1997). Maternal permissiveness and the evolution of non-maternal caretaking among anthropoid primates. *Behavioral Ecology and Sociobiology* **40**, 213–20.

Menard, N. and Vallet, D. (1993). Population dynamics of *Macaca sylvanus* in Algeria: an 8-year study. *American Journal of Primatology* **30**, 101–18.

Nash, L.T. (1983). Reproductive patterns in Galagos (*Galago zanzibaricus* and *Galago garnetti*) in relation to climatic variability. *American Journal of*

Primatology **5**, 181–96.

van Noordwijk, M.A., Hemelrijk, C.K., Herremans, L.A. and Sterck, E.H.M. (1993). Spatial position and behavioural sex differences in juvenile long-tailed macaques. In *Juvenile Primates: Life History, Development and Behaviour*, ed. M.E. Pereira and L.A. Fairbanks, pp. 57–74. Oxford: Oxford University Press.

O'Brien, T.G. and Robinson, J.G. (1993). Stability of social relationships in female wedge-capped capuchin monkeys. In *Juvenile Primates: Life History, Development and Behaviour*, ed. M.E. Pereira and L.A. Fairbanks, pp. 197–210. Oxford: Oxford University Press.

Pagel, M. and Harvey, P.H. (1993). Evolution of the juvenile period in mammals. In *Juvenile Primates: Life History, Development and Behaviour*, ed. M.E. Pereira and L.A. Fairbanks, pp. 528–37. Oxford: Oxford University Press.

Peters, R.H. (1983). *The Ecological Implications of Body Size*. Cambridge: Cambridge University Press.

Pianka, E.R. (1970). On r- and K-selection. *American Naturalist* **104**, 592–7.

Promislow, D.E.L. and Harvey, P.H. (1990). Living fast and dying young: a comparative analysis of life history variation among mammals. *Journal of Zoology, London*, **220**, 417–37.

Purvis, A. (1995). A composite estimate of primate phylogeny. *Philosophical Transactions of the Royal Society, London, Series B* **348**, 405–21.

Purvis, A. and Harvey, P. (1995). Mammal life history evolution: a comparative test of Charnov's model. *Journal of Zoology, London*, **237**, 259–83.

Purvis, A. and Rambaut, A. (1995). Comparative analysis by independent contrasts (CAIC): an Apple Macintosh application for analysing comparative data. *Computer Applications for the Biosciences* **11**, 247–51.

Rasamimanana, H.R. and Rafidinarivo, E. (1993). Feeding behaviour of *Lemur catta* females in relation to their physiological state. In *Lemur Social Systems and their Ecological Basis*, ed. P.M. Kappeler and J.U. Ganzhorn, pp. 123–34. New York: Plenum Press.

Rasmussen, D.T. (1985). A comparative study of breeding seasonality and litter size in eleven taxa of captive lemurs (*Lemur* and *Varecia*). *International Journal of Primatology* **6**, 501–17.

Rhine, R.J. (1997). Criteria of reproductive success. *American Journal of Primatology* **41**, 87–101.

Richard, A.F. (1985). *Primates in Nature*. New York: W.H. Freeman.

Richard, A.F., Ratotomanga, R. and Schartz, M. (1991). Demography of *Propithecus verreauxi* at Beza Mahafaly, Madagascar: sex ratio, survival, and fertility, 1984–1988. *American Journal of Physical Anthropology* **84**, 307–22.

Robinson, J.G. (1988). Demography and group structure in wedge-capped capuchin monkeys, *Cebus olivaceus*. *Behaviour* **104**, 202–32.

Ross, C. (1988). The intrinsic rate of natural increase and reproductive effort in primates. *Journal of Zoology, London* **214**, 199–219.

Ross, C. (1992a). Environmental predictability and the intrinsic rate of natural increase in primates. *Oecologia* **90**, 383–90.

Ross, C. (1992b). Life history patterns and ecology of macaque species. *Primates* **33**, 207–15.

Ross, C. and MacLarnon, A. (1995). Ecological and social correlates of maternal expenditure on infant growth in haplorhine primates. In *Motherhood in Hu-*

man and Nonhuman Primates: Biological and Social Determinants, ed. C. Pryce, R.D. Martin and D. Skuse, pp. 37–46. Basel: Karger.

Ross, C. and MacLarnon, A. (1996). The evolution of non-maternal care in haplorhine primates. (Abstract.) *Proceedings of the XVIth Congress of the International Primatological Society.*

Rowell, T.E. and Richards, J.M. (1979). Reproductive strategies of some African monkeys. *Journal of Mammalogy* **60**, 58–69.

Sade, D.S. (1990). Intrapopulation variation in life history parameters. In *Primate Life History and Evolution*, ed. C. Jean DeRousseau, pp. 181–84. New York: Wiley-Liss.

Sauther, M.L. (1991). Reproductive behaviour of free-ranging *Lemur catta* at Beza Mahafaly Special Reserve, Madagascar. *American Journal of Physical Anthropology* **84**, 43–58.

Sauther, M.L. and Sussman, R.W. (1993). A new interpretation of the social organisation and mating system of *Lemur catta*. In *Lemur Social Systems and their Ecological Basis*, ed. P.M. Kappeler and J.U. Ganzhorn, pp. 123–34. New York: Plenum Press.

van Schaik, C.P. and van Noordwijk, M.A. (1985). Interannual variability in fruit abundance and the reproductive seasonality in sumatran long-tailed macaques (*Macaca fascicularis*). *Journal of Zoology, London* **206**, 533–49.

Sigg, H., Stolba, A., Abegglen, J.J. and Dasser, V. (1982). Life history of hamadryas baboons: physical development, infant mortality, reproductive parameters, and family relationships. *Primates* **23**, 473–87.

Smith, R.J. and Jungers, W.L. (1997). Body mass in comparative primatology. *Journal of Human Evolution* **32**, 523–9.

Smuts, B.B., Cheney, D.L., Seyfarth, R.M., Wrangham, R.W. and Struhsaker, T.T. (eds.) (1987). *Primate Societies*. Chicago: University of Chicago Press.

Smuts, B.B. and Nicolson, N. (1989). Reproduction in wild olive baboons. *American of Journal Primatology*, **19**, 229–46.

Soini, P. (1982). Ecology and population dynamics of the pygmy marmoset, *Cebuella pygmaea*. *Folia Primatologica* **39**, 1–21.

Stearns, S.C. (1977). The evolution of life history traits: a critique of the theory and a review of the data. *Annual Review of Ecological Systematics* **8**, 145–71.

Stearns, S.C. (1983). The influence of size and phylogeny on patterns of co-variation among life history traits in mammals. *Oikos* **41**, 173–87.

Stearns, S.C. (1992). *The Evolution of Life Histories*. Oxford: Oxford University Press.

Strier, K.B. (1991). Demography and conservation of an endangered primate, *Brachyteles arachnoides*. *Conservation Biology* **5**, 214–18.

Strier, K.B. (1993). Growing up in a patrifocal society: sex differences in the spatial relations of immature muriquis. In *Juvenile Primates: Life History, Development and Behaviour*, ed. M.E. Pereira and L.A. Fairbanks, pp. 138–47. Oxford: Oxford University Press.

Strier, K.B., Mendes, F.D.C., Rimoli, J. and Rimoli, A. (1993). Demography and social structure of one group of muriquies (*Brachyteles arachnoides*). *International Journal of Primatology* **14**, 513–26.

Sussman, R.W. (1991). Demography and social organisation of free-ranging *Lemur catta* in the Beza Mahafaly Reserve, Madagascar. *American Journal of Physi-*

cal Anthropology **84**, 43–58.

Sugiyama, Y. and Ohsawa, H. (1982). Population dynamics of Japanese monkeys with special reference to the effect of artifical feeding. *Folia Primatologica* **39**, 238–63.

Wasser, L.M. and Wasser, S.K. (1995). Environmental variation and developmental rate among free-ranging yellow baboons (*Papio cynocephalus*). *American Journal of Primatology* **35**, 15–30.

Watts, D.P. (1989). Infanticide in mountain gorillas: new cases and a reconsideration of the evidence. *Ethology* **81**, 1–189.

Watts, D.P. (1991). Mountain gorilla reproduction and sexual behavior. *American Journal of Primatology* **24**, 211–25.

Watts, D.P. and Pusey, A.E. (1993). Behavior of juvenile and adolescent great apes. In *Juvenile Primates: Life History, Development and Behaviour*, ed. M.E. Pereira and L.A. Fairbanks, pp. 148–67. Oxford: Oxford University Press.

Western, D. (1979). Size, life history and ecology in mammals. *African Journal of Ecology* **17**, 185–204.

Western, D. and Ssemakula, J. (1982). Life history patterns in birds and mammals and their evolutionary interpretation. *Oecolgia* **54**, 281–90.

Wilson, R.S. (1979). Twin growth: initial deficit, recovery, and trends in concordance from birth to nine years. *Annals of Human Biology* **6**, 205–20.

Winkler, P., Loch, H. and Vogel, C. (1984). Life history of hanuman langurs (*Presbytis entellus*). Reproductive parameters, infant mortality and troop development. *Folia Primatologica* **43**, 1–23.

Wootton, J.T. (1987). The effects of body mass, phylogeny, habitat, and trophic level on mammalian age at first repoduction. *Evolution* **41**, 732–49.

Wright, P. (1995). Demography and life history of free-ranging *Propithecus diadema edwardsi* in Ranomfana National Park, Madagascar. *International Journal of Primatology* **16**, 835–54.

5 Comparative ecology of postnatal growth and weaning among haplorhine primates

PHYLLIS C. LEE

Introduction

Why are the problems of growth interesting? Somatic growth relates to individual survival; growth failure or faltering is associated with an increased likelihood of mortality (rhesus: Small and Smith, 1986; humans: Skuse *et al.*, 1995), and weight for age can be used as a proxy for morbidity or mortality risks (humans: Tomkins, 1994). Maternal mass and body composition can influence reproductive potential and efficiency in terms of fetal growth and lactation capacity (Martin, 1984; Lee and Bowman, 1995). Growth is also a variable with a time component and as such it links with a suite of life history traits within a species, and has been used to describe variation in the timing of events within a life history strategy (Harvey, Clutton-Brock and Martin, 1987; Ross, 1988; Lee, 1996). Growth thus raises two distinct questions in primate comparative biology. The first concerns the underlying selected mechanisms for attaining metabolic and reproductive efficiencies. The second concerns whether and how such growth strategies are achieved at the population or individual level.

Growth can be partitioned into separate periods, each with possibly independent rates and trajectories, as well as unique problems to be solved. These distinct periods may have repercussions on subsequent growth stages, as well as an influence on behaviour and reproductive ability (Bercovitch, 1987; Altmann, 1991). The three major periods are fetal growth, postnatal growth during the period of lactation, and finally growth between weaning and the attainment of a terminal mass or stature, often associated with the onset of reproduction among females (see Leigh, 1994a; Leigh and Shea, 1996).

Most of the scant data we have on normal patterns of growth among the non-human primates rely on small numbers of individuals from captive populations, with only a few samples from free-ranging populations. Most data are also restricted to mass growth rather than statural or skeletal growth, and little is known about the growth of internal organs other than the brain. There are important exceptions, such as dental development (Smith, 1992; Smith, Crummett and Brandt, 1994). While generally we can but lament the inadequacies of the sample, there are sufficient data available to use for a comparative analysis that may at least point to future issues for investigation.

An attempt is made in this chapter to describe patterns of mass growth, specifically weight gain, and to make comparisons between haplorhine primate species. The comparisons should highlight the extent of possible variation in growth, and may illuminate the sources of that variation – those behavioural and ecological factors that affect growth. Because patterns of fetal growth are strongly influenced by maternal mass allometry (Martin, 1996), this is initially examined. One problem specific to weaning is presented: the need to attain a threshold weaning size (Lee, Majluf and Gordon, 1991). The weaning weight hypothesis suggests that growth to weaning is under a metabolic constraint rather than reflecting life history variation in a temporal sequence of events. Lee *et al.* argued that a weanling must attain approximately four times its neonate mass before it is metabolically able to sustain itself through independent feeding, and that this mass could be attained rapidly or slowly as a function of the maternal ability to support tissue growth through lactation.

The chapter thus emphasises the difference between mass growth and the time over which that growth occurs. It is suggested that primates in general are trading-off a short-term reduction in energy costs against an increase in time costs in relation to postnatal growth and weaning. Furthermore, this primate growth strategy can either be a function of the energy available to the mother to fuel growth through lactation or, alternatively, may be necessitated by extrinsic uncertainty or risk in the environment, and these issues are explored. Proximate mechanisms such as hormonal competence and the role of growth factors will not be considered, although these systems are thought to act as control mechanisms underlying patterns of growth (see Tanner and Preece, 1989).

Methods

Data sources and their problems

There is a vast literature on body weights among adult primates, and the selection of a species-typical mass is fraught with problems (see Smith and Jungers, 1997). For a small sample such as that used here, variation of the order of a few hundred grams is probably relatively unimportant. The term 'weight' is used as synonymous with mass, because for terrestrial species there is unlikely to be discordance between the measured weight and the assumed mass. In the selection of the original data base (Lee *et al.*, 1991), priority was given to populations for which as much information as possible was available. This will, indeed, result in some skew in the data, as many growth data are derived from captive sources, which tend to be heavier and develop more rapidly than wild populations (Leigh, 1994b). Mass data are presented for adult females, neonates and weanlings. Weaning weights were calculated from growth curves, and taken as the weight for age at the end of the major period of lactation.

Lactation duration was defined as the population-specific (where weights were derived from a population) interbirth interval – gestation time. This measure incorporates an average duration of lactational anovulation, when the majority of the infant's nutrition is derived from milk, plus some cycling time. As primates frequently lactate into the next pregnancy, defining weaning as the period of major nutritional input that affects the mother's subsequent reproduction should standardise the term across species, and thus reflect a process rather then an endpoint (Lee, 1996). Data on gestation length are 'species-average' because variation due to infant sex (Clutton-Brock, Albon and Guinness, 1989) or ecology (Silk, 1986) is in the order of days rather than months. Other data presented as species averages are adult brain mass and neonate brain mass. These have been taken from Harvey *et al.* (1987). Data on M1 eruption age were taken from Smith *et al.* (1994), and used in these analyses because previous work (reviewed by Smith, 1992) suggests M1 eruption age is an excellent life history measure.

Two 'measures' of environmental risk were incorporated in the analyses. The first was that defined by Ross (1988) in relation to predictability of the environment. Ross' measure assessed resource type, its productivity and seasonality, without relying on specific dietary variables (see also Chapter 4). Diet may be related to evolved digestive capacities, and thus more subject to phylogenetic error than are general habitat parameters. The second measure categorises potential mortality risk due to predation pressure, and is derived from a population-level assessment of predator presence, contacts between primates and predators, antipredator behaviour and observed predations (see Hill and Lee, 1998). These qualitative categories were used in preference to more quantitative vari-ables, as a means of discriminating between gross habitat qualities.

Data were limited to haplorhine species bearing single young, in order to eliminate any possible 'grade' effects or biases from including strepsirhines, and to reduce the variation introduced in individual growth for a litter size greater than one – factors recently addressed in detail by Garber and Leigh (1997). Only a small number of species is represented ($n = 41$) and data used are presented in the appendix to this chapter.

All data were natural log transformed to facilitate comparisons between variables and to normalise data for parametric tests. This was done in order to use multiple regression statistics, and to compute residuals from observed relationships. Residual analysis has been widely used in life history studies as a means of 'removing' any allometric effects and thus to explore variation remaining after controlling for autocorrelation. Here, multiple regressions are also used to assess autocorrelation between size variables. As such, the results are presented as least squares regression slopes, which may be less accurate and produce a lower slope. It should be noted that when comparisons are made between slopes and constants, at least some of the differences may be simply a result of the two techniques for calculating the slopes rather than demonstrating any biological difference.

As noted above, much of the analysis was conducted at the level of the species, without phylogenetic correction techniques. There are two reasons for this: firstly, to explore the variation within closely related taxa (see below); and, secondly, because many allometric studies find no effect of phylogeny on the overall variation (Martin, 1996). Rogers and Cashdan (1997) note that when closely related species experience different rates of selection, phylogenetic subtraction techniques may statistically obscure species trends.

Analysis of phylogenetic effects on growth variables

In these data, 41 species from 23 genera and 12 subfamilies were also partitioned into a higher node, which was established to explore the potential existence of 'grades' within haplorhines. The data were separated into time variables (gestation length, duration of lactation, interbirth intervals, and growth rates) and mass data (maternal mass, neonate mass, weight at weaning).

Several different techniques were used to attempt to assess the effect of phylogeny on the variance within these categories of data. The first was that of phylogenetic regression, using dummy variables to assess the distribution of variation across taxonomic levels (see Harvey and Pagel, 1991). Species were not included because, with the exception of *Macaca* and *Papio*, only one or two species were represented within a genus and little variance could be statistically partitioned at this level. 'Genus' in this

Table 5.1. *Results of phylogenetic regression analysis on mass, time and
residual variables used in the comparative growth analyses*

	Percentage of variance accounted by		
	'Grade'	Subfamily	Genus
Mass variables			
Maternal mass (MM)	55.8	30.8	9.0
Neonate mass (BM)	67.1	24.6	7.0
Weight at weaning (WW)	52.6	35.4	10.2
Time variables			
Gestation	52.5	32.8	11.8
Duration of lactation	31.5	38.0	15.8
Mean interbirth interval	43.3	28.9	18.4
Range of interbirth intervals	38.6	22.3	20.4
Mass residual variables			
Z BM/MM[a]	21.4	31.7	23.6
Z WW/BM[a]	7.3	41.1	24.3
Z Gestation/MM	48.8	17.2	25.9
Z Duration of lactation/MM[a]	35.3	27.8	21.3
Z interbirth interval/MM	36.6	15.4	30.3
Z Growth rate/MM[a]	10.6	37.6	28.9

[a]Entered into discriminant analysis.

sample thus approximates the level of variation found between the most
closely related taxa, while the 'grade' effectively partitions the considerable
variance between the small platyrrhines, the large hominoids and the
medium-sized colobids and cercopithecids in mass. For time variables,
more closely related taxa account for double the variance in comparison to
mass (Table 5.1). The same analysis on 'size-corrected' variables (using the
standardised residuals from species regression lines) suggests that even
more of the variance exists between closely related taxa when the
phylogentically constrained mass effects are removed (Table 5.1).

How much, then, does phylogeny determine clusters of similarity with
these mass-controlled variables? A discriminant analysis was used to see if
residuals (zgrowth rate, zduration of lactation, zbirth mass and zwean
weight) in Table 5.1 were classified correctly by 'grades' (61%), subfamily
(50%), or genus (64%), entering residuals into a single matrix function. It is
interesting that the genus with the greatest number of species (*Macaca*) was
the group with the lowest prediction from the observed function, with only
14.3% classified correctly. The next most specious group (*Papio*), with
three species, had only a 33% prediction.

Phylogeny is obviously a constraint on the potential for variation in these data; no platyrrhine can approach the size of a gorilla (although gibbons are similar in size to atelines). As such, it could be argued that the bias in these data comes not from the inclusion of similar, closely related, non-independent species, but rather from the significant 'biological' skew inherent in primate evolutionary phylogeny. Small primates may be simply different from large primates, and thus the inclusion of extremes (hominoids and several platyrrhines) within otherwise relatively homogenous groups (colobids and cercopithecids) will exaggerate the statistical partitioning of phylogeny.

'Grades' correctly predicted in the discriminant analyses of residuals may represent biological differences in the extent of variance that is possible, once size has been accounted for. One suggestion is that, irrespective of size, the potential for growth rates to vary is constrained by other factors such as seasonality, infant mortality and energy balance. These results also suggest that the residual, mass-corrected values are worth investigation as individual species data points, although the possibility of a '*Macaca*' or other lineage skew cannot be eliminated in this analysis. It was notable in the subfamily discriminant analysis that the 13 species of papionines were classified correctly in only 23% of cases, atelines in 33%, and cercopithecines and colobines in 75%. There does appear to be both phylogenetic clustering in growth traits and some interesting variance observed at the lowest taxonomic levels. As such, the remainder of the analyses concentrate on the species level, with the caveat that *p* valves may be inflated due to some statistical dependence. Other phylogenetic effects are discussed further below with respect to specific analyses.

Maternal mass and postnatal growth relations

As has been demonstrated in a number of previous studies (Table 5.2), maternal mass is a significant predictor of both prenatal and postnatal growth parameters. These relationships are proposed to result from maternal ability to metabolise resources for infant growth: species with high growth costs need additional energetic investment (see below). In this sample, platyrrhine mothers tend to be small and hominoids tend to be large, while the weanlings are of similar relative weights (Fig. 5.1).

Maternal mass is highly correlated with neonate mass, and is also associated with gestation length and neonate brain mass (Table 5.2). Gestation length may be more of a function of the relations between birth and brain mass than due to maternal mass alone (Leutenegger, 1973, 1979; Martin, 1996). Indeed, neonate brain and body masses are closely linked in

Table 5.2. *Comparison of growth relationships with maternal mass among primates from literature*

r^2	Slope	α	n	Comments (source)
Gestation length				
0.73	0.13		66	All primates (1)
0.77	0.10		44	Simian primates only (1)
0.55	0.13	1.9		All primates (2)
0.49	0.11	4.9	39	Single litter haplorhines (3)
0.50	0.12	1.2	42	All primates (4)
0.66	0.09	1.9	31	Haplorhines (4)
Neonate mass				
0.87	0.68	0.12	37	Single litter haplorhines (3)
0.87	0.66	0.13	32	Single litter haplorhines (5)
0.93	0.85		74	All primates (1)
0.97	0.77		51	Simian primates only (1)
0.94	0.93	0.12		All primates (2)
0.98	0.70	0.10	26	Haplorhines (6)
0.94	0.83	0.25	42	All primates (4)
0.97	0.73	0.62	31	Haplorhines (4)
Duration of lactation or weaning age				
0.35	0.35	6.2	32	Single litter haplorhines (5)
0.47	0.47	4.7	39	Single litter haplorhines (3)
0.83	0.56	2.71		All primates (2)
0.82	0.50	− 0.84	23	Strepsirhines and haplorhines (7)
Postnatal growth rate				
0.89	0.33		44	Haplorhines (8)
0.52	0.43	1.69	35	Single litter haplorhines (3)
0.66	0.35	1.53	32	Primates (9)
	0.75	1.57		Theoretical (10)
	0.42			Theoretical – primates (13)
Interbirth interval				
0.74	0.37	29.1		All primates (2)
0.54	0.34	10.4	39	Single litter haplorhines (3)
Neonate brain mass				
0.90	0.93	0.014		All primates (2)
0.93	0.77	10.7	13	Single litter haplorhines (3)
0.92	0.83	28.8	30	Primates (11)
Weaning mass				
0.88	0.81	0.406	30	Single litter haplorhines (5)
0.85	0.87	0.332	35	Single litter haplorhines (3)
	0.75	0.25		Theoretical (12)

(1) Martin and MacLarnon (1985). (2) Harvey *et al.* (1987). (3) This study. (4) Ross (1988). (5) Lee *et al.* (1991). (6) Leutenegger (1979). (7) Schwartz and Rosenblum (1983). (8) Ross and MacLarnon (1995). (9) Pontier *et al.* (1989). (10) Kirkwood (1985). (11) Martin (1983). (12) Charnov (1991). (13) Charnov (1993).

Where known, r^2 (calculated as square of correlation coefficients when not provided), slopes, constant (α), and sample size are given. Estimates for slopes from references 1, 2, 4 and 11 are derived from major axis regression; others are from least squares.

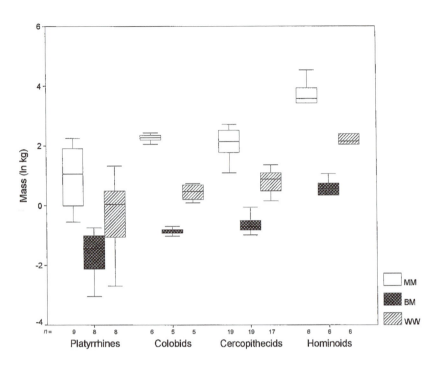

Fig. 5.1   Mean maternal mass (MM), neonate mass (BM) and weaning mass (WW) for the sample of haplorhine primates analysed.

this sample ($r = 0.99$, $n = 14$, $p < 0.001$), and gestation length was most strongly influenced by neonate brain mass ($r^2 = 0.75$, df = 12, slope = 0.13). The inclusion of maternal mass accounted for all residual variation (25.5%, slope = $- 0.32$) in gestation length.

Once the infant has been born, maternal mass appears to pace infant growth, but not to determine the ultimate weaning mass. Previous studies have found relationships between weaning age and maternal mass, and between postnatal growth rates and maternal mass for a variety of primate and non-primate species (Table 5.2). Weight at weaning, however, was more strongly affected by neonate mass ($r^2 = 0.88$, df = 34, $p < 0.001$) than by maternal mass (r^2 change = 1.6%, n.s.). Growth rates were a function primarily of maternal mass (Table 5.2). This might be the result of autocorrelation between time variables, because growth rates to weaning are expressed as g/day. Time variables are interrelated, with 12 correlations of 0.4 or greater out of a possible 15 (Table 5.3). However, when maternal mass is controlled using partial correlations, only seven of the original correlations remain, and three new correlations appear. The maternal

Table 5.3. *Correlations between time variables*

	Duration of lactation (n = 40)	Growth rate	M1 age	Brain growth	Mean interbirth interval
Growth (g/day)	—				
(n = 36)	**− 0.52**				
Age M1 eruption	0.83	0.64			
(n = 16)	**0.73**	—			
Brain growth (g/day)	—	0.67	0.68		
(n = 14)	**(− 0.39)**	—	—		
Mean interbirth interval	0.81	0.43	0.80	—	
(n = 40)	**0.62**	—	**(0.46)**	**− 0.50**	
Gestation length	0.79	0.41	0.86	(0.47)	0.76
(n = 40)	**0.58**	—	**0.83**	—	**0.50**

Partial correlations controlling for maternal mass are indicated in bold. *n* varies for pairwise comparisons (trends for $p < 0.10$).

mass-independent correlations suggest that the duration of lactation is negatively linked to growth rates for body mass and M1 eruption age, while gestation is positively related to M1 eruption age (Table 5.3). Interestingly, interbirth interval correlates negatively with brain growth when maternal size is removed, suggesting that brain growth may be costly in terms of fertility because it requires additional maternal energy and time investment.

Mean interbirth interval varies considerably across the primates (mean = 23.8 months ± 13.6; range 9–72 months), with a consistent effect of maternal mass ($r = 0.73$, $n = 40$, $p < 0.001$). What may be of greater significance is that the reported ranges of maximum and minimum interbirth interval are also strongly positively associated with maternal mass ($r = 0.70$, $n = 38$, $p < 0.001$). Larger mothers appear to have the potential for greater variation in interbirth interval. The extent of this variation is strongly influenced by two factors: the age of M1 eruption (74.8% of variance) and maternal mass (16% of variance).

Thus, an extended range of interbirth intervals is a function of the mass of the mother: larger mothers have a greater potential to extend interbirth intervals, although this may be constrained by the programme (genetically determined) for somatic growth. Indeed, it appears that mothers with greater residual variation in interbirth interval for their mass have infants with relatively more change in total brain mass ($r = 0.63$, $n = 14$, $p = 0.017$). Furthermore, relatively greater total change in brain mass is also associated with relatively later M1 eruption ($r = 0.86$, $n = 9$, $p = 0.003$).

Table 5.4. *Correlations between standardised residuals of mass-controlled time variables (as Table 5.3)*

	Zges	Zdurlac	Zgro	Zbraingro
Zdurlac (n = 40)	0.58			
Zgro (n = 36)	n.s.	− 0.54		
Zbraingro (n = 14)	n.s.	− 0.60	n.s	
ZM1 (n = 16)	n.s.	(0.42)	(− 0.43)	n.s.

Brackets indicate trends for $p < 0.10$.
Zdurlac = duration of lactation for maternal mass; Zgro = growth rates to weaning age for maternal mass; Zges = gestation for maternal mass; Zbraingro = brain growth rate for maternal mass; ZM1 = eruption age for maternal mass.

Correlations between mass-specific residuals (Table 5.4) suggest a distinction between prenatal and postnatal growth rates. Relative gestation length is unrelated to most postnatal growth residuals, although there is some 'time' concordance. Relatively long gestation is associated with relatively long lactation but, somewhat surprisingly, not with a relatively larger neonate ($r = 0.29, n = 38$, n.s.). Slowly investing species may, irrespective of their mass, simply maintain a trend in time allocation, but show no rate concordance in energy allocation after birth.

The weaning weight hypothesis

The original analysis was carried out on 88 mammal species from 13 families and five orders. Primates alone have been considered here. The least squares regression of weight at weaning to neonate mass gives a slope of 1.25 ($\pm$ 0.076), within the confidence limits of the original slope (1.135 $\pm$ 0.178) and a constant of 4.9 compared to 4.6. Interestingly, the primates appear to have relatively high weights at weaning for their birth weight compared to some other taxa, suggesting that extending a life history or spreading the costs of growth over longer periods (e.g. Janson and van Schaik, 1993) can effectively produce a larger weanling with a lower chance of mortality.

This analysis demonstrates two important points. Firstly, the result appears relatively robust even when further taxa are added or data are refined. Secondly, the effect of phylogenetic groups is minimal in the underlying relationships between birth and weaning weight (Lee, 1996),

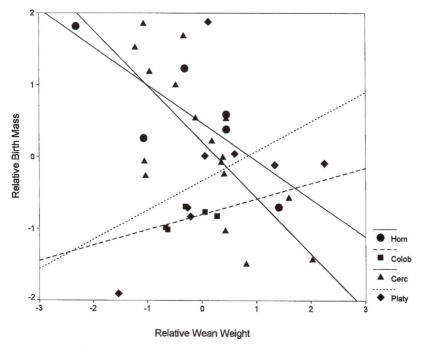

Fig. 5.2 Residuals of weight at weaning for birth mass plotted against residuals of birth mass for maternal mass. Lines fitted for each major taxonomic group separately. Groups defined as in Figure 5.1.

although an interesting trend appeared. When residuals of weights at weaning for birth weight were compared with residuals for birth to maternal mass (Fig. 5.2), a negative correlation was found overall ($r = -0.347$, $n = 36$, $p = 0.038$), such that relatively large neonates tended to be relatively small weanlings. However, there appear to be two distinct strategies, at least in this sample. Among the hominoid and cercopithecid primates, this negative trend was observed, whereas among colobids and platyrrhines, relatively larger neonates were also relatively larger weanlings. The differences between slopes were significant (ANCOVA, $F_{1,4}$ (taxa) $= 2.92$, $p = 0.05$). Thus, several different strategies may exist, both in the mass attained at weaning and in the way in which this mass gain is attained, even in this restricted sample.

Intraspecific variation

Relatively few studies have examined variation in birth mass, growth rates and age at weaning within a species, and made comparisons with the patterns predicted from the interspecific trends. Here, intraspecific patterns in growth are briefly reviewed (see also Bowman and Lee, 1995; Lee and Bowman, 1995).

Across primate species, infant mass at birth is a constant proportion of maternal mass, but the relative mass of the neonate (as a proportion of maternal mass) declines with maternal size. Larger species give birth to a proportionately smaller neonate. Within rhesus macaques, larger individual mothers give birth to larger neonates (Bowman and Lee, 1995) and larger mothers appear to be more efficient at sustaining fetal growth as gestation length varies little. However, as found in interspecific comparisons (see also Reiss, 1985), the infants of larger macaque mothers were a relatively lower proportion of their mothers' mass. The patterns are similar between and within species, although with lower slopes, as has been predicted (Kozlowski and Weiner, 1997).

A threshold weight of 1335 g (range = 1000–1600 g) was found for an intraspecific sample of rhesus macaques (Bowman and Lee, 1995), but with a constant of 3.2 (assuming exponent = 1) rather than 4. Correcting for the observed exponent of 1.25 produces a constant of 3.4. For the smaller vervet monkey, the observed threshold is lower, at 1170 g, again at about 3.4 times birth mass. These results suggested that, while a threshold weight at weaning can be identified within a species, variation in the constant may be a function of interindividual rearing strategy. A number of infants in the rhesus growth sample were not weaned within the first year of life. These infants had rapid early growth by comparison to their weaned counterparts, and reached a larger weight for age. We suggested (Bowman and Lee, 1995) that mothers were making behavioural decisions about the termination of investment in the current infant in the light of that infant's relative growth, the social risks present in its environment, and the demands of the infant for nutrition and support. It would thus appear that, while a metabolic threshold is at least theoretically likely at the level of a species, individual growth and investment decisions may produce different values for weight at weaning around that threshold.

Although the duration of lactation shows considerable variance, between species it is still somewhat related to maternal size and growth rates during lactation. Multiple regression analysis suggests that whereas 52% of variance in growth rates $(F_{1,34} = 38.2, p < 0.001)$ is accounted for by the positive allometry with maternal mass, a further negative effect of duration of lactation is found (13% of variance, $F_{2,33} = 31.6$, $p < 0.001$, $B = -0.45$). Although over 88% of the variation in weight at weaning is predicted by birth mass, there is again some residual variation between species. An increased time cost to attaining a relatively larger weaning

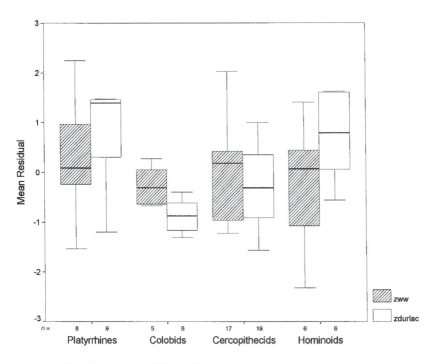

Fig. 5.3 Mean (+ 2SD) of relative weight at weaning (zww) and relative duration of lactation for maternal mass (zdurlac), plotted for each group separately.

weight might be expected. However, in this sample there was considerable variation between the species groups in this prediction of a relationship between relative weaning weight and relative duration of lactation (Fig. 5.3). Relatively long lactations were associated with relatively slow growth, low rates of energy transfer, and relatively high weaning weights for the hominoids and platyrrhines, but there was a negative trend for colobids and no relationship for the cercopithecids.

If infant growth is a function of maternal lactation capacity, it should be sensitive to both maternal size and condition, and thus influenced by the local ecology and social context of the mother. The interspecific variation in weaning trends was mirrored within the rhesus macaque sample. It was argued that behavioural components, suckling and the probability of reconception, contributed to these patterns. The duration of lactation was unrelated to maternal size in the intraspecific rhesus sample, but was related to maternal status (Bowman and Lee, 1995). Alpha-ranking mothers had slowly growing infants which attained the threshold weaning weight later. This was especially true for male infants (males being some-

Table 5.5. Intraspecific variation in growth parameters

	Reference	MM	BM	WW	Wean weight predicted by			GR g/d	Lact (mo)
					GR2	BM2	MM2		
Rhesus macaque		**8.5**	**0.47**	**1.3**		**1.9**	**2.1**	**4.5**	**6.4**
Captive (n = 31)	1	7.5	0.46	1.3	1.4	2.1	1.9	4.2	7.4
Cayo (n = 38)	2	8.2	0.45	1.5c	1.5	1.8	2.1	5.3	6.4
Mountain	3	6.5	0.43				1.7	2.1e	20.5
Japanese macaque		**9.2**	**0.50**	**2.4**		**2.1**	**2.3**	**3.5**	**6.4**
Provisioned + Density (n = 7)	4	8.1	0.50	2.0	2.0	2.1	2.1	4.1	12
High density (n = 15)	4	7.2	0.47	2.3	2.3		1.8	2.5	24
Long-tail macaque		**4.3**	**0.38**	**1.7**		**1.5**	**1.2**	**5.8**	**7.6**
Captive	5			1.6				3.65	11
Wild	5			1.9				2.8	18.4
Yellow Baboon		**11.0**	**0.71**	**3.0**		**3.2**	**2.7**	**5.2**	**15**
Wild (n = 10)	6	11.9	0.60	3.1	3.2		2.9	3	29
Garbage (n = 24)	6	16.7	0.82	3.5	4.6		3.8	5.5	23
Captive (n = 42)	7	14.8	0.89	4.2	3.6	4.2	3.5	7.9	15
Olive baboon		**13.9**	**0.95**	**3.8**		**4.6**	**3.3**	**4.8**	**20**
Wild (n = 52)	8	14.7	0.76	3.4	3.4		3.4	4.6	19.2
Raiders (n = 39)	8	15.6	0.79	3.8m	4.2		3.6	6.8	16.8
Captive (n = 91)	9	16.9	0.95	3.9	3.6	4.6	3.9	7.35	12
Chimpanzee		**31.0**	**1.75**	**8.5**		**9.8**	**6.6**	**4.0**	**56**
Captive (n = 22)	10	45.0	1.77	9.5	9.2	10.0	9.1	6.9	36
Howler		**6.0**	**0.32**	**1.1**		**1.2**	**1.6**	**1.6**	**16.5**
Wild	11	5.7	0.40				1.5	2.6e	13.7

	50.0	**2.90**	**9.2**		**18.5**	**10.0**	**6.8**	**31**
Human								
Dobe !kung	12	40.1	2.73	9.2	9.2	17.2	8.1	5.4	39.8
Gambia	13	51.0	2.90	7.7	7.9	18.5	9.8	13.2	12
Edinburgh	13	55.0	3.40	9.0	9	22.6	10.4	20.7	9
Australia	14	62.0					11.5		18

References: (1) Bowman and Lee (1995). (2) Johnson and Kapsalis (1995). (3) Mori (1979). (4) Melnick and Pearl (1987). (5) Janson and van Schaik (1993). (6) Altmann and Alberts (1987); Altmann et al. (1993); (7) Glassman et al. (1984). (8) Bercovitch (1987); Strum (1991); Smuts and Nicolson (1989). (9) Coelho (1985). (10) Leigh and Shea (1996). (11) Fedigan and Rose (1995). (12) Howell (1979); Truswell and Hansen (1976). (13) Prentice et al. (1986). (14) Short et al. (1991).

m = males only; c = estimated from regression; e = growth rate necessary to wean at weight predicted by MW for reported duration of lactation. Data are presented on maternal mass (MM), wean weights (WW), interbirth intervals (Lact mo) and growth rates (GR g/d); species averages are given in bold. Italicised data are calculated from allometric equations when no data were available. Lactation duration for captive species is given in italics as the average minimum for the species range when not specified for populations. Weaning weights predicted by the regression equations for growth rate (GR2), birth mass (BM2) and maternal mass (MM2) are presented for comparison. All units of mass are in kilograms.

what more costly in terms of their growth rates). Suckling patterns, which influence reconception through the mechanism of lactational anovulation, appear to underlie the differences in growth rates to the weaning threshold and thus the duration of lactation. High rates of suckling in both rhesus and vervet monkeys are associated with rapid growth, but a lower likelihood of conception, while low rates of suckling are associated with slower growth and reconception (Lee and Bowman, 1995). Mothers appear to balance the costs of ensuring growth for a current infant against their ability to reconceive – trading-off rapid but prolonged offspring growth with no reconception, against slower offspring growth to a minimal weaning threshold with reconception. The trade-off here concerns both maternal reproduction and infant survival, mediated by growth rates to the threshold. A study of rhesus infants (Johnson and Kapsalis, 1995) found that growth was a function of maternal body mass index (a condition index), as well as of infant sex. This is strong evidence for an effect of maternal condition on infant growth, also apparent in food-limited baboon populations (Strum, 1991; Altmann *et al.*, 1993) and in captive vervets (Fairbanks and McGuire, 1996).

Trends in growth to a threshold mass within species are similar to those predicted from interspecific comparisons (Table 5.5). What is most striking is that local ecology affects maternal size and either growth rates or their correlate – duration of lactation – most dramatically, while weight at weaning is predicted to vary relatively little. However, while suggesting a general mammalian trend in the attainment of weaning weight from which the primates as a group deviate relatively little, the intraspecific data do not yet confirm the hypothesis. Further tests are needed, based on studies of variation between individuals within species. By determining sources of variance in growth to weaning between individuals, we can better identify the constraints and selective pressures acting on patterns of growth as a life history variable.

The Martin maternal energy hypothesis

The maternal energy hypothesis, as elaborated by Martin (1996), suggests that the need for a mother to sustain brain growth lies at the heart of the time course of a reproductive event, from conception to weaning. Thus, while the mother's body size and condition determine some of her ability to allocate milk resources for growth, the growth requirements are twofold and possibly separate: firstly, that of somatic growth to a metabolic weaning weight (as noted above); and, secondly, that of brain growth during gestation and lactation.

Assuming that the majority of growth from neonate brain mass to adult brain mass occurs during lactation – a model that fits most of the *non-human* primates (Martin, 1983) and may apply to humans when an average duration of lactation is four years (Dettwyler, 1995) – then the change in brain mass should be a function of maternal energy capacity via the mother's metabolic rate and body mass relations (Martin, 1996). A relationship between brain mass and maternal mass has been noted for a variety of taxa (Martin, 1981; Gittleman, 1994), with further effects of brain mass on other life history parameters (e.g. age at first reproduction; Harvey *et al.*, 1987) for primates.

As predicted, adult brain mass scales with the duration of a reproductive event from conception to weaning among the primates in this sample. Larger-brained primates have longer reproductive events ($r = 0.77, n = 38$, $p < 0.001$). However, when the effects of maternal body mass are removed from the duration of a reproductive event and from the change in brain mass between birth and adult mass, the residuals show only a weak relationship ($r = 0.42, n = 14, p = 0.13$)! The underlying factor in the relationship appears to be simply that of maternal mass. In part, this is to be expected because mothers must ensure brain growth through lactation, and milk energy is itself a function of maternal mass (Martin, 1984; Oftedal, 1984). But, as noted above, the residual variation in interbirth intervals does correlate with residual brain mass change. Thus, primates with greater change in brain mass than predicted for their maternal mass also tend to have more potential to prolong interbirth intervals than is expected for size alone. This finding provides additional support for both the maternal energy hypothesis and the metabolic constraints hypothesis, in that expensive growth processes can be more effectively supported when time variables are labile and thus sensitive to local ecology, individual condition or social status.

As noted above, more expensive brain growth is associated with relatively higher somatic growth costs (relative neonate mass and relative wean mass), suggesting some concordance or rate constraints in growth. A further trend appears in relation to brain growth, which might be unrelated to its metabolic costs. The strong association between relative M1 eruption and relative brain growth (e.g. Smith, 1992), also noted here, suggests that, despite the tiny sample size, when maternal size is removed, brain and tooth growth appear to covary. I would suggest that one expense of growth is that of maintaining a suite of rate-limited traits that all contribute to the energy burden on the mother.

The risky environments hypothesis

In a comprehensive review of mammalian development, Eisenberg (1981) suggested links between reproductive duration, encephalisation and predation risks. Studies of different mammalian groups have emphasised the importance of environmental risks in patterns of infant development and associated life history traits (pinnipeds: Trillmich 1989, 1996; carnivores: Gittleman and Oftedal, 1987). Promislow and Harvey (1990) clearly demonstrated the importance of early mortality as a pacer of life history events in mammals. In particular, they noted the association between juvenile mortality and gestation length, duration of maternal investment (age at weaning) and growth rate to weaning independent of body size. Theoretical analyses by Charnov (1991; Charnov and Berrigan, 1993) emphasise the importance of mortality in determining the switch between individual investment in growth and investment in reproduction. These 'strategies for maturation' (e.g. Pagel and Harvey, 1993) appear to be sensitive to the resources available for growth and the risks of mortality, and also may select for adult body size itself, an argument recently developed by Kozlowski and Weiner (1997).

Janson and van Schaik (1993) have explored the concept of ecological risk in relation to the evolution of primate juvenile periods through mortality induced by food shortages imposed by the physical or social environment. Although their hypotheses relate specifically to the post-weaning period, these illustrate the importance of ecological variation affecting growth rates among primates and touch on the issue of predation avoidance as a specific problem for the smaller-bodied immatures.

There are, as yet, few data to test the relationship between growth, weaning ages, environmental variability and mortality among primates. The original study is that of Ross (1988; see also Chapter 4), which notes the influence of ecological predictability on reproductive rates and mortality among primates. Further support for the concept that ecological variation mediates species growth patterns has been provided by Leigh's (1994a) finding that folivorous primates have rapid growth over a short duration in comparison to frugivores. Among New World primates, several studies have noted the sensitivity of interbirth intervals (Fedigan and Rose, 1995) and maternal investment and allocare (Tardif, 1994; Ross and MacLarnon, 1995) to ecological variability or risk. Growth rates among small New World monkeys are unrelated to foraging risks but may be sensitive to predation risk, at least for squirrel monkeys (Garber and Leigh, 1997). Differences in growth rates among apes, again, may be a function of the degree of ecological risk or of energy derived from the diet (Leigh and Shea,

1996). In general, the overall pattern appears to suggest that risky environments, either through predation or unpredictable variation in food supply, should maximise growth rates at some cost to the mother in terms of future reproduction.

Estimates of infant mortality were not available for the interspecific sample examined here, but when a general predation risk category was assigned (Hill, 1995; Hill and Lee, 1998), relative birth mass was greater under high risks (ANOVA, $F = 3.27$, df $= 1$, $p = 0.05$), while relative weaning mass tended to be lower (ANOVA, $F = 2.8$, df $= 1$, $p = 0.075$) (Fig. 5.4A). However, neither relative growth rates nor relative duration of lactation were associated with predation risk. Thus, a mass strategy may play a greater role in predation risk aversion than does a time strategy.

When growth variables were compared between Ross' (1988) dichotomy of predictable or unpredictable environments, there were some slight effects of environment on the residual growth rate ($F_{1,34} = 2.0$, $p = 0.17$) and residual brain growth rate ($F_{1,9} = 3.9$, $p = 0.08$; Fig. 5.4B), with higher relative growth rates when environments were unpredictable. There was also an interaction with predation risk for relative brain growth ($F_{2,9} = 9.44$, $p = 0.013$), with rapid relative brain growth in unpredictable environments with high predation, and low relative brain growth in predictable environments with low predation. The sample size, however, is too small to reach any definitive conclusions. Relative growth rate also appeared to be lowest in predictable environments with low predation risks, and highest in unpredictable environments with high predation risks (Interaction $F_{1,2} = 2.45$, $p = 0.102$; Fig. 5.4C). Although crude, grossly oversimplistic, and of low significance, there is at least some suggestion that environmental risk affects primate growth strategies – with predation influencing mass, while environmental quality affects the tempo of events, as suggested by Leigh (1994a).

Conclusions

Postnatal growth strategies among primates are interdependent, and the hypotheses examined briefly above are probably intercorrelated. Indeed, the maternal energy hypothesis underlies all other effects of environments or metabolic requirements because infants can only grow when mothers meet those needs through milk (Martin, 1984; Oftedal, 1984; Tilden and Oftedal, 1997). When environmental risk is high, mothers may need to sustain additional relative costs of growth. Risky environments are associated with relatively high birth mass, high growth rate and lower

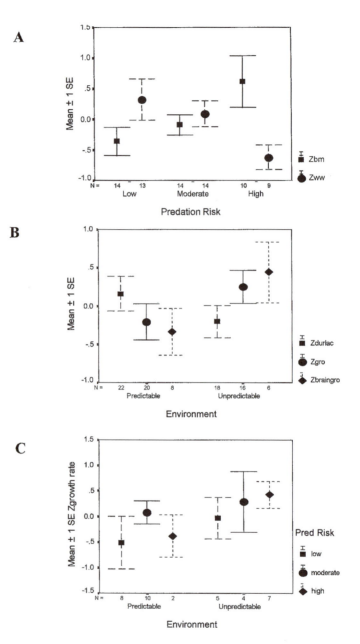

Fig. 5.4   (A) Mean relative birth mass (Zbm) and relative weaning weight (Zww) by predation risk category. (B) Mean residual duration of lactation (Zdurlac), residual growth rate (Zgro) and residual brain growth rate (Zbraingro) for each category of environmental quality. (C) Mean residuals of growth rate (Zgro) for each predation category within the two quality of environment categories.

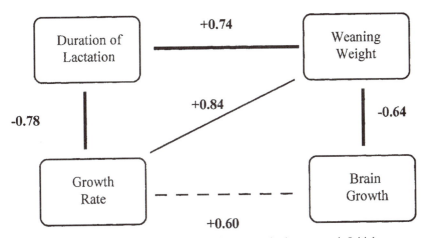

Fig. 5.5 Path of partial correlations in postnatal primate growth. Initial correlations were determined with maternal and birth mass controlled, subsequent correlations with each set of independent variables controlled.

relative weaning weights. In such environments, mothers may be investing in fetal and postnatal growth over a short period as a strategy to minimise the vulnerability of small neonates to either predation or variation in food supply. Leigh (1994a) suggested that folivorous species sustain higher rates of growth to adult size than do frugivores, which is unexpected if folivores occupy relatively predictable, low-predation environments. However, the differences between these suggestive results may be a consequence of the difference in relative weaning weights: if low-risk environments promote large weaning size, then the attainment of an adult body mass will be generally quicker. The question to be addressed concerns the difference in growth rates prior to and after weaning between these groups.

As a summary of the major relationships between postnatal growth variables, Figure 5.5 presents the paths of highest correlation when maternal and neonate masses are removed, and when each correlation also removes the other independent variables. High weaning weights are associated with high growth rates, longer lactations and, independently, with slower rates of brain growth during the lactation period. Lactation length is most strongly associated with rates of growth, although negatively. This suggests that lactation length and growth rates can be decoupled, resulting in variation in subsequent weight attained at weaning. Finally, higher rates of brain growth tend to be found in species with high rates of postnatal growth.

While these trends may be a consequence of the limited data available here, two patterns emerge. One is wean small and early, but with a high energy cost to the mother in terms of maintaining growth and especially that due to sustaining rapid brain growth. This can be considered the energy-expensive, but time-cheap option. The other pattern is to sustain slow growth over a long period, producing a hefty weanling, although with a lower rate of brain growth and potentially lower brain maturity. This is the time-expensive strategy, while possibly less energy expensive in the short term.

It might be argued that these strategies should differ between large and small mothers, due to the relationship between maternal mass and the efficiency of sustaining growth. However, these trends are derived from mass-controlled correlations. Gordon (1989) suggests that when the costs of tissue maintenance are taken into account, the energy expense of a reproductive event is similar across mammals of different sizes. The evolution of trends in postnatal growth may thus reflect more about energy balance among primates than about selection for body size. Another important consideration is the robust trend for M1 eruption age to relate to a number of other time variables. As Smith (1992) has shown, M1 eruption appears to be a sensitive measure of time in life histories, although results here suggest that it is less related to attained mass strategies. We can thus ask if growth varies between organs, with teeth and brains more sensitive to a rate-limited temporal sequence, while mass reflects ecological and behavioural variation. Until sufficient data are available for growth patterns of different organs such as guts, as well as for brains and teeth, we will be unable to produce a comprehensive perspective on growth costs or constraints for primates.

The period of infancy remains one of crucial significance: the major mortality occurs within this time, adult size and hence reproductive ability are established, and lifetime reproductive output is affected by the maternal trade-off between sustaining current infant growth and investing in subsequent reproduction. Are primate mothers simply opting for one mammalian strategy as a group – the slow lane of growth (e.g. Charnov and Berrigan, 1993), as the classical r/K dichotomy would propose – or can the strategies suggested above, and indeed those within species, be used to explain behaviour in relation to social systems and local ecology? Comparative analyses point to interesting differences between species and habitats, reflecting recent evolutionary history as well as ecological niche. Investigation of growth patterns among the strepsirhines is needed to test further for 'primate' patterns, and to assess the extent of variation within the order. Species with larger litters and those with extensive allocare represent groups with different strategies for growth (Tardif, 1994; Ross and MacLarnon, 1995; Garber and Leigh, 1997). Haplorhines with single infants are but one part of this adaptive array, and even they appear to be able to utilise a variety of postnatal growth strategies. These strategies appear to depend on absolute size, metabolic relationships, ecological risk and mortality patterns, as well as varying between individuals as a function of social and ecological constraints. The alternative reproductive strategies available to primate mothers are set within the bounds of evolved species life histories, yet with this potential for variation. Infant growth is one important component of such a strategy, and while a metabolic efficiency may be necessary for weanling survival, how mothers invest to attain that threshold provides the key to life history differences.

Acknowledgements

I thank J. Bowman for a long-term fruitful collaboration, R.D. Martin and R. Barton for stimulating discussions, and R. Foley, M. Lahr, S. Leigh, A. MacLarnon, R. Martin and C. Ross for comments on the manuscript.

References

Altmann, J. and Alberts, S. (1987). Body mass and growth rates in a wild primate population. *Oecologia* **72**, 15–20.

Altmann, J., Schoeller, D., Altmann, S.A., Murithi, P. and Sapolsky, R.M. (1993). Body size and fatness of free-living baboons reflect food availability and activity levels. *American Journal of Primatology* **30**, 149–61.

Altmann, S.A. (1991). Diets of yearling female primates (*Papio cynocephalus*) predict lifetime fitness. *Proceedings of the National Academy of Sciences, USA* **88**, 420–3.

Bercovitch, F.B. (1987). Female weight and reproductive condition in a population of olive baboons (*Papio anubis*). *International Journal of Primatology* **12**, 189–95.

Bowman, J.E. and Lee, P.C. (1995). Growth and threshold weaning weights among captive rhesus macaques. *American Journal of Physical Anthropology* **96**, 159–75.

Charnov, E.L. (1991). Evolution of life history variation among female mammals. *Proceedings of the National Academy of Sciences, USA* **88**, 1134–7.

Charnov, E.L. (1993) *Life History Invariants: Some Explorations of Symmetry in Evolutionary Ecology*. Oxford: Oxford University Press.

Charnov, E.L. and Berrigan, D. (1993). Why do female primates have such long lifespans and so few babies? or life in the slow lane. *Evolutionary Anthropology* **1**, 191–4.

Clutton-Brock, T.H., Albon, S.D. and Guinness, F.E. (1989). Fitness costs of gestation and lactation in wild mammals. *Nature* **337**, 260–2.

Coelho, A.M. Jr (1985). Baboon dimorphism: growth in weight, length and adiposity from birth to eight years of age. *Monographs on Primatology* **6**, 125–9.

Dettwyler, K.A. (1995). A time to wean: the hominid blueprint for a natural age of weaning in modern human populations. In *Breastfeeding: Biocultural Perspectives*, ed. P. Stuart-Macadam and K.A. Dettwyler, pp. 39–73. New York: Aldine.

Eisenberg, J.F. (1981). *The Mammalian Radiations*. Chicago: University of Chicago Press.

Fairbanks, L.A. and McGuire, M.T. (1996). Maternal condition and the quality of maternal care in vervet monkeys. *Behaviour* **132**, 733–54.

Fedigan, L.M. and Rose, L.M. (1995). Interbirth interval variation in three sympatric species of neotropical monkey. *American Journal of Primatology* **37**, 9–24.

Garber, P.A. and Leigh, S.R. (1997). Ontogenetic variation in small-bodied New World primates: implications for patterns of reproduction and infant care. *Folia Primatologica* **68**, 1–22.

Gittleman, J.L. (1994). Female brain size and parental care in carnivores. *Proceedings of the National Academy of Sciences, USA* **91**, 5495–7.

Gittleman, J.L. and Oftedal, O.T. (1987). Comparative growth and lactation energetics in carnivores. *Symposium of the Zoological Society of London* **57**, 41–77.

Glassman, D.M., Coelho, A.M. Jr, Carey, K.D. and Bramblett, C.A. (1984). Weight growth in savannah baboons: a longitudinal study from birth to adulthood. *Growth* **48**, 425–33.

Gordon, I.J. (1989). The interspecific allometry of reproduction: do larger species invest relatively less in their offspring? *Functional Ecology* **3**, 285–8.

Harvey, P.H., Clutton-Brock, T.H. and Martin, R.D. (1987). Life histories in comparative perspective. In *Primate Societies*, ed. B.B. Smuts, D.L. Cheney, R.M. Seyfarth, R.W. Wrangham and T.T. Struhsaker, pp. 181–96. Chicago: University of Chicago Press.

Harvey, P.H. and Pagel, M.C. (1991). *The Comparative Method in Evolutionary Biology*. Oxford: Oxford University Press.

Hill, R.A. (1995). Modelling the effects of predation on the evolution of primate social systems and morphology. M.Phil. Thesis, University of Cambridge.

Hill, R.A. and Lee, P.C. (1998). Predation risk as an influence on group size in cercopithecoid primates: implications for social structure. *Journal of Zoology* **245**, 447–56.

Howell, N. (1979). *Demography of the Dobe !Kung*. New York: Academic Press.

Janson, C.H. and van Schaik, C.P. (1993). Ecological risk aversion in juvenile primates: slow and steady wins the race. In *Juvenile Primates*, ed. M.E. Pereira and L.A. Fairbanks, pp. 57–74. Oxford: Oxford University Press.

Johnson, R.L. and Kapsalis, E. (1995). Determinants of postnatal weight gain in infant rhesus monkeys: implications for the study of interindividual differences in neonatal growth. *American Journal of Physical Anthropology* **98**, 343–53.

Kirkwood, J.K. (1985). Patterns of growth in primates. *Journal of Zoology, London* **205**, 123–36.

Kozlowski, J. and Weiner, J. (1997). Interspecific allometries are a by-product of body size optimisation. *American Naturalist* **149**, 352–80.

Lee, P.C. (1996). The meanings of weaning: growth, lactation and life history. *Evolutionary Anthropology* **5**, 87–96.

Lee, P.C. and Bowman, J.E. (1995). Influence of ecology and energetics on primate mothers and infants. In *Motherhood in Human and Nonhuman Primates*, ed. C.R. Pryce, R.D. Martin and D. Skuse, pp. 47–58. Basel: Karger.

Lee, P.C., Majluf, P. and Gordon, I.J. (1991). Growth, weaning and maternal investment from a comparative perspective. *Journal of Zoology, London* **225**, 99–114.

Leigh, S.R. (1994a). The ontogenetic correlates of folivory in anthropoid primates. *American Journal of Physical Anthropology* **94**, 499–522.

Leigh, S.R (1994b). The relations between captive and non-captive weights in anthropoid primates. *Zoo Biology* **13**, 21–44.

Leigh, S.R. and Shea, B.T. (1996). Ontogeny of body size variation in African apes. *American Journal of Physical Anthropology* **99**, 43–65.

Leutenegger, W. (1973). Maternal–foetal weight relationships in primates. *Folia Primatologica* **20**, 280–93.

Leutenegger, W. (1979). Evolution of litter size in primates. *American Naturalist* **114**, 525–31.

Martin, R.D. (1981). Relative brain size and basal metabolic rate in terrestrial vertebrates. *Nature* **293**, 57–60.

Martin, R.D. (1983). *Human Brain Evolution in an Ecological Context*. New York: American Museum of Natural History.

Martin, R.D. (1984). Scaling effects and adaptive strategies in mammalian lactation. *Symposia of the Zoological Society of London* **51**, 87–117.

Martin, R.D. (1996). Scaling of the mammalian brain: the maternal energy hypothesis. *News in Physiological Sciences* **11**, 149–56.

Martin, R.D. and MacLarnon, A.M. (1985). Gestation period, neonatal size and maternal investment in placental mammals. *Nature* **313**, 220–3.

Melnick, D.J. and Pearl, M.C. (1987). Cercopithecines in multi-male groups: genetic diversity and population structure. In *Primate Societies*, ed. B.B. Smuts, D.L. Cheney, R.M. Seyfarth, R.W. Wrangham and T.T. Struhsaker, pp. 121–34. Chicago: University of Chicago Press.

Mori, A. (1979). Analysis of population changes by measurements of body weight in the Koshima troop of Japanese monkeys. *Primates* **20**, 371–97.

Oftedal, O.T. (1984). Milk composition, milk yield and energy output at peak lactation: a comparative review. *Symposia of the Zoological Society of London* **51**, 33–85.

Pagel, M.D. and Harvey, P.H. (1993). Evolution of the juvenile period in mammals. In *Juvenile Primates*, ed. M.E. Pereira and L.A. Fairbanks, pp. 28–37. Oxford: Oxford University Press.

Pontier, D., Gaillard, J.M., Allaine, D., Trouvilliez, J., Gordon, I. and Duncan, P. (1989). Postnatal growth rate and adult body weight in mammals: a new approach. *Oecologia* **80**, 390–4.

Prentice, A.M., Paul, A., Prentice, A., Black, A, Cole, T. and Whitehead R.G. (1986). Cross-cultural differences in lactational performance. In *Human Lactation 2*, ed. M. Hamosh and A.S. Goldman, pp. 13–44. New York: Plenum Press.

Promislow, D.E.L. and Harvey, P.H. (1990). Living fast and dying young: a comparative analysis of life history variation among mammals. *Journal of Zoology, London* **220**, 417–38.

Reiss, M.J. (1985). The allometry of reproduction: why larger species invest relatively less in their offspring. *Journal of Theoretical Biology* **113**, 529–44.

Rogers, A.R. and Cashdan, E. (1997). The phylogenetic approach to comparing human populations. *Evolution and Human Behaviour* **18**, 353–8.

Ross, C. (1988). The intrinsic rate of natural increase and reproductive effort in primates. *Journal of Zoology, London* **214**, 199–219.

Ross, C. and MacLarnon, A. (1995). Ecological and social correlates of maternal expenditure on infant growth in haplorhine primates. In *Motherhood in Human and Nonhuman Primates*, ed. C.R. Pryce, R.D. Martin and D. Skuse, pp. 37–46. Basel: Karger.

Schwartz, G.G. and Rosenblum, L.A. (1983). Allometric influences on primate mothers and infants. In *Symbiosis in Parent–Offspring Interactions*, ed. L.A. Rosenblum and H. Moltz, pp. 215–48. New York: Plenum Press.

Short, R.V., Lewis, P.R., Renfree, M.B. and Shaw, G. (1991). Contraceptive effects of extended lactational amenorrhea: beyond the Bellagio Consensus. *Lancet* **337**, 715–17.

Skuse, D., Wolke, D., Reilly, S. and Chan, I. (1995). Failure to thrive in human infants: the significance of maternal well-being and behaviour. In *Motherhood in Human and Nonhuman Primates*, ed. C.R. Pryce, R.D. Martin and D. Skuse, pp. 162–70. Basel: Karger.

Silk, J.B. (1986). Eating for two: behavioural and environmental correlates of gestation length. *International Journal of Primatology* **7**, 583–602.

Small, M.F. and Smith, D.G. (1986). The influence of birth timing upon infant growth and survival in captive rhesus macaques (*Macaca mulatta*). *International Journal of Primatology* **7**, 289–304.

Smith, B.H. (1992) Life history and the evolution of human maturation. *Evolutionary Anthropology* **1**(4), 134–42.

Smith, B.H., Crummett, T.L. and Brandt, K.L. (1994). Ages of eruption of primate teeth: a compendium for ageing individuals and comparing life histories. *Yearbook of Physical Anthropology* **37**, 177–231.

Smith, R.J. and Jungers, W.L. (1997) Body mass in comparative primatology. *Journal of Human Evolution* **32**, 523–59.

Smuts, B.B. and Nicolson, N. (1989). Reproduction in wild female olive baboons. *American Journal of Primatology* **19**, 229–46.

Strum, S.C. (1991). Weight and age in wild olive baboons. *American Journal of Primatology* **29**, 219–37.

Tanner, J.M. and Preece, M.A. (eds.) (1989). *The Physiology of Human Growth*. Cambridge: Cambridge University Press.

Tardif, S.D. (1994). Relative energetic cost of infant care in small bodied neotropical primates and its relation to infant-care patterns. *American Journal of Primatology* **34**, 133–43.

Tilden, C.D. and Oftedal, O.T. (1997). Milk composition reflects patterns of maternal care in prosimian primates. *American Journal of Primatology* **41**, 195–211.

Tomkins, A. (1994). Growth monitoring, screening and surveillance in developing countries. In *Anthropometry: the Individual and the Population*, ed. S.J. Ulijaszek and C.G.N. Mascie-Taylor, pp. 108–16. Cambridge: Cambridge University Press.

Trillmich, F. (1990). The behavioural ecology of maternal effort in fur seals and sea lions. *Behaviour* **114**, 3–20.

Trillmich, F. (1996). Parental investment in pinnipeds. *Advances in the Study of Behaviour* **25**, 533–77.

Truswell, A.S. and Hansen, J.D.L. (1976). Medical research among the !Kung. In *Kalahari Hunter–Gatherers*, ed. R.B. Lee and I. DeVore, pp. 166–95. Harvard: Harvard University Press.

Appendix 5.1
Species and values used in the analysis

Species	Subfamily	Group	Predation risk[a]	Environment	Maternal weight (kg)	Birth weight (kg)	Gestation (kg)	Duration of lactation (months)[b]	Mean interbirth interval (months)	Minimum interbirth interval (months)	Maximum interbirth interval (months)	M1 age[c]	Wean weight (kg)
Tarsius bancanus	tar	Tar	M	P	0.1	0.024	178	2.5	11.2	5.5	14.0	4.32	0.256
Aotus trivirgatus	aot	Platy	L	P	1.0	0.098	133	2.5	9.0	4.0	14.0	4.40	0.479
Saimiri sciureus	ceb	Platy	H	P	0.6	0.146	170	8.0	22.0	14.0	26.0	13.8	1.00
Cebus apella	ceb	Platy	M	P	2.6	0.232	154	8.7	26.4	14.0	55.0		1.35
Cebus capucinus	ceb	Platy	M	P	2.9	0.239	162	17.0	24.0	13.0	50.0		2.00
Ateles paniscus	atel	Platy	L	P	7.7	0.480	230	27.0	37.4	17.0	48.0		3.79
Ateles geoffroyi	atel	Platy	L	P	6.8	0.426	229	25.0	18.3	13.0	26.0		1.10
Alouatta palliata	atel	Platy	L	P	6.0	0.318	186	16.5	33.8	10.5	44.0		
Brachyteles arachnoides	atel	Platy	L	P	9.5		233	28.5		26.0			
Callimico goeldii	callim	Platy	H	U	0.6	0.048	151	2.2	9.0	6.0	12.0		0.067
Miopithecus talapoin	cerco	Cerco	H	P	1.1	0.180	165	6.5	12.0	12.0	24.0		0.042
Erythrocebus patas	cerco	Cerco	H	U	5.6	0.625	167	8.5	12.0	7.3	23.1		2.40
Cercopithecus aethiops	cerco	Cerco	H	U	3.0	0.430	163	6.7	12.0	10.0	36.0	9.96	1.17
Cercopithecus neglectus	cerco	Cerco	M	P	4.0	0.260	168	14.0	12.0	12.0	42.0		1.64
Lophocebus albigena	papio	Cerco	M	P	7.5	0.500	186	7.0	33.0	18.0	48.0		2.20
Macaca nemestrina	papio	Cerco	M	P	7.8	0.473	171	7.8	14.0	8.0	36.0	16.32	1.32
Macaca arctoides	papio	Cerco	M	P	8.0	0.489	178	13.1	19.0	13.0	36.0	16.44	2.30
Macaca fuscata	papio	Cerco	L	U	9.2	0.503	173	6.4	14.0	12.0	36.0	18.00	2.40
Macaca fascicularis	papio	Cerco	M	P	4.3	0.375	167	7.6	13.0	11.0	24.0	16.50	1.70
Macaca thibetana	papio	Cerco	L	U	12.8	0.500	170	18.7	24.0				2.40
Macaca cyclopis	papio	Cerco	L	U	6.2	0.402	162	6.8	15.4	10.5	24.0		
Macaca mulatta	papio	Cerco	M	U	8.5	0.473	164	6.4	12.0	8.7	36.0	16.14	1.33
Macaca sylvanus	papio	Cerco	L	U	13.3	0.450	165	7.0	22.0	8.0	60.0		2.42
Mandrillus sphinx	papio	Cerco	M	P	11.5	0.613	220	10.1	17.3	11.0	20.0		3.00
Papio cynocephalus	papio	Cerco	H	U	11.0	0.710	175	15.0	23.0	15.0	29.0	19.98	3.03
Papio anubis	papio	Cerco	H	U	13.9	0.950	180	20.0	24.5	9.0	37.0	20.04	3.80
Papio ursinus	papio	Cerco	H	U	15.2	0.600	187	29.0	29.0	20.0	48.0		
Papio hamadryas	papio	Cerco	H	U	12.0	1.00	170	18.7	24.0	12.0	36.0		3.10
Theropithecus gelada	papio	Cerco	L	U	13.6	0.465	180	18.0	24.0	12.0	30.0		3.90
Colobus polykomos	colob	Colob	M	P	9.0	0.400	170	7.2	22.0	12.0	32.0		1.24
Colobus guereza	colob	Colob	M	P	9.3	0.445	170	11.0	20.0	8.0	32.0		1.60
Semnopithecus entellus	colob	Colob	H	U	11.4	0.500	184	8.3	16.7	10.6	76.4		2.10
Trachypithecus phayrei	colob	Colob	L	P	10.5		165	10.2	15.0				
Trachypithecus vetulus	colob	Colob	L	P	7.8	0.360	200	9.0	31.5	15.0	48.0		1.10
Nasalis larvatus	nasal	Colob	L	P	10.0	0.450	166	7.0	18.0	12.0	24.0		2.00
Hylobates lar	hylob	Hom	L	P	5.3	0.400	205	24.5	30.0	24.0	60.0	21.00	1.07

Pan paniscus	pan	Hom	L	P	32.0	1.40	240	36.0	48.0	36.0	84.0		8.50
Pan troglodytes	pan	Hom	M	U	31.0	1.75	240	56.0	60.0	39.0	90.0	38.94	8.50
Gorilla gorilla	pan	Hom	L	P	93.0	2.11	285	30.0	47.0	36.0	84.0	42.00	19.80
Pongo pygmaeus	pongo	Hom	M	U	40.0	1.73	244	42.0	72.0	36.0	108.0	42.00	11.00
Homo sapiens	hom	Hom	L	U	50.0	2.9	270	31.0	36.0	15.0	100.0	74.88	9.2

[a]Predation risk assigned by body size, crypsis and defence in combination to known vulnerability to arboreal, aerial or terrestrial predators within habitats. M = medium; L = low; H = high.
[b]Duration of lactation = reported duration of lactational anovulation; where not reported, calculated for seasonal breeders as 12 months gestation, for non-seasonal breeders = average inter-birth – gestation.
[c]Mandibular M1 eruption in months.
P = predictable; U = unpredictable.

6 Some current ideas about the evolution of the human life history

NICHOLAS BLURTON JONES, KRISTEN HAWKES AND
JAMES F. O'CONNELL

Introduction

Adolph Schultz's famous diagram (Fig. 6.1) aged well: human life history is characterised by a long juvenile period (weaning to reproductive maturity), and a long post-reproductive lifespan in females. Table 6.1 compares human and great ape life history parameters and invariants of Charnov (1993). Notable here are low adult mortality, high fertility squeezed into a reproductive span similar in length to that of *Pan* (thus, shorter interbirth intervals), late age at first reproduction and great length of the juvenile period.

How do we explain these differences between our nearest relatives and ourselves? This chapter summarises some recent attempts to use life history models on data from contemporary hunter–gatherers, and other non-contracepting populations with little access to modern medicine (see also Borgerhoff Mulder, 1991; Hill, 1993; and for a comprehensive review of hunter–gatherer research, Kelley, 1995).

Trade-off between numbers and care of offspring

Hill and Hurtado (1996) examine interbirth interval and the trade-off between increased fertility and increased infant and child mortality among Ache foragers in Paraguay. As in other populations, after controlling for early death of a previous infant, mother's age, and mother's weight, shorter interbirth intervals are accompanied by higher infant and child mortality. But, in contrast to Blurton Jones (1986; for discussion of Harpending's (1994) critiques see Blurton Jones, 1994; for Hill and Hurtado's results, see Blurton Jones, 1996), they found that this effect was much too weak to render the observed intervals optimal. Ache values predict that the optimal interbirth interval would be much shorter than is observed. Hill and Hurtado discuss possible reasons, suggesting that there must be costs to

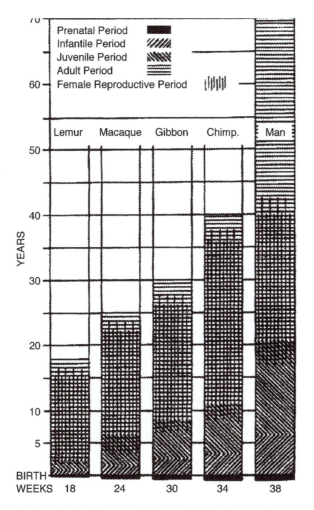

Fig. 6.1   Schultz's diagram of primate life histories. (Modified from Schultz, 1969.)

very short intervals that we have yet to appreciate. They also emphasise the effect of mother's size, bringing the study of human fertility closer both to the concept of productivity in Charnov's life history model and to primate literature (e.g. Lee, Majluf and Gordon, 1991; Lee and Bowman, 1995). But the source of Blurton Jones and Sibly's (1978) predictions (tested by Blurton Jones, 1986, 1987), maternal provisioning of juveniles, which so markedly distinguishes humans from other primates, will come back into the picture below.

Table 6.1. *Average values for selected life history variables*

	Average adult life span (years) ELC method (1/M)	Age at maturity[b] (years)	Age at weaning[c] (years)	α[d] (years)	αM	Weaning weight/adult weight[e]	Daughters/ year[f]	αb
Orangutan	17.9	14.3	6.0	8.3	0.46	0.28	0.06	0.52
Gorilla	13.9	9.3	3.0	6.3	0.45	0.21	0.126	0.79
Chimpanzee	17.9	13.0	4.8	8.2	0.46	0.27	0.087	0.70
Human	32.9	17.3	2.8	14.5	0.44	0.21	0.142	2.05

[a]The method described in Charnov (1993, caption to Figure 5.6, p. 104) is used to estimate average adult life span (1/M) from maximum observed life spans (T_{max}): $1/M = 0.4T_{max} - 0.1$. Values for orangutans: Leighton et al. (1995); gorillas: Stewart, Harcourt and Watts (1988); chimpanzees: Nishida, Takasaki and Takahata (1990). The human value is estimated from Howell's (1979) oldest observed !Kung individual (age 88), and Hill and Hurtado's (1996) oldest observed (forest living) Ache individual (age 77).

[b]Age at first birth minus gestation. Orangutans: Leighton et al. (1995); gorillas: Stewart et al. (1988); chimpanzees: mean of means from Wallis (1997) for Gombe, Nishida et al. (1990) for Mahale, and Sugiyama (1984) for Bossou; humans: mean of the mode for !Kung in Howell (1979) and Ache in Hill and Hurtado (1996).

[c]Orangutans: Galdikas and Wood (1990); gorillas: Stewart et al. (1988); chimpanzees: the mean of the estimate from Goodall (1986) for Gombe and from Nishida et al. (1990) for Mahale; humans: the mean of the median for !Kung in Howell (1979) and Ache in Hill and Hurtado (1996).

[d]Defined as the period of independent growth, from weaning to maturity.

[e]Data from Lee et al. (1991) for the great apes. Maternal size for orangutans is estimated to be 40 kg, gorillas 93 kg, chimpanzees 40 kg. In that data set, δ for humans is 0.16 with maternal size at 55 kg (the upper end of the range for modern foragers who are generally smaller than contemporary non-foragers and pre-Mesolithic moderns). The mean of the !Kung (Howell, 1979) – who are at the upper end – is used to represent humans.

[f]Great ape data from Galdikas and Wood (1990), who reappraise birth spacing in all species in the same way. Medians calculated therein (for closed intervals) plus two months are used to approximate the mean interval; then divide by 2 to get the rate for daughters. Galdikas and Wood use the Gainj, a population of horticulturalists in highland Papua New Guinea, to represent humans, for which $b = 0.132$. The mean of the !Kung (Howell, 1979) and the Ache (Hill and Hurtado, 1996) are used by the authors.

Belsky, Steinberg and Draper (1991) – taking their lead from Draper and Harpending (1982) and Pennington and Harpending (1988) – and Chisholm (1993) interpret variation in parental care and reproductive strategies in human societies as responses to high and uncontrollable mortality. Poverty and unpredictable environments were proposed to go with higher investment in fertility and less care, and earlier maturation, resulting from an early developmental shift triggered by parental care as an indicator of future reproductive opportunities. Empirical studies provide mixed support for earlier maturation among poorer or more stressed girls (e.g. Surbey, 1990).

Blurton Jones (1993) suggested that in humans parental fitness can be enhanced by effort allocated not only to numbers and survivorship of offspring but also to a variety of ways to enhance offsprings reproductive success, such as accumulating and endowing wealth and status, caring for grandchildren, teaching, arranging mates. Variation in returns to effort directed to such ends should predict variation in the amount of effort directed towards them. We might then be able to account for extensive variation in patterns of parental care. (The distinction between depreciable and non-depreciable care would play an important role in this enterprise.) For example, anthropologists have noted the scarcity of direct teaching – we would suggest that most mothers have more fitness-enhancing demands upon their time, such as housing and feeding themselves and their children. Where added effort directed towards shelter and food yields little increase in fitness, we expect to see parents taking more time to train and educate children. Kaplan (1996) developed a much more advanced model of the allocation of parental resources that links together issues as diverse as the demographic transition, education and health-seeking behaviour, socioeconomic status and labour markets, in what must be the most ambitious (and promising) invasion of the social sciences by evolutionary thinking yet undertaken.

Age at first birth

Hill and Hurtado (1996) report their intensive investigation of demography of the Ache hunter-gatherers of Paraguay by means of carefully interlinked interviews conducted between 1977 and 1995. The information covers a 'pre-contact' or 'forest' period, which ended in 1971, a brief but decimating (from 544 in 1970 to 338 in 1976) period of epidemics during the first years of settlement, and a subsequent period during which most people spent most of their time at settlements, and the population rapidly grew to pre-contact levels – 537 people in 1989.

Hill and Hurtado examined women's weight as a predictor of fertility. After controlling for age, and whether the previously born child was dead or alive, they found a significant contribution of body weight to predicting time to the next birth. Heavier women had children after a shorter interval than lighter women. Thus, delaying maturation to grow bigger yields greater fertility. Delaying too long cuts into the time available to reproduce, and lowers the probability of surviving from birth until reproduction begins. Fitting real-life parameters to equations summarising this trade-off predicts an optimal age and weight at first birth. The observed Ache age at first birth (17.5 years) fits well with the prediction, but the model predicts maturity at greater weight than observed. Hill and Hurtado then fit !Kung weight and mortality data from Howell (1979) (but use the Ache regression of fertility on weight) to the equations and show that observed age at first birth fits the slightly higher age predicted (19 years). Growth data predicted the much younger age at menarche observed in the USA. Analyses of individual variation among fast-growing and slow-growing Ache girls provide further support for the generality of the model. Faster-growing Ache girls have their first baby earlier than slow-growing girls. Apparently, neuroendocrine maturation mechanisms do their job flexibly and efficiently.

Hill and Hurtado point out that variation in adult mortality rates within humans makes little difference to the calculated optimal age at maturity, but that lowering mortality to chimpanzee levels predicts maturity at 14 years – close to the 13 years noted in Table 6.1.

Like Stearns and Koella (1986), Hill and Hurtado sought the strategy that maximised fitness, measured as r (intrinsic rate of increase). Stearns (1992: p. 148) points out that Kozlowski and Wiegert (1987) examined maximisation of $\Sigma l_x m_x$ and obtained the prediction that faster growth predicted later maturity. This point, apparently commonplace to life history theorists, is noted as a warning to us 'end-users' that the field is complex, in flux, and there is as yet no single 'right' model.

Hill and Hurtado make an interesting excursion into male life histories: among Ache men who have completed their growth, fertility relates to size but follows an inverted U shape. They suggest that this is because very heavy men have lower hunting success than men of intermediate weight (their Fig. 11.12 is reminiscent of Lee's (1979) observation that taller !Kung had lower hunting success than shorter men). Using male mortality, growth rate, and age-specific and weight-specific fertility, Hill and Hurtado sought the optimal age at cessation of growth. The result shows that optimal male age and size at cessation of growth are greater than for females, as observed among Ache and almost every other population. The model predicts body weight dimorphism ratio (male/female) of 1.03, rather than the observed 1.11. Hill and Hurtado draw several interesting further implications, for example that variance in body weight should be lower for men than for women, and among Ache it is.

They link these observations of developing hunting success to the idea that humans mature late because this allows them more time to acquire fitness-enhancing skills (Lancaster and Lancaster, 1983; Bogin, 1990; Lancaster, 1997). Kaplan's interest in human capital theory (1994), often used to discuss the costs and benefits of staying on at school, may lead to a more explicit version of this view. These authors have argued in another direction (Blurton Jones, Hawkes and O'Connell, 1997; Hawkes, O'Connell and Blurton Jones, 1997) and do so below.

Zero population growth?

Hill and Hurtado make a suggestion that runs counter to much that has been believed about human population dynamics. The zero population growth observed among the !Kung, and the density-dependent effects shown by Wood and Smouse (1982) for the Gainj were taken to confirm that in prehistoric times our species conformed to the expectable steady state, density-dependently controlled, level population size. Population growth would thus result slowly from expansion into new habitats, and from technological advances. Our currently high rates of population growth were then regarded as some errant outcome of modernity arriving in the Third World.

Hill and Hurtado point out that the accumulated data suggest otherwise. Mortality levels (especially adult mortality) in populations without access to modern medicine (they list Ache, Batak, Yanomamo, !Kung, and the authors would add Hadza – Blurton Jones *et al.*, 1992) are high by modern standards but are very low compared to our primate relatives, including other great apes. Human fertility also – whether you take the low-fertility !Kung (total fertility rate, TFR 4.7: Howell, 1979; or 5.05; Blurton Jones, 1994) or the high-fertility Ache (TFR 8.3) or the intermediate Hadza (TFR 6.2) – is high by the standards of our nearest relatives (as shown in Table 6.1). These figures suggest extremely rapid rates of increase, as are actually displayed by the Ache and the Hadza. Hill and Hurtado suggest that a saw-tooth pattern may have characterised much of human evolutionary history. !Kung zero population growth may owe its existence to sexually transmitted diseases of the reproductive tract (STDs) that greatly increased in frequency with the arrival of larger numbers of herders in the 1950s (Harpending, 1994).

It may not be unusual to observe short-term imbalance of fertility and mortality in animal populations and we should be cautious about Hill and Hurtado's suggestion. But simulations suggest that the 400–1000-person forager populations can give reliable demographic data (Jenike, personal communication), so Hill and Hurtado's argument cannot be dismissed as due to unreliable data. Furthermore, Hill and Hurtado point out that fertility and mortality schedules are part of the coherent pattern that links life history parameters together. If our mortality schedule was 'really' chimp-like, there could have been no selection pressure to produce individuals that senesce so much later than chimps. Nor would humans be the size they are, nor mature at the age they do. Hill and Hurtado point out that 67% pre-adult mortality would be needed to give a constant population with a 48-month interbirth interval (the upper extreme observed among foragers; shorter interbirth intervals require even higher mortality), and comment 'Such high mortality has never been observed in any traditional population'.

Thus, Hill and Hurtado imply that a long-standing feature of human biology may have been not just the Malthusian possibility, but actual rapid increase, and a saw-tooth population history. They suggest that this may influence our ability to assess accurately the significant selective forces responsible for our biology and behaviour: 'Perhaps trade-offs were not detected, and menopause not favored by kin selection, because the Ache were in a period of resource abundance'. There must also be implications for archaeologists' ideas about invention and 'intensification' as a response to population pressure.

Helpers

Humans are expert at recruiting and distributing help. Hill and Hurtado suggest that many costs and benefits of alternative ways of behaving, growing and reproducing may be rendered unmeasurable by the ability of the individual who 'made an error' to cover it by recruiting help from kin, who also benefit from helping to remedy the miscalculation. Thus, a grandmother gains more fitness by going to go to help her daughter who bore too many children too fast (and thus risks being unable to feed them) than to her daughter who has few (depending on why the latter has few!). Grandmother's decision then reduces the effect of her daughter's 'miscalculation'. If instead the first daughter has more children because she already has helpers and better circumstances, it may pay the mother to go to help the second daughter, if her slow reproduction is due to lack of help or other soluble hardship. But it also follows that individuals who can recruit more help have different optima from those with no helpers. A woman who has a co-resident mother and a co-resident childless aunt can well afford to bear babies faster than her orphan friend.

Despite Turke's (1988) early work on daughters (see also Blurton Jones *et al.*, 1997), the study of men and grandmothers as 'helpers' has dominated recent work. Hawkes (1990, 1991, 1993) has argued against the widely held view that hunting is best understood as paternal investment, and that this was the major force in human evolution. We will continue to investigate hunting less as 'bringing home the bacon' and more as acquiring and distributing political 'pork'. An even more uniquely human class, that of grandmothers, has attracted more attention.

Hill and Hurtado (1991, 1996: p. 433) tested the classic version of the grandmother hypothesis of menopause: do the benefits of help given to daughters outweigh the benefits of continued child bearing? Ache demographic values show that, even under the most favourable assumptions, the effect of help on kin would have to be massively increased to give menopause a selective advantage. Ache women provide only 13% of the calorie income, and Hill and Hurtado note that the analysis needs to be repeated in populations in which women's contribution is much greater. Nonetheless, Hill and Hurtado (and Rogers, 1993) show that the effect would have to be almost unrealistically large.

Appealing to Schultz (Fig. 6.1) for initial support, Hawkes *et al.* (1997) have suggested that the derived character awaiting an explanation is post-reproductive life, not 'early cessation of fertility'. Chimpanzees cease to reproduce by 40, and their somatic senescence coincides with this. Human females cease to reproduce at about the same age but remain active and strong, and survive, even without access to modern medicine, for many years after this – women's life expectancy at 45 is 22 years for Ache (see Table 6.1 in Hill and Hurtado, 1996), 19.9 years for !Kung, and 21.3 years for Hadza (Howell, 1979; Blurton Jones *et al.*, 1992, from best fitting stable population models in Coale and Demeny, (1983) – but is close to zero for Gombe chimpanzees. The adaptation question then becomes: what could give rise to a selective advantage for delayed somatic senescence in an animal that is no longer reproducing? Natural selection cannot act on a post-reproductive individual unless there is a way in which its survival can influence the spread of its genes even after it has finished reproducing. Help given to descendants could have just this effect.

Hadza women past the childbearing years forage as efficiently as

Table 6.2. *Some return rates for human foragers*

Species	Part	Users	kcal/h on site	kcal/h with travel	kcal/h by child on site[c]	Daily yield[a] 7-hour day	People fed at 2000 kcal/day[d]
Adansonia digitata	Fruit	Hadza	1745–2328	1745–2328	600	12 215–16 296	6.1–8.1
Ricinodendron rautanenii	Nut	!Kung	2325–2738	727 +		5089–7000	2.5–3.5
Vigna frutescens	Tuber	Hadza	1615–1955	1423–1723	200	9964–12 062	4.98–6.03
Grewia retinervis	Berry	!Kung	187–973	113–335		791–2345 (1353–4024/ 12-hr day)	0–1
Grewia bicolor	Berry	Hadza	497–1130				
Cordia gharaf (undushi)[b]	Berry	Hadza	2771		2223		
Salvadora persica (tafabe)[b]	Berry	Hadza	1281		964		
Coccinea rehmannii (Tan)	Root	!Kung	2886	?	?	?	
Vigna dinteri (Sha)	Root	!Kung	3043	?	?	?	
Eminia antenullifera (makalita)	Tuber	Hadza		2000	500	14000	7
Panjuko	Root	Hadza		1600	500	11 200	5.6

The authors sought two kinds of information, mostly from data presented in Hawkes *et al.* (1989, 1994b, 1997) and in Blurton Jones *et al.* (1989, 1995, 1997). (1) Return rate on encounter: if an individual encounters a food plant or patch, how many kcal does it get per minute of picking, digging or whatever it takes to acquire the resource? For some of the foods that humans target, processing costs are small (e.g. some roots). For others (Baobab fruit, Mongongo nuts), processing costs are high. These were reported where available. (2) Estimated returns per day: some of the foods humans target are distributed through the environment in very localised patches; Mongongo nuts are an extreme example. Thus, the time people have to spend to get to the patch is an important issue, and it reduces the daily return rate considerably. These figures are quite approximate. They may only roughly bracket seasonal, annual and local variation.

[a]Day yield includes travel time, which is very large for *Ricinodendron*, and sometimes very large for *Cordia* and *Salvadora*.

[b]No data exist for other primates using these very high-return berries.

[c]Figures for human dietary requirements are widely used, and are believed to err on the generous side. The authors use a figure of 2000 kcal/day for an active forager-sized woman, and another 2000 for each child (suppose an assortment from 5 to 15 years old). They divide the daily return rates shown in Table 6.3, for a seven-hour day, by 2000 to find the number of people a woman could feed (including herself).

[d]Return rates for Hadza children picking undushi and tafabe berries are from Hawkes *et al.* (1995). The figures for Baobab, //ekwa, makalita and panjuko are those for a 10-year-old child, read from the plots of returns against age in Blurton Jones *et al.* (1997).

younger women, and work slightly longer hours, and so bring home more food (Hawkes, O'Connell and Blurton Jones, 1989). Hawkes *et al.* (1997) show that the growth of weaned children of women with a suckling infant correlates with their grandmother's work hours, not with their mother's. Thus, older women's foraging may increase the fitness of their young descendants.

The opportunity to help depends upon the use of highly productive resources that are difficult to acquire or process, such as tubers and nuts (Table 6.2). Hawkes *et al.* (1997) suggest: (a) that use of these resources both allows and gives rise to the need to provision juveniles, who can acquire or process them only inefficiently if at all (see Hawkes, O'Connell and Blurton Jones, 1995, and Blurton Jones *et al.*, 1994a and 1994b, for analyses of the variety of combinations of mother and juvenile foraging displayed by Hadza and !Kung); and (b) that the need to provision juveniles provides an opportunity for helpers to make a difference to the fitness of close relatives.

The main point about Hawkes *et al.*'s grandmother hypothesis is the interesting set of links to other life history parameters, which are now outlined. One of the most successful attempts to offer a unified theory of life histories is that of Charnov (1993) – CM. CM shows that relationships among certain variables remain approximately 'invariant' across wide ranges in body size. These relationships account for the correlations long recognised among life history variables and body size, and also for correlations among life history variables when body size is removed. Primates, for example, have 'slow' life histories compared with other mammals of similar size (Harvey, Promislow and Read, 1989). This can be explained as a result of a difference in the 'production function' for primates.

CM assumes that 'production' can be allocated either to growing oneself or to growing offspring (Charnov, 1991, 1993; Charnov and Berrigan, 1993). This simple model divides growth into two periods: (1) conception to weaning, in which growth is set by the mother's production (a function of her size); and (2) weaning to maturity (α), in which growth is a function of an individual's own body size. At maturity, production previously allocated to growth is redirected to babies. Growth rates are a function of body mass (W), a characteristic 'production coefficient' (A), and an allometric exponent (c). Individual growth rates take the form: $dW/dt = AW^c$, where production energy at time t for an individual of body mass W equals the production coefficient times body mass to the c power (usually assumed to be 0.75). Adult size for a given period of independent growth (α) and the production available to funnel into offspring for a given adult size both depend on A. If A is large, the result is faster growth and production (for a given size) to funnel into babies. Primates have a very low A, averaging less than half that of other mammals. This accounts for small size at a given age of maturity and low fecundity for size compared with other mammals (Charnov and Berrigan, 1991, 1993).

CM assumes that, given adult life span, selection sets the evolutionarily stable strategy for duration of independent growth (α) according to the trade-off between (1) the benefits of growing longer, and so being larger with more production to put into offspring, and (2) beginning to reproduce sooner, and so having a greater chance to reproduce before dying. CM assumes that key features of the mortality schedule can be modelled as an early burst of high mortality followed by a drop to adult mortality levels before age of first reproduction. Because production put into babies is a function of maternal size (W_α), that production increases with age of maturity. The time available to use those gains depends on adult life span, the inverse of the instantaneous mortality rate (M). As adult life span increases (adult mortality rate falls), selection favours delayed maturity to reap the gains of larger size. Thus, both a and M vary widely but inversely. Their product (αM) is approximately invariant.

There is another constraint in this model: the size at which babies are weaned is a function of adult body size. For a sample of mammals (and for primates separately), the ratio of size at independence (weaning) to adult size ($W_o/W_\alpha = \delta$) is approximately constant (Lee *et al.*, 1991; Charnov, 1993). Because δ scales almost isometrically with body size while production scales up more slowly (the growth allometry is a power of 0.75), the size of weanlings goes up faster with maternal size than does the production the mother can put into them. Thus, annual fecundity, the number of daughters produced per year (b), goes down as age at maturity (α) goes up. Larger mothers produce larger but fewer babies, making αb another approximate invariant.

These 'assembly rules' for mammalian life histories seem quite robust. The general fit of empirical patterns to the predictions of CM (since confirmed on other, larger data sets: Purvis and Harvey, 1995) suggests that the model identifies key trade-offs that shape mammalian life histories. The invariant relationships reveal scaling rules: some life history variables are adjusted to others. In this model, fecundity depends on age at maturity, and age at maturity is adjusted to adult life spans. If ancestral human life spans increased due to grandmothering, then that increase should have distinctive effects on the age at maturity, time or size at weaning, and fertility.

Grandmothering, age at maturity, interbirth intervals, and fecundity

In CM, αM is approximately invariant because longer life spans favour more advanced age at maturity. More time to accrue the benefits of increased production associated with growing longer before reproducing offsets the cost of delay. If gains from growing longer continue to pay off *after* menopause, as the grandmother hypothesis proposes, then α should be adjusted accordingly. It is. These authors found that the late age at maturity for humans (high α) combined with our long life spans (low M) result in an αM similar to that of the other great apes. The delay in maturity for humans is as predicted if the gains from growing longer before reproducing pay off throughout adulthood, during *both* childbearing and grandmothering years.

If the grandmother hypothesis is correct, it also implies that childbearing women must be producing babies faster than otherwise expected because of grandmothers' contribution to that production. Human interbirth intervals *are* smaller than those of any other great ape (Galdikas and Wood, 1990). Grandmothers could affect the growth of infants in one of two ways: (1) they might contribute to nursing mothers directly, and so add to the production that goes through them to infants, who would then grow to the size of independence faster than otherwise expected; or (2) mothers might wean infants before they reach the expected size of independence because they can pass the youngsters on to grandmothers, who supply the necessary nutrient stream.

The second alternative is suggested by the contribution grandmothers make to the nutrition of their weaned grandchildren (Hawkes *et al.*, 1997). In the light of the Hadza pattern, the authors hypothesised that grandmothers allow mothers to wean infants earlier than they otherwise might. If so, weaning would mark a shift to grandmother's support, not feeding independence, and so δ (the ratio of weaning weight to adult weight) would be lower for humans than for other apes.

The ratio of offspring size at weaning to maternal size ($W_o/W_\alpha = \delta$) is approximately 0.33 across the mammals generally, and primates in particular (Lee *et al.*, 1991; Charnov, 1993). All the great apes (Lee *et al.*, 1991) have δs lower than the order mean (see Table 6.1), which illustrates the slight negative relationship between δ and maternal size in both primates and other mammalian taxa (Lee *et al.*, 1991; Charnov, 1993; Purvis and Harvey, 1995). *Within* the apes there is no relationship between δ and adult body size. The authors used the mean of two ethnographic hunter–gatherer values (Ache: Hill and Hurtado, 1996; and !Kung: Howell, 1979) to represent δ for humans (see Table 6.1). That value is not lower than all the other great apes, although it ties with that of gorillas for lowest place, a result that is at best equivocal for our expectation that humans wean at a relatively small size.

It is possible that our estimate for human δ errs on the conservative side (i.e. against the hypothesis). There appears to be a much greater variation in adult than in weanling sizes in humans (Lee *et al.*, 1991), so a larger sample of human populations might dilute the large effect of the small size of !Kung adults in our sample. The δ value for humans in Lee *et al.* (1991), for which the sample is not restricted to foragers, *is* much lower than all the other apes (0.16 compared to 0.21 for gorillas, 0.27 for chimpanzees, 0.28 for orangutans). Ethnographically known hunter–gatherers are generally smaller than pre-Mesolithic human populations, so perhaps the δ for Paleolithic moderns was closer to the Lee *et al.* value.

Our short interbirth intervals, perhaps made possible by a small δ, are especially striking given the negative scaling of fecundity with age at maturity across the mammals. Because b (daughters per year) scales approximately with α, the direct comparison of interbirth intervals underestimates the relatively high fertility of humans. Later age at maturity is usually associated with reductions in b. But if later maturity in humans is due to grandmothering, then grandmothers' contribution to production will have important countervailing effects that *increase b*.

Across the primates and mammalian taxa generally (of all sizes), αb is approximately 1.7 (Charnov, 1993). For all the other great apes, the αb number is substantially lower than this, less than 1.0. Humans, however, have an αb value greater than than 2.0, more than double the other large-bodied apes. The grandmother hypothesis predicts just this: αb *should* be high because it incorporates the production of *both* mothers *and* grandmothers. The baby production of the entire life span is concentrated in the childbearing years.

The grandmother hypothesis combined with CM accounts for several distinctive features of human life history, including long life spans after menopause, late age at maturity, short interbirth intervals, and high fertility. Other hypotheses have been offered to explain each of these individually (Smith and Tompkins, 1995), but 'grandmothering' can explain them all at once.

Criticisms of the grandmother hypothesis

The grandmother hypothesis has attracted several critical comments, among them:

1. Old Hadza women help young kin, but is this a general pattern (Gibbons, 1997)? Hill and Hurtado (1991) show an effect of Ache grandmothers upon their daughters' reproduction, and !Kung informants of Blurton Jones, Hawkes and Draper (1994a) claimed that older women, less distractable and tougher, are the most productive foragers.

2. What about meat obtained by males, which often forms around 50% of the food intake in hunter–gatherer populations? CM implies that attention to female life histories is sufficient to account for major taxonomic variation. But as long as females can access the meat, it has an effect on female economics – the debate is not concluded!

3. Some chimpanzees use resources juveniles cannot process (Table 6.3), they sometimes give food to juveniles, and some live beyond their apparent reproductive span. Is grandmothering really distinctly human? Great ape populations are so small that meaningful comparative data take years to accumulate. The chimpanzee nut-cracking data are very important; they support the view that hard-to-process food is a context in which adults transfer food to juveniles. If nuts were the staple food of these populations, we might expect to see benefits to delayed senescence. But the nuts are used for only a third of the year, are not available in some years and, most intriguing, are only exploited for about an hour a day (Sugiyama and Koman, 1979a, 1979b; Boesch and Boesch, 1984).

4. Current evidence suggests patrilocality as a conservative hominoid trait (Foley and Lee, 1989). Would grandmothering work as well in patrilocal groups as in the matrilocal groups it has been studied in? The authors would rank grandmother's fitness benefit accruing from help to children of beneficiaries as: greatest from help to daughter > sister > niece > son > nephew because of relatedness and paternity uncertainty. (Grandmother as mate guard is a role we have not examined. It might have interestingly different implications for life history.) For grandmothering to win over continued births, the only hopeful candidate would be help to daughter; help to others is too heavily discounted. But help to any of these might be adequate to select for delayed senescence. Thus, the grandmother hypothesis could apply to either matrilocal or patrilocal settings. But grandmothers benefit most by living with their daughters and should favour matrilocality. Provisioning enables older juveniles to help their mother, offering another advantage to keeping daughters at home. (Blurton Jones *et al.*, 1997, suggest that teenage sons pursue interests that conflict with efficient provisioning of younger siblings.) If men gain by kin co-operation in hunting, or the defence of females, they should favour patrilocality.

Table 6.3. *Returns on food patches for primate species and human foragers, and an estimate of the number of individuals that could potentially be supported by a foraging individual*

Species	Resource	kcal/hour in patch	kcal/day	Expend/ day	Number that can be supported
Papio cynocephalus[a]	Various	297	1013	626	1.62
Cercopithecus aethiops[b]	Acacia flowers	311	1244	505	2.46
M. fuscata[c]	Seeds	115	717	529	1.36
Pan troglodytes[d]	Coula nuts	1408–1508	2616 (potential maximum 9856–10 556)	1767	1.48 5.6–5.97
Pan troglodytes[e]	Figs	1200	?	1767	?
Homo sapiens	Baobab fruit	1700	11 900	2400	4.96

[a]Baboons (Stacey, 1986): feeding efficiency seems to show 'on-site' returns of 20.7 kJ per min = 1242 kJ/h = 297 kcal/h. Stacey's 'foraging efficiency' includes moving during the day, which seems equivalent to the authors' foraging returns with travel time and processing included. Stacey gives 9.4 kJ/min (564 kJ/h), which is 135 kcals/h; multiply this by 7 hours' work = 945 kcal/day. Food intake is 385 kJ/kg per day (57 kcal). Females weigh 11 kg = 4235 kJ per female per day. Energy expenditure calculated at 238 kJ/kg per day. Females of 11 kg need 2618 kJ per day. Ratio of intake/expenditure = 1.51–1.62.

[b]Vervets: feeding returns = 311 kcal/h (Lee, personal communication), with 4-hour foraging day = 1244 kcal/day intake. Food requirement calculated as 2 × BMR, with 65% digestible energy. Vervet female weight = 3.15 kg; ratio of intake over requirements = 2.46 individuals (self plus 1.46 others).

[c]Japanese macaque (Saito, 1996): intake rate up to 8 kJ/min on seeds Zelkova and Carpinus = 480 kJ/h/4.18 = 115 kcal/h. Return rate × feeding time up to 3000 kJ/day = 717 kcal/day. Weight of female *M. fuscata* is 9.2 kg (Lee *et al.*, 1991). Using Stacey's 238 kJ/kg per day expediture = 2189.6 kJ/day per female (or 529 kcal/day). A female *M. fuscata* can support 717/529 = 1.35 individuals (i.e. self and 0.35 more) on these berries. Assuming a 7-hour foraging day, 115 kcal/h × 7 = 805 kcal/day. This calculation suggests she can support 805/529 = 1.52 individuals.

[d]Chimpanzee (Boesch and Boesch, 1984): female weight is 31 kg = 1767 kcal/day expenditure. If we take Boesch's 'mean intake per individual and per day represents 2616 kcal', then a female chimp can only support 2616/1767 individuals = 1.48. But the 2616 kcal are obtained in only 1.5 h at an anvil site. This represents 1744 kcal/h. Other figures in Boesch suggest 1408–1508 kcal/h. These figures are comparable to human return rates. Not surprisingly, if a chimp stayed at the anvil for a 7-hour day, it could support more than five individuals.

[e]Chimpanzee (Wrangham *et al.*, 1993): midpoint in plot of return rates in fig trees = 1200 kcal/h. If the supply was not depleted over 7-hours foraging, in one tree it could obtain 8400 kcal. 1767 kcals are required per day and 4.75 individuals could be supported in the tree.

The resulting conflict of interest should give the kind of variability Ember (1975) reports for human hunter–gatherers, and is in line with ideas about animal group transfer patterns (Greenwood, 1980; Pusey and Packer, 1995).

5. Isn't it too easy (cheating?) to give up assessing the trade-off between care for kin and another birth (even though evidence against it is accumulating), and switch attention to selection for delayed somatic senescence? We have long thought of menopause as a special adaptation, an endocrine modification that brings reproduction to an early end. But evidence for such a view is scant. The prime mover in menopause appears to be the loss of oocytes. Other mammals show a similar rapid loss of oocytes as they age. Experimental removal of ovarian tissue early in life hastens a later decline in reproductive rates (see Adams, 1984). Finch (1990) concludes his survey of the literature 'The oviprival syndromes of reproductive senescence seem to occur widely in mammals'. Given the high attrition rate of oocytes, extending the reproductive span may require prohibitively costly increases in investment in reproduction during prenatal development (Wood, 1994). Evidence compatible with such a trade-off between early investment and later reproduction is offered by Cresswell *et al.* (1997), who found in longitudinal data that early menopause was associated with indicators of retardation in the woman's own prenatal growth. A few mammals appear able to extend the reproductive span as long or longer than the human span: elephants, perhaps, and the long-finned pilot whale *Globicephala melaena* (Marsh and Kasuya, 1986), in contrast to the short-finned pilot whale (*G. macrorhynchus*) with 25% of adult females as post-reproductive. A selectionist examination of the peculiar development of the mammalian female germ line might be rewarding. If it can be interpreted as another manifestation of the contrast between male pay-offs from quantity and female pay-offs from quality (quantity being limited by gestation), then we must suppose repeated mitosis (found in sperm production throughout the individual male's life but in females contributing to oocyte production only before birth) poses a greater threat to quality than the suspended, resting state of the primary oocyte. Some data on mutation rates appear to support this suggestion (Drost and Lee, 1995).

6. Why would any animal take on the burden of provisioning juveniles? Are we not suggesting evolution into a disadvantage from which the animal is rescued only by the simultaneous, improbable and rapid arrival of grandmothers? Table 6.2 show s that staple foods used by !Kung and Hadza women allow them to provide for several dependents,

Table 6.4. *A South American example: plant foods of Ache and* Cebus apella

Species	Part	User	kcal/h
Arecastrum romanzolfianum	Growing shoot	Ache women	1584
	Fibre starch	Ache women	2246
	Fibre and shoot	Ache	2436
Casimiroa sinesis	Fruit	Ache	4181
Philodendron sellam	Unripe fruit	Ache	2708
	Ripe fruit	Ache	10078
Campomanesia zanthocarpa	Fruit	Ache	6417
Rheedia brasilense	Fruit	Ache	3245
Acromia totai	Nut	Ache	2243
Astrocaryum murumuru	Flower	*Cebus apella*	56.5
	Seeds	*C. apella*	191.3
Scheelia	Fruit	*C. apella*	127.2
	Frond pith	*C. apella*	63.4
Quararibea cordata	Nectar	*C. apella*	148.4
Strychnos asperula	Fruit	*C. apella*	174.2
Ficus kilipii	Fruit	*C. apella*	80.8

Ache plant foods from Table 2 in Hill *et al.* (1987; fruits for which $n < 10$ excluded). *Cebus apella* from Table 1 in Janson (1985). Return rates for Ache children are currently not available. The preponderance of fruit is discussed in the text. Nuts and palm parts require processing to render them accessible.

and that juveniles exploit these resources only inefficiently. Table 6.3 suggests that, with the important exception of chimpanzees, other primates acquire resources at a rate that does not generate a surplus. Thus, exploiting tubers and nuts, even though they must be provided for juveniles, appears likely to at least match the alternative strategy.

Table 6.4 offers a South American example and an important puzzle. This table compares plant foods of Ache foragers – forest dwellers among whom women acquire only 17% of the calories, in contrast to !Kung (58%) and Hadza (about 50%) – and *Cebus apella*. Notable here are the absence of tubers and the presence of just one nut in the Ache diet. Yet, the return rates Ache derive from fruit far exceed those of *Cebus*. Skill and agility seem unlikely to account for this difference. More likely candidates may be found in body size, gut transit times, or in the possession of baskets in which to stash fruit instead of waiting until the previous mouthful has been chewed. The low kcal/h return rate of primate foraging may result from a limit to the speed at which food can be processed, not a limit to the rate at which it could be acquired. Thus, it might be that many primates could acquire food fast enough (if they were going to give it away and thus avoid their lengthy preprocessing) to produce a surplus and provision offspring. But the chimpanzee data, which suggest this most strongly, are in-patch acquisition rates, which leave us ignorant about how much of the day these rates could be sustained, and how much of the year. Important aspects of the rich but hard-to-acquire resources used by humans may be that their total biomass is very great, their seasons very long, and because they are difficult and time consuming to extract, competition takes the form of comparative skill and strength at extraction more than immediate interindividual competition for access, and hence depletion is slower. These considerations emphasise the role of juvenile inefficiency in the evolution of provisioning and show that the grandmother hypothesis may guide forager researchers towards greater participation in comparative research on primate and hominid diets.

7. Does provisioning juveniles really give a greater opportunity for helpers to enhance their fitness? Could helpers not help any primate mother just as much? Non-human primates provisioned in captivity maintain interbirth intervals half as long as in the wild. Any mammalian mother could gain from a helper, but helpers are found mostly among carnivores (including non-reproducing but not post-reproductive helpers) and humans. As suggested above, perhaps not all mammals can produce a surplus. But we also questioned this interpretation of the foraging data, and another argument may be considered. If we look at potential losses of fitness when the mother is temporarily incapacitated, the provisioning mother (who lives in an ecology in which her juvenile offspring cannot efficiently forage for themselves) appears at risk of losing the products of great portions of her reproductive career (dependent juveniles), while the non-provisioner appears at risk of losing only the current infant, or a delay to the next birth. Thus, the long-term contribution of a helper may be much greater when there is a series of still-dependent weaned offspring.

8. Is it right to divorce the evolution of human life history from the evolution of a large brain? Many authors have attempted to link the advantages of learning to large brains, and particularly to late maturity (Bjorklund, 1997, is a recent and unusually thorough example). Childhood is seen as a time for learning; more time spent learning improves adult competitive ability and the delay is repaid. Blurton Jones *et al.* (1997) suggested juvenile life is primarily waiting time, which could be filled with fitness-enhancing activities such as learning, if not too costly. Let us continue our devil's advocacy. Individuals who acquire skills

more rapidly are presumably at an advantage over those who acquire them slowly – they benefit from the skills without losing by delayed reproduction. Thus, a skill-based subsistence should select strongly for rapid learning (and the rapid construction of a brain that performs such rapid learning – most brain growth is accomplished within the first five years of life, why then would having a large brain account for another 12 pre-reproductive years?). If the length of life span renders cultural transmission useful (Boyd and Richerson, 1985, show cultural transmission is most useful at intermediate rates of environmental change; rate is relative to generation time), there may be an increased pay-off to rapid forms of learning. (We would draw the causal arrows from longer adult span, to late maturity, to long generation time, to increased pay-off from cultural transmission, to advantages in observational and 'hearsay' learning, to increased brain size.) Even if we concede that there is 'lots to learn' to be a successful forager, it is easier for us to believe that this selected for faster mechanisms of learning (such as observational learning) than that it selected for later maturity. A selective advantage to faster learning, employing mechanisms of learning not widely found in other species, seems more likely to account for the evolution of a large brain than does an assumption that because there is so much to learn we need to take longer to learn it. There seems to be ample experimental evidence that mechanisms of learning seen in other species would, indeed, require a long time (many trials) to acquire many skills and much cultural or topographical knowledge. 'Skill theory' proponents need to decide whether they wish to explain big brains or late maturity! If humans were acquiring more knowledge with a monkey's brain, it would indeed take many years to learn what young foragers know! But what we see among hunter–gatherers are individuals with the capacity to observe an older individual prepare a skin, cut climbing pegs, winnow Baobab flour, perform sacred dance steps, etc., and immediately repeat the activity themselves. These authors suggest that the human brain, while in a sense just more of the same, is not providing merely more space to store more information, nor merely more of the same learning mechanisms, but rather exhibits processes weakly evidenced in other species but startlingly efficient in our species.

If causal arrows are to be drawn from Machiavellian learning to big brains, and (separately, because brain volume reaches close to adult levels long before maturity) to late maturity, we need to spell out the components of Machiavellian intelligence that can be added with a larger brain and are absent from other primates, and we need to show whether these things take a long time to learn. Because many aspects of social interaction must be shaped by real interests and power differences (including size and weaponry, and the interests of allies), it is likely that the scheduling of development of juvenile social intelligence is not determined by difficulty or time needed to learn but by the need to keep out of unnecessary trouble.

9. If grandmothering is such a good idea, why do we not see menopause or post-reproductive life in more species? Vervet grandmothers contribute to the fitness of their daughters and granddaughters (Fairbanks, 1988) but reproduce until death. If it is not costly to produce enough oocytes to last five or ten years, the trade-off that we should consider for the vervet may be between help to kin, and continued births (the 'classic' trade-off examined by Hill and Hurtado, 1991, and Rogers, 1993). This is even less likely to pay when: (a) the help given is non-depreciable, thus less impaired by the addition of offspring, and (b) the grandmother has already transferred some of her status to her daughters, in which case the marginal difference made by the grandmother is unlikely to out-weigh the benefit to her of continued reproduction.

We should ask a parallel question about Belsky *et al.* (1991) and Chisholm's (1993) view of differences in parental strategy, and perhaps even of Kaplan's (1994) view of demographic transition: why do high-ranking primates not have fewer offspring than low-ranking ones? The distinction between depreciable and non-depreciable care is critical here, too. Kaplan (1996) attends to highly depreciable care – money, direct instruction, social–intellectual interaction – that improves the child's chances in the competitive labour market. The nearest analogy in other animals is rank order, but its maintenance is non-depreciable. High-ranking vervets gain nothing by distributing this care among fewer offspring.

Discussion and conclusions

This chapter concentrates on efforts to apply perspectives from life history theory in biology to data on contemporary hunters and gatherers. Topics include the allocation of resources between offspring number and fitness, age at first reproduction, lack of balance of fertility and mortality schedules, the special significance of helpers in human adaptation, and contrasting ideas about the evolution of post-reproductive life. These investigations share an interest in ecological contexts that may have shaped the life history parameters. Each attaches primary importance to rich resources that are difficult for juveniles to acquire (meat, deep roots and difficult nuts). The significance of such foods has been frequently discussed (e.g. Hatley and Kappelman, 1980). There are disagreements about the relative importance of hunting and the exploitation of difficult-to-acquire but rewarding plant foods, and about the way in which the development of the ability to exploit either resource accounts for the human life history.

The opportunities for meaningful comparisons of humans and other primates are greatly increased by the addition of more quantitative studies of hunters and gatherers. The Ache and Hadza studies add significantly to the landmark studies of the !Kung San, showing us both variability and commonality. Quantitative study of the use of plant foods (and hunting methods) among farmers and pastoralists would also provide useful information for comparative and evolutionary debate. More long-term studies of primate demography would also be very useful.

From the Hadza and San studies (e.g. Lee, 1979), we are beginning to build knowledge of African savanna economics to add to the comparative studies of Peters and O'Brien (1981), Peters (1987) and McGrew, Baldwin and Tutin (1988). Despite the obvious relevance of understanding savanna foraging ecology to the rational discussion of human origins, many data have yet to be collected. We need more data on the returns obtainable from such foods in different savanna habitats (enough traditional use of wild plants persists for useful work to be done), and the assumed association of roots and nuts with savanna habitats needs testing by quantitative analysis of availability in different kinds of forest (Vincent, 1985; Peters, 1987; McGrew *et al.*, 1988). Foraging return data from Hadza and !Kung and other users of wild plants can suggest which plants critically define forager habitats. Data on primate foraging that are easier to reconcile with foraging theory would be very useful. The authors were surprised by the difficulty of finding and comparing foraging data from primates. Comparative rainfall data for chimp habitats are made usefully available by Moore (1992). At 1717 mm/year, the mean rainfall is about five times (and even the driest localities are some three times) the rainfall in !Kung and Hadza country. Foley (1982) showed that prehistoric hunters and gatherers probably occupied areas with rainfall and productivity higher than the habitats of contemporary foragers. But we do not know the consequences of such differences for the real-life economic decisions of foragers. Peters' (1987) maps suggest that productive, hard-to-exploit plant resources are found in all these habitats.

Our species is often described as the ultimate omnivore. But the use of hard-to-acquire but highly rewarding plant foods predicts the narrow diet breadth actually observed among hunters and gatherers. While human informants can give a list of edible plants as long as the known diets of other primates, relatively few of these foods are taken regularly. Many days following Hadza on foraging trips, and many additional days weighing the food they brought home, have given us a list of nine species regularly used in a year (//ekwa, makalita, panjuko, shumuko, tafabe, undushibe, emberipe, baobab, tamarind). Rather fewer days following !Kung foraging in locations chosen by us (many chosen to see how little food they produce – thus likely to expand recorded diet breadth) nonetheless gave a similarly short list. Hill *et al.* (1987) list ten species taken often enough by Ache to yield foraging data. People primarily exploit a small number of 'staple' foods. In contrast, McGrew *et al.* (1988) observed chimpanzees using 43 species and suspect the use of an additional 41 species, but still describe this as a narrow diet breadth relative to other primates. Tutin and Fernandez (1993) report chimpanzees using 111 plant species (and observed them eating 67% of these, i.e. 74 species). Given the very high returns for human foods shown in Table 6.2, diet breadth theory would lead us to expect a narrower diet than accompanies the low returns obtained by other primates. While the global human catalogue of plant foods may far exceed that of other primates, local human populations are likely to have often been characterised by a narrow diet spectrum.

The authors think it is also now clear that female provisioning of weaned offspring (dependent on the use of rewarding but hard-to-acquire plant foods) challenges hunting as a productive explanation of the path of human evolution and the distinctiveness of our species.

Acknowledgements

The authors wish to thank the Tanzanian Commission on Science and Technology for permission to conduct research in Tanzania. They also thank several hundred individual Hadza for their patience and good spirits, their field assistants Gudo Mahiya and the late Sokolo Mpanda for their expertise and collegiality, and David Bygott and Jeannette Hanby for providing a home away from home and vital logistic facilities, Professor C.L. Kamuzora of the University of Dar es Salaam, and numerous citizens of Mbulu district for help and friendship. The research was funded by the National Science Foundation Anthropology Program, the Swan Fund, B. Bancroft, the University of Utah, and the University of California Los Angeles.

References

Adams, C.E. (1984). Reproductive senescence. In *Reproduction in Mammals 4: Reproductive Fitness*, ed. C.R. Austin and R.V. Short, pp. 210–33. Cambridge: Cambridge University Press.

Belsky, J., Steinberg, L. and Draper, P. (1991). Childhood experience, interpersonal development, and reproductive strategy: an evolutionary theory of socialization. *Child Development* **62**, 647–70.

Bjorklund, D.F. (1997). The role of immaturity in human development. *Psychological Bulletin* **122**, 153–69.

Blurton Jones, N.G. (1986). Bushman birth spacing: a test for optimal interbirth intervals. *Ethology and Sociobiology* **7**, 91–105.

Blurton Jones, N.G. (1987). Bushman birth spacing: direct tests of some simple predictions. *Ethology and Sociobiology* **8**, 183–204.

Blurton Jones, N.G. (1993). The lives of hunter–gatherer children: effects of parental behavior and parental reproducive strategy. In *Juvenile primates*, ed. M. Perreira and L. Fairbanks, pp. 309–26. Oxford: Oxford University Press.

Blurton Jones, N.G . (1994). A reply to Harpending. *American Journal of Physical Anthropology* **93**, 391–7.

Blurton Jones, N.G. (1996). Too good to be true? Is there really a trade-off between number and care of offspring in human reproduction? In *Human Nature, a Critical Reader*, ed. L. Betzig, pp. 83–6. Oxford: Oxford University Press.

Blurton Jones, N., Hawkes, K. and O'Connell, J. (1989). Modeling and measuring costs of children in two foraging societies. In *Comparative Socioecology*, ed. V. Standen and R. Foley, pp. 367–390. Oxford: Blackwell Scientific Publications.

Blurton Jones, N.G., Hawkes, K. and Draper, P. (1994a). Differences between Hadza and !Kung children's work: original affluence or practical reason?. In *Issues in Hunter Gatherer Research*, ed. E.S. Burch, pp. 189–215. Oxford: Berg.

Blurton Jones, N.G., Hawkes, K. and Draper, P. (1994b). Foraging returns of !Kung adults and children: why didn't !Kung children forage? *Journal of Anthropological Research* **50**, 217–48.

Blurton Jones, N.G., Hawkes, K. and O'Connell, J.F. (1997). Why do Hadza children forage? In *Uniting Psychology and Biology: Integrative Perspectives on Human Development*, ed. N. Segal, G.E. Weisfeld and C.C. Weisfeld, pp. 279–313. Washington, DC: American Psychological Association.

Blurton Jones, N. and Sibly, R.M. (1978). Testing adaptiveness of culturally determined behaviour: do bushman women maximise their reproductive success by spacing births widely and foraging seldom? In *Human Behaviour and Adaptation*, ed. N. Blurton Jones and V. Reynolds, pp. 135–58. Society for Study of Human Biology Symposium No. 18. London: Taylor & Francis.

Blurton Jones, N.G., Smith, L., O'Connell, J., Hawkes, K. and Kamuzora, C.L. (1992). Demography of the Hadza, an increasing and high density population of savanna foragers. *American Journal of Physical Anthropology* **89**, 311–18.

Boesch, C. and Boesch, H. (1984). Possible causes of sex differences in the use of natural hammers by wild chimpanzees. *Journal of Human Evolution* **13**, 415–40.

Bogin, B. (1990). The evolution of human childhood. *Bioscience* **40**, 16–24.

Borgerhoff Mulder, M. (1991). Human behavioral ecology. In *Behavioural Ecology:*

an Evolutionary Approach, ed. J.R. Krebs and N.B. Davies, pp. 69–98. Oxford: Blackwell Scientific Publications.

Bowman, J.E. and Lee, P.C. (1995). Growth and threshold weaning weights among captive rhesus macaques. *American Journal of Physical Anthropology* **96**, 159–75.

Boyd, R. and Richerson, P.J. (1985). *Culture and the Evolutionary Process.* Chicago: University of Chicago Press.

Charnov, E.L. (1991). Evolution of life history variation in female mammals. *Proceedings of the National Academy of Sciences, USA* **88**, 1134–7.

Charnov E.L. (1993). *Life History Invariants.* Oxford: Oxford University Press.

Charnov, E.L. and Berrigan, D. (1991). Dimensionless numbers and the assembly rules for life histories. *Philosophical Transactions of the Royal Society*, Series B **33**, 241–8.

Charnov, E.L. and Berrigan, D. (1993). Why do female primates have such long lifespans and so few babies? Or life in the slow lane. *Evolutionary Anthropology* **1**, 191–4.

Chisholm, J.S. (1993). Death, hope and sex: life history theory and the development of reproductive strategies. *Current Anthropology* **34**, 1–24.

Coale, A.J. and Demeny, P. (1983). *Regional Model Life Tables and Stable Populations.* New York: Academic Press.

Creswell, J.D., Egger, P., Fall, C.H., Osmond, C., Fraser, R.B. and Barker, D.J. (1997). Is the age of menopause determined in utero? *Early Human Development* **49**, 143–8.

Draper, P. and Harpending, H. (1982). Father absence and reproductive strategy: an evolutionary perspective. *Journal of Anthropological Research* **38**, 255–73.

Drost, J.B. and Lee, W.R. (1995). Biological basis of germline mutation: comparisons of spontaneous germline mutation rates among *Drosophila*, mouse, and human. *Environmental and Molecular Mutagenesis* **25**, (suppl. 26), 48–64.

Ember, C.R. (1975). Residential variation among hunter–gatherers. *Behavior Science Research* **10**, 199–227.

Fairbanks, L. (1988). Vervet monkey grandmothers: effects on mother–infant relationships. *Behaviour* **104**, 176–88.

Finch, C.E. (1990). *Longevity, Senescence, and the Genome.* Chicago: University of Chicago Press.

Foley, R.A. (1982). A reconsideration of the role of predation on large mammals in tropical hunter–gatherer adaptation. *Man* (N.S.) **17**, 383–402.

Foley, R.A. and Lee, P.C. (1989). Finite social space, evolutionary pathways, and reconstructing hominid behavior. *Science* **243**, 901–6.

Galdikas, B. and Wood, J. (1990). Birth spacing patterns in humans and apes. *American Journal of Physical Anthropology* **63**, 185–91.

Gibbons, A. (1997). Ideas on human origins evolve at anthropology gathering. *Science* **276**, 535–6.

Goodall, J.(1986). *The Chimpanzees of Gombe: Patterns of Behavior.* Cambridge, Mass: Harvard University Press.

Greenwood, P.J. (1980). Mating systems, philopatry and dispersal in birds and mammals. *Animal Behavior* **28**, 1140–62.

Harpending, H.C. (1994). Infertility and forager demography. *American Journal of Physical Anthropology*, **93**, 385–90.

Harvey, P.H., Promislow, D.E.L. and Read, A.F. (1989). Causes and correlates of life history differences among mammals. In *Comparative Socioecology*, ed. V. Standen and R. Foley, pp. 305–18. Oxford: Blackwell Scientific Publications.

Hatley, T. and Kappelman, J. (1980). Bears, pigs, and Plio-Pleistocene hominids: a case for the exploitation of belowground food resources. *Human Ecology* **8**, 371–87.

Hawkes, K. (1990). Why do men hunt? Benefits for risky choices. In *Risk and Uncertainty in Tribal and Peasant Economies*, ed. E. Cashdan, pp. 145–66. Boulder, Col.: Westview Press.

Hawkes, K. (1991). Showingoff: tests of an hypothesis about men's foraging goals. *Ethology and Sociobiology* **12**, 29–54.

Hawkes, K. (1993). Why hunter–gatherers work: an ancient version of the problem of public goods. *Current Anthropology* **34**, 341–61.

Hawkes, K., O'Connell, J.F. and Blurton Jones, N.G. (1989). Hardworking Hadza grandmothers. In *Comparative Socioecology*, ed. V. Standen and R. Foley, pp. 341–66. Oxford: Blackwell Scientific Publications.

Hawkes, K., O'Connell, J. and Blurton Jones, N.G. (1991). Hunting income patterns among the Hadza: big game, common goods, foraging goals and the evolution of the human diet. *Philosophical Transactions of the Royal Society* Series B, **334**, 243–51.

Hawkes, K., O'Connell, J.F. and Blurton Jones, N.G. (1995). Hadza children's foraging: juvenile dependency, social arrangements, and mobility among hunter–gatherers. *Current Anthropology* **36**, 688–700.

Hawkes, K., O'Connell, J.F. and Blurton Jones, N.G. (1997). Hadza womens time allocation, offspring provisioning, and the evolution of long post-menopausal lifespans. *Current Anthropology* **38**, 551–77.

Hill, K. (1993). Life history theory and evolutionary anthropology. *Evolutionary Anthropology* **2**,78–88.

Hill, K. and Hurtado, A.M. (1991). The evolution of reproductive senescence and menopause in human females. *Human Nature* **2**, 315–50.

Hill, K. and Hurtado, A.M. (1996). *Ache Life History: the Ecology and Demography of a Foraging People*. New York: Aldine de Gruyter.

Hill, K., Kaplan, H., Hawkes, K. and Hurtado, A.M. (1987). Foraging decisions among Ache hunter–gatherers: new data and implications for optimal foraging models. *Ethology and Sociobiology* **8**, 1–36.

Howell, N. (1979). *Demography of the Dobe Area !Kung*. New York: Academic Press.

Janson, C. (1985). Aggressive competition and individual food consumption in wild brown capuchin monkesy (*Cebus apella*). *Behavioral Ecology and Sociobiology* **18**, 125–38.

Kaplan, H. (1994). Evolutionary and wealth flows theories of fertility. Empirical tests and new models. *Population and Development Review* **20**, 753–91.

Kaplan, H. (1996). A theory of fertility and parental investment in traditional and modern human societies. *Yearbook of Physical Anthropology* **39**, 91–135.

Kelley, R.L. (1995). *The Foraging Spectrum: Diversity in Hunter–Gatherer Lifeways*. Washington DC: Smithsonian Institution Press.

Kozlowski, J. and Wiegert, R.G. (1987). Optimal age and size at maturity in annuals and perennials with determinate growth. *Evolutionary Ecology* **1**, 231–44.

Lambert, J.E. (1997). Digestive strategies, fruit processing, and seed dispersal in the chimpanzees (*Pan troglodytes*) and redtail monkeys (*Cercopithecus ascanius*) of Kibale National Park, Uganda. PhD thesis, University of Illinois, Urbana-Champaign.

Lancaster, J.B. (1997). The evolutionary history of human parental investment in relation to population growth and social stratification. In *Feminism and Evolutionary Biology*, ed. P.A. Gowaty, pp. 466–88. New York: Chapman and Hall.

Lancaster, J.B. and Lancaster, C. (1983). Parental investment: the hominid adaptation. In *How Humans Adapt: a Biocultural Odyssey*, ed. D.J. Ortner, pp. 33–56. Washington DC: Smithsonian Institution Press.

Lee, P.C., Majluf, P. and Gordon, I.J. (1991). Growth, weaning and maternal investment from a comparative perspective. *Journal of Zoology* 225, 99–114.

Lee, R.B. (1979). *The !Kung San*. Cambridge: Cambridge University Press.

Leighton, M., Seal, U.S., Soemarna, K. *et al.* (1995). Orangutan life history and VORTEX analysis. In *The Neglected Ape*, ed. R.D. Nadler, B.F.M. Galdikas, L.K. Sherran and N. Rosen, pp. 97–107. New York: Plenum Press.

Marsh, H. and Kasuya, T. (1986). Evidence for reproductive senescence in female Cetaceans. *Report of the International Whaling Commission, Special Issue* 8, 57–74.

McGrew, W.C., Baldwin, P.J. and Tutin, C.E.G. (1988). Diet of wild chimpanzees (*Pan troglodytes verus*) at Mt. Assirik, Senegal: 1. composition. *American Journal of Primatology* 16, 213–26.

Moore, J. (1992). 'Savanna' chimpanzees. In *Topics in Primatology* Vol. 1: *Human Origins*, ed. T. Nishida, W.C. McGrew, P. Marler, M. Pickford and F.B.M. de Waal, pp. 99–118. Tokoyo: University of Tokyo Press.

Nishida, T., Takasaki, H. and Takahata, Y. (1990). Demography and reproductive profiles. In *The Chimpanzees of the Mahale Mountains: Sexual and Life History Strategies*, ed. T. Nishida, pp. 63–97. Tokyo: University of Tokyo Press.

Pennington, R. and Harpending, H.C., (1988). Fitness and fertility among Kalahari !Kung. *American Journal of Physical Anthropology* 77, 303–19.

Peters, C.R. (1987). Nut-like oil seeds: food for monkeys, chimpanzees, humans, and probably ape-men. *American Journal of Physical Anthropology* 73, 333–63.

Peters, C.R. and O'Brien, E.M. (1981). The early hominid plant-food niche: insights from an analysis of plant exploitation by *Homo*, *Pan*, and *Papio* in Eastern and Southern Africa. *Current Anthropology* 22, 127–40.

Purvis, A. and Harvey, P.H. (1995). Mammalian life history evolution: a comparative test of Charnov's model. *Journal of Zoology* 237, 259–83.

Pusey, A.E. and Packer, C. (1995). Dispersal and philopatry. In *Primate Societies*, ed. B. Smuts, D.L. Cheney, R. Seyfarth, R. Wrangham and T. Struhsaker, pp. 250–66. Chicago: University of Chicago Press.

Rogers, A.R. (1993). Why menopause? *Evolutionary Ecology* 7, 406–20.

Saito, C. (1996). Dominance and feeding success in female Japanese macaques, *Macaca fuscata*: effects of food patch size and inter-patch distance. *Animal Behavior* 51, 967–80.

Schultz, A. (1969). *The Life of Primates*. New York: Universe Books.

Stacey, P.B. (1986). Group size and foraging efficiency in yellow baboons. *Behavioral Ecology and Sociobiology* 18, 175–87.

Smith, H. and Tompkins, R.L. (1995). Toward a life history of the Hominidae.

Annual Reviews in Anthropology **24**, 257–79.

Stearns, S.C. (1992). *The Evolution of Life Histories.* Oxford: Oxford University Press.

Stearns, S.C. and Koella, J. (1986). The evolution of phenotypic plasticity in life history traits: predictions for norms of reaction for age- and size-at-maturity. *Evolution* **40**, 893–913.

Stewart, K. J., Harcourt, A.H. and Watts, D.P. (1988). Determinants of fertility in wild gorillas and other primates. In *Natural Human Fertility: Social and Biological Determinants,* ed. P. Diggory, M. Potts and S. Teper, pp. 22–38. New York: MacMillian Press.

Sugiyama, Y. (1984). Population dynamics of wild chimpanzees at Bossou, Guinea, between 1976 and 1983. *Primates* **25**, 381–400.

Sugiyama, Y. and Koman, J. (1979a). Social structure and dynamics of wild chimpanzees at Bossou, Guinea. *Primates* **20**, 323–39.

Sugiyama, Y. and Koman, J. (1979b). Tool-using and -making behavior in wild chimpanzees at Bossou, Guinea. *Primates* **20**, 513–24.

Surbey, M.K. (1990). Family composition, stress, and the timing of human menarche. In *Socioendocrinology of Primate Reproduction,* ed. T.E. Ziegler and F.B. Bercovitch, pp. 11–32. New York: Wiley-Liss.

Turke, P.W. (1988). Helpers at the nest: childcare networks on Ifaluk. In *Human Reproductive Behavior: a Darwinian Perspective,* ed. L. Betzig, M. Borgerhoff Mulder and P. Turke, pp. 173–88. Cambridge: Cambridge University Press.

Tutin, C.E.G. and Fernandez, M. (1993). Composition of the diet of chimpanzees and comparisons with that of sympatric lowland gorillas in the Lope Reserve, Gabon. *American Journal of Primatology* **30**, 195–211.

Vincent, A.S. (1985). Wild tubers as a harvestable resource in the East African savannas: ecological and ethnographic studies. PhD dissertation, University of California, Berkeley.

Wallis, J. (1997). A survey of reproductive parameters in the free-ranging chimpanzees of Gombe National Park. *Journal of Reproduction and Fertility* **109**, 297–307.

Wood, J.W. (1994). *Dynamics of Human Reproduction.* New York: Aldine de Gruyter.

Wood, J. and Smouse, P. (1982). A method for nalyzing density-dependent vital rates with an application to the Gainj of Papua New Guinea. *American Journal of Physical Anthropology* **58**, 403–11.

Wrangham, R.W., Conklin, N.L., Etot, G. *et al.* (1993). The value of figs to Chimpanzees. *International Journal of Primatology* **14**, 243–56.

7 The evolutionary ecology of the primate brain

R O B E R T   B A R T O N

Introduction

Comparative studies of the brain have long intrigued and frustrated researchers. They have intrigued because of what they might reveal about the evolutionary forces shaping cognitive attributes. They have frustrated because the results have been so difficult to interpret, leading to a raft of further questions. What is the cognitive significance of differences in brain size? Do correlations between brain size and ecology indicate selection pressures on specific-information processing abilities, or merely some kind of general constraints on brain size? The ratio of speculation to hypothesis testing in this area has been so far above unity as to question its scientific credibility. Yet there are hypotheses to be tested. Is variation in brain size associated with the differential expansion of specific neural systems? If so, does the differential expansion of such systems correlate with relevant behavioural or ecological variables? Or is brain size simply determined by biological constraints, like metabolic rate? Tests of predictions arising from such questions are described in this chapter, based on the data in Appendices 7.1 and 7.2. Some of the methodological principles and problems associated with these tests are emphasised, starting with a discussion (parallel to Purvis and Webster's, Chapter 3) of the importance of controlling for phylogeny when testing for correlated evolution among brain traits and other variables.

Phylogeny and brain evolution

The comparative method can be used to test hypotheses about adaptation. Indeed, 'adaptation is an inherently comparative idea' (Harvey and Pagel, 1991: p. 13). Purvis and Webster (Chapter 3) explain why comparative studies must take phylogeny into account when testing for correlated evolution between biological, behavioural and ecological traits (Harvey and Pagel, 1991). Hypothesising that variable Y is adaptively linked to variable X implies that selection has caused the correlation between them. Species cannot be assumed to represent statistically independent data points in testing such hypotheses, because similarities may stem from common ancestry rather than from independent evolution under similar selection pressures.

The principle is illustrated clearly by considering the evolution of brain size. Within the primates, there is a fundamental taxonomic division between the haplorhines (monkeys and apes) and strepsirhines (lemurs and lorises). Haplorhine primates tend to have larger brains relative to body size than do strepsirhine primates (Fig. 7.1). This 'grade shift' affects the consideration of both the scaling of brain size to body size, and the patterns of correlated evolution between brain size and ecological variables. Each of these is considered in turn.

Brain allometry

Several authors have analysed brain to body size relationships in primates (e.g. Jerison, 1973; Martin, 1981). In theory, the slope of the best-fit line on a log-log plot of the two variables represents the rate at which brain size has to increase in order to maintain functional equivalence as body size increases. Early comparative analyses of mammals using individual species as data points gave slopes of around 0.67 (Jerison, 1973), whereas later studies put the slope at 0.75 (Martin, 1981). In each case, a fundamental biological reason for these slopes was suggested. The 0.67 slope suggested a connection with surface-to-volume ratios, because the surface area of a solid increases to the 2/3 power of its volume. The biological reason for this might be that numbers of sensory receptors and motor effectors must increase in direct proportion to the area of the body surfaces over which they are distributed. The more recently accepted 0.75 slope suggested a link with basal metabolic rate, because, for unknown reasons, the latter also scales to body mass with a 0.75 slope.

Scaling relationships are, however, potentially confounded by grade shifts, as clearly explained by Martin (1996). Fitting a single line across the species values in Figure 7.1 yields an artificially high slope because two grades are present, each having a similar slope that is lower than the one calculated across the whole order. The least-squares regression slope across suborders is 0.77, close to the theoretical 0.75 value. The slopes within suborders, however, are 0.66 and 0.70, closer to the 0.67 value implied by surface–volume geometry, and with 95% confidence intervals that barely include 0.75 (0.58–0.73 for strepsirhines and 0.66-0.75 for haplorhines). This reduction in slope values does not appear to be the result of

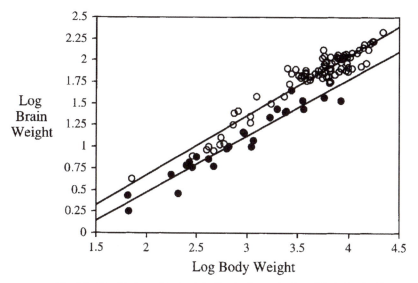

Fig. 7.1   Double-logarithmic plot of brain size on body weight in haplorhines (open circles) and strepsirhines (filled circles). The elevations of the least-squares regression slopes for the two sub-orders are significantly different ($t = 8.6$, df $= 115$, $p < 0.0001$).

the tendency for least-squares regression to underestimate slopes with lower correlations (correlation coefficients are about 0.96 in both across-order and within-order analyses). One should not, however, be tempted to conclude that the slopes within suborders necessarily represent the true line of functional equivalence, because further grade shifts may be present within suborders. The average slope obtained by analysing the species values in Figure 7.1 separately for individual primate families is 0.56 (range $= 0.45$–0.67).

A grade shift is, of course, simply a specific case of the general tendency for more closely related species to be more similar. The divergence of two taxa that produces separate grades is thus an evolutionary event like any other, and not qualitatively different from the smaller-scale adaptive differentiation between two species. As such, grade shifts should be grist to the mill of comparative analysis. The point is, however, that no single evolutionary event, such as the divergence of strepsirhines and haplorhines, should unduly weight a comparative analysis. The method of independent contrasts (see Chapter 3) is one way of determining scaling relationships free of phylogenetic bias. This method treats the grade shift in Figure 7.1 as just one of the many relevant evolutionary changes in brain and body size. Using independent contrasts, Harvey and Krebs (1990) obtained a brain–body size slope of 0.69 across a broad sample of mam-

Table 7.1. *Ecological correlates of brain size*

Variable	Standardised coefficient	t value	t value
Body weight	0.91	19.98	< 0.0001
Percentage fruit in diet	0.15	2.53	0.014
Social group size	0.16	2.77	0.007

The figures are based on a multiple regression (through the origin) carried out on independent contrasts in brain weight, body weight, percentage of fruit in the diet and social group size. The contrasts were generated using the CAIC programme (Purvis and Rambaut, 1995) and primate phylogeny (including branch lengths) in Purvis (1995), applied to the data in Appendix 7.1. All variables were log-transformed prior to analysis. For overall regression, adjusted $r^2 = 0.86$, $n = 67$, $F = 141.2$, $p < 0.0001$.

mals. Within primates, using Purvis's Comparative Analysis by Independent Contrasts (CAIC) programme and his composite primate phylogeny (see Chapter 3) gives a regression slope of 0.55 ($r^2 = 0.81$, $n = 106$, 95% confidence intervals = 0.49–0.60). Because error variance has a relatively large effect on contrasts calculated between closely related species (Purvis and Rambaut, 1995; Chapter 3), the author repeated the analysis excluding nodes dated in Purvis' composite phylogeny at younger than 5 million years (approximately half the nodes). This gave a higher slope of 0.64 ($r^2 = 0.84$, $n = 50$, 95% confidence intervals = 0.57–0.72). In fact, it is well known that, even using species as data points, brain–body weight slopes increase with taxonomic level, a phenomenon for which various biological and statistical explanations have been advanced (e.g. Lande, 1979; Pagel and Harvey, 1988a, 1989). Whilst this variability suggests we should not attach too much biological significance to any particular brain size scaling exponent, however calculated, analyses that take phylogeny into account appear to give values closer to 0.67 than to 0.75. The 0.75 scaling exponent for basal metabolic rate is apparently more robust, in that it survives independent contrasts analysis and does not vary significantly with taxonomic level (Harvey and Pagel, 1991; Martin, 1996). Hence, the claim that similar scaling exponents for brain size and for metabolic rate imply some kind of causal relationship is questionable.

Adaptive variation in brain size

Phylogenetic bias also complicates the use of comparative data for inferring adaptive associations between brain size and other traits. Strepsirhines, for example, not only have smaller brains than haplorhines, they also tend to have lower metabolic rates (Martin, 1996), shorter gestations, a different type of placentation and smaller neonates (Martin, 1990), and tend to be more nocturnal and live in smaller groups (Smuts *et al.*, 1987; Kappeler and Heymann, 1996). Interspecific analysis of any of these variables with brain size would risk finding a spurious correlation as a result of the overall grade differences between the two suborders. The word 'spurious' is used here to mean a correlation that does not reflect a general adaptive association. In order to infer that two traits have such an association, it is necessary (though not sufficient) to show that they exhibit correlated evolution: that is, they can be shown to have covaried in a consistent way across multiple evolutionary events. As an example, an interspecific analysis of relative brain size and activity timing (diurnal versus nocturnal habits) in primates shows a statistically significant difference: with the effects of body weight partialled out, diurnal species have significantly larger brains ($t = 4.3$, df $= 115$, $p < 0.0001$). There should not, however, really be 115 degrees of freedom in this analysis, because there have only been a few evolutionary changes in activity timing against which to match changes in brain size. The result could largely reflect the fact that strepsirhines are small brained and predominantly nocturnal, whereas haplorhines are relatively large brained and predominantly diurnal. Indeed, using the BRUNCH procedure for categorical variables in Purvis' implementation of the independent contrasts method (see Chapter 3), there is no significant trend for activity timing and relative brain size to evolve together ($t = 1.38$, df $= 4$, $p = 0.24$). This remains true regardless of the allometric exponent used to generate residual brain size values. Other ecological variables do, however, correlate with overall brain size. A multiple regression analysis on 68 independent contrasts shows that, when body size is taken into account, brain size shows independent positive correlations with the percentage of fruit in the diet and social group size (Table 7.1). This confirms the well-known correlation in primates between diet and relative brain size (Clutton-Brock and Harvey, 1980; Foley and Lee, 1992). The size of one particular brain structure, the neocortex, is also correlated with both percentage frugivory and social group size (Barton, 1996). The neocortex size analysis controlled for the size of the rest of the brain, thus examining relative neocortical expansion rather than size relative to body size. Hence, the ecological correlates of relative neocortical expansion and of overall brain size are similar. This might be because brain size differences are, at least partly, a consequence of selection operating specifically on neocortical processing systems, a possibility explored further below.

Ecological correlates of brain and neocortex size have been interpreted as evidence for the idea that specific aspects of lifestyles have selected for specific cognitive abilities. The correlation with frugivory led to the suggestion that large brains are needed to memorise and integrate information on the location of fruit trees distributed patchily in large home ranges (Clutton-Brock and Harvey, 1980; Milton, 1988; see Eisenberg and Wilson (1978) for a similar explanation of large brains in frugivorous bats). Correlations between neocortex size and social group size have been taken as evidence for selection on social intelligence (Dunbar, 1992; Sawaguchi, 1992; Barton and Dunbar, 1997). Whereas diet and social group size have sometimes been seen as alternative explanations for the evolution of brain size, Table 7.1 suggests that both are independently important. Some authors, however, are more sceptical that brain size can be related to the specific information-processing demands of different ecological niches. Their ideas are explored in the next section.

Life histories, maternal energetics and brain size

Perhaps differences in brain size reflect overall life history strategies or biological constraints (Sacher, 1959; Hofman, 1984; Shea, 1987; Harvey *et al.*, 1987; Parker, 1990; Allman, McLaughlin and Hakeem, 1993; Martin, 1996), rather than ecologically related neural specialisation. One suggestion has been that brain size is linked to life span (Sacher, 1959; Hofman, 1984; Allman *et al.*, 1993; see also Harvey and Read (1988) for a discussion). Sacher (1959) found that brain size and life span were more strongly correlated than either were with body size. Similarly, Allman *et al.* (1993) found that primate life spans and brain sizes were positively correlated even after the effects of body weight had been removed from each. Harvey *et al.* (1987) suggested that brain size is more directly related to age at maturity than to life span, because age at maturity reflects the amount of postnatal brain development and learning during the juvenile phase (see also Joffe, 1997). Others have suggested that energy constraints may play a key role in limiting brain size (Martin, 1981; Armstrong, 1983; Foley and Lee, 1992). Martin's ideas (1981, 1996) have been particularly influential and stimulating. He suggested that variation in adult brain size is linked to the metabolic turnover of the mother during gestation and lactation. The reason why frugivores have comparatively large brains might be simply that their energy-rich diets allow them to have higher basal metabolic rates (Martin, 1996: p. 153). As mentioned above, this idea originated in part from the apparent similarity of the allometric exponents for brain size and basal metabolic rate (Martin, 1981). Subsequently, Martin (1996) included gestation length with basal metabolic rate as components of maternal energy expenditure. He analysed the partial correlations between brain size, basal metabolic rate, gestation length and body weight in 53 species of placental mammals, finding significant correlations between brain weight and both metabolic rate and gestation length, with the effect of body weight partialled out.

The analyses described above suffer from two problems. The first is that they were carried out using means for species, or for higher taxonomic groupings such as genera, as independent data points, and thus are susceptible to the type of phylogenetic bias discussed above. The second is the 'Economos problem' (see Harvey and Krebs, 1990). Economos (1980) pointed out that body weight is an intrinsically 'messy' variable, subject to a high degree of intraspecific variability related to nutrition, disease and genetics. Life span and brain size, or metabolic rate and brain size, may be more highly correlated with each other than with body weight merely because they are more accurate indices of a species' body size than is weight. Using regression to remove the effects of body weight from both variables of interest (e.g. Allman *et al.*, 1993) would only compound the problem, because this adds the same error variance in the body size measurements to both variables. If we then correlate the 'size-corrected' variables with each other, there is a relatively high chance of finding a spurious positive correlation, because of the correlated error added to each variable. To make this clearer, imagine the (admittedly unlikely) case in which the true correlations between brain size and body size and between life span and body size were perfect. Residual values for brain size and life span after regressing them on mass would be entirely determined by the measurement errors in each variable (because, in theory, all points should lie on the line, making their true residual values zero). If mass was substantially more error prone than brain size and life span, then the size-corrected values for each of the latter variables would be strongly and similarly affected by the error in body size, leading to positive correlations even in the absence of any biological relationship. A possible solution to mass errors is to estimate size from separate samples of individuals for each variable being investigated (Harvey and Krebs, 1990), or perhaps to use size measures other than weight (Mace and Eisenberg, 1982). The extent to which this theoretical problem is a practical reality has not yet been assessed, and therefore the results of studies in which two or more variables corrected for body size are correlated with each other should be treated with caution. In mitigation, Allman *et al.* (1993) showed that the size of organs other than the brain did not correlate with life span in the same way, which would have been expected if the original correlation was a statistical artefact.

If we ignore the theoretical Economos problem for the moment, what does phylogenetic re-analysis of the links between brain size and life histories reveal? Life-history variables such as age at first reproduction and life span are so closely linked to each other (e.g. Harvey et al., 1987; Charnov, 1991; Purvis and Harvey, 1995) that it may not make sense to treat them as different variables within a single comparative analysis. Thus, separate multiple regressions were carried out for each with 55 independent contrasts. Body weight was accounted for in these analyses by entering it into the multiple regressions ($p < 0.001$ in each case). There is no significant association between brain size and life span ($t = 1.58, p = 0.12$), nor between brain size and age at first reproduction ($t = -0.42, p = 0.67$). These tests are, however, quite strongly affected by a few outlying contrasts calculated at younger nodes in the phylogeny. As noted earlier and in Chapter 3, error variance tends to be amplified in contrasts at younger nodes. Excluding nodes younger than 5 million years gives significant results (with 38 contrasts, $t = 3.27$, $p = 0.002$ for life span, $t = 2.24$, $p = 0.03$ for age at first reproduction).

In the previous section it was shown that ecological variables (percentage frugivory and social group size) are correlated with brain size. How can this be reconciled with the results for life history variables? Are ecology and life histories confounded, such that only one of them is the true correlate of brain size? Multiple regressions, again on contrasts calculated at older nodes (> 5 million years), suggest that both ecological and life history variables are separately correlated with brain size. Brain size is significantly positively correlated with age at first reproduction ($t = 2.34$, $p = 0.03$), group size ($t = 2.61$, $p = 0.01$) and percentage frugivory ($t = 2.33$, $p = 0.03$). In a separate multiple regression, brain size is positively related to life span ($t = 2.79$, $p = 0.01$), group size ($t = 1.98$, $p = 0.05$) and frugivory ($t = 2.42, p = 0.02$). These analyses therefore support both ecological specialisation and life histories as correlates of brain size.

Next, the maternal energy hypothesis (Martin, 1996) is re-examined, following Martin in considering the joint effects of two maternal energy variables, basal metabolic rate and gestation length. The problem of phylogenetic bias is very apparent here. For example, correlating basal metabolic rate with brain size across species gives a statistically significant result (multiple regression controlling for body weight, $t = 2.7$, df = 2,18, $p = 0.01$). However, this appears to be largely because, relative to their body sizes, haplorhines have significantly higher metabolic rates ($t = 2.6$, df = 2,19, $p = 0.02$) and brain sizes ($t = 8.1$, df = 2,116, $p < 0.001$) than do strepsirhines. When the original multiple regression is re-run, controlling for suborder membership by entering this as a dummy variable, brain size is uncorrelated with basal metabolic rate ($t = 1.6$, df $= 3,17$, $p = 0.13$). Similarly, there are no significant associations between brain size and maternal energy variables when the independent contrasts method is used (multiple regression on 18 sets of contrast values, controlling for body size: $t = -0.21$, $p = 0.84$ for basal metabolic rate, and $t = 1.19$, $p = 0.25$ for gestation length). The contrasts were nearly all made at older nodes, and excluding the one node younger than 5 million years does not make the results significant (multiple regression on 17 contrasts, $t = 0.47$, $p = 0.65$ for basal metabolic rate, and $t = 1.17$, $p = 0.26$ for gestation length), nor does the additional removal of another contrast that appeared as a relatively large outlier in scatterplots (with 16 contrasts, $t = 0.47$, $p = 0.65$ for basal metabolic rate and $t = 1.14$, $p = 0.27$ for gestation length). Perhaps life history variables are confounding relationships with the maternal energy variables? To test this, the author incorporated age at first reproduction into the analysis. Because this makes the sample size rather small ($n = 11$) relative to the number of test variables, a stepwise regression was used to select the variables significantly associated with brain size, taking body size into account by entering it first. This showed a significant positive correlation for age at first reproduction ($p = 0.02$), but not for basal metabolic rate or gestation length ($p > 0.1$ in each case).

These results cast doubt on a link between maternal energy and brain size in primates. Results of other comparative studies also lend little support. Among mammals generally, neonatal brain size is not correlated with basal metabolic rate (Pagel and Harvey, 1988b). Although neonatal brain size *is* correlated with gestation length (Pagel and Harvey, 1988b), species with small brained neonates are not small-brained as adults, because they compensate by greater postnatal brain growth (Harvey and Read, 1988). While frugivores have large brains, they do not seem to have high basal metabolic rates (Elgar and Harvey, 1987; Ross, 1992), and carnivory appears to be the only dietary correlate of basal metabolic rate that is independent of phylogeny (Elgar and Harvey, 1987). Within the order Carnivora, the more frugivorous species actually have the lowest metabolic rates (McNab, 1995). Because several of these studies did not control for phylogenetic bias, further work is needed to establish the validity of their conclusions.

If substantiated by further work, the lack of correlations between brain size and metabolic rate does raise the question of how large brains are accommodated in metabolic terms. Because brains are metabolically expensive (Martin, 1981, 1996; Foley and Lee, 1992; Aiello and Wheeler, 1995), the extra energy for brain expansion must come from somewhere.

One possibility is a trade-off between the size of the brain and of other organs, such as the liver and gut (Aiello and Wheeler, 1995). A folivorous diet creates more demands on the digestive system than does a frugivorous diet, requring a larger, more metabolically active gut. Thus, frugivory may simultaneously select for large brains and for small guts, leading to a trade-off between the two (Aiello and Wheeler, 1995). Presumably, the implication of such a trade-off is that the need for large brains and the relaxation of the need for large guts is a happy coincidence.

Ontogenetic constraints on brain specialisation?

Finlay and Darlington (1995) proposed that mammalian brain evolution has been highly constrained by ontogenetic mechanisms limiting variability in the size of specific brain regions. They stated that 'the most likely brain alteration resulting from selection for any behavioural ability may be a coordinated enlargement of the entire non-olfactory brain'. If true, this means that individual brain regions have not been free to vary in size independently of the whole, and that selection has acted primarily on overall brain size. The idea fits most easily with theories of brain size that invoke general life history strategies or metabolic constraints, rather than adaptive specialisation in response to the sensory and cognitive demands of particular niches. Is it correct?

Finlay and Darlington's claims are based largely on the apparent predictability of the size of major brain structures from overall brain size across three orders of mammals. They showed that the size of each individual structure is highly correlated with overall brain size, explaining 96% of the variance in the size of any given structure. Only the olfactory bulb and associated structures deviated from this predictability to any extent, and a two-factor model comprising the effects of overall brain size and olfactory bulb size accounted for 99% of the variance in the size of individual brain structures. Brain structures that develop late in ontogeny, such as the neocortex, were found to increase relatively rapidly in size with interspecies increases in brain size, leading to the suggestion that evolutionary changes in brain size occur by simple developmental shifts, such as an increase in the period of brain growth. An increase in this period would result in all structures growing larger, but late-developing structures would be most affected.

Finlay and Darlington's results seem to provide strong support for their conclusion that individual brain systems have not, in general, evolved independently of the whole, and therefore that adaptive specialisation has

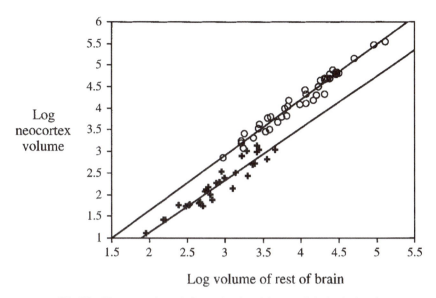

Fig. 7.2   Neocortex size, relative to the size of the rest of the brain, in primates (circles) and insectivores (crosses). It is clear that primates' larger brains are associated with relative expansion of the neocortex.

been minimal. There are, however, a number of reasons to doubt this conclusion.

First, Finlay and Darlington's analysis used species values for primates, bats and insectivores as independent data points. Their analysis of the regression slopes of structure size on overall brain size depends on the validity of these slopes, but, as explained in the section on phylogeny and brain evolution above, the slopes are likely to have been biased by grade shifts between orders. For example, they pointed to the 'explosive nature of the change in neocortex size' with increasing brain size. The apparently explosive scaling of neocortex size is, however, partly a product of adaptive grade shifts between orders and suborders (Fig. 7.2). Primates have larger neocortices relative to brain size than do insectivores, and slope values calculated across orders will therefore be inflated by this grade shift. Additional grade shifts in neocortex size are present within orders (e.g. Barton, 1996). Other structures, such as the cerebellum, exhibit similar grade shifts (see Barton, in press). Although a re-analysis by Darlington (personal communication) suggests that between-order grade shifts have a small effect on slopes, the fact that they exist at all of course contradicts the idea that the size of each brain structure is strictly tied to overall brain size by a universal growth law. Instead, such grade shifts indicate adaptive specialisation in brains.

Second, the regression slopes relating the size of each structure to overall brain size are affected by autocorrelation. Brain structures that make up a relatively large proportion of total brain size (such as the neocortex and cerebellum) are inevitably more strongly correlated with it than are smaller structures. Large structures will therefore tend to exhibit higher slopes on total size than do small structures, even in the absence of any real biological effect, because least-squares regression slopes vary with the strength of the correlation (the higher the correlation, the greater the slope). Thus, the differences in regression slopes may indicate differences in the extent of autocorrelation as much as they do differences in ontogenetically driven scaling laws.

Third, the conclusion that only a trivial amount of variance in brain structure size remains unexplained by Finlay and Darlington's two-factor model depends on some questionable assumptions. Much of the variance they attributed to the overall brain size factor may, in fact, be associated with body size. They argued that the first factor in their principal components analysis was most usefully construed as simply overall brain size, rather than body size. One reason for this was that the correlation with body weight was relatively low. As explained above, however, this may be simply because of the greater error in body size data. A slightly different pattern is revealed by performing separate principal components analyses for each order, first removing the effects of body size, and using independent contrasts rather than species as data points.[1] The first factor accounts for 49%, 62% and 67% of the variance among bats, insectivores and primates respectively, which are quite high, but substantially less than the 96% in Finlay and Darlington's analysis. As in Finlay and Darlington's analysis, the first factor was a 'global' factor, in that it loaded moderately to strongly on all structures except for, in primates and insectivores, the olfactory bulb. Ontogenetic constraints are, however, only one of a number of possible explanations for this global factor. Another is failure to remove all body size-related variance owing to the use of error-prone weight measurements. The second factor extracted in this author's analyses differed between orders: in primates and insectivores it loaded most highly on the olfactory bulb and another olfactory structure, the piriform lobe, in broad agreement with Finlay and Darlington's two-factor model. Again functional specialisation is suggested because olfactory systems are correlated with specific ecological factors (Barton, Purvis and Harvey, 1995). Bats showed a different pattern. A second and a third factor were extracted, the former loading on the cerebellum, medulla and mesencephalon, the latter on the striatum and neocortex (which have major functional and anatomical links – see Keverne, Martel and Nevison, 1996). These differences between orders are not predicted by strict ontogenetic constraints, but accord with the idea of selection affecting separate but functionally linked brain structures.

A difficulty with interpreting the results of principal components analysis is that the cut-off between what constitutes interesting variance and what constitutes trivial residual variance is quite arbitrary; a variance proportion that superficially seems trivially small may nevertheless be evolutionarily and neurally highly significant. For example, when the author extracted a third factor for primates, this loaded on both schizocortex and hippocampus, in accord with the fact that these two structures have strong anatomical and functional connections and so may be expected to have co-evolved under appropriate selection pressures. The point is reinforced by examining partial correlations between all brain structures. This shows that many significant correlations are found when variation in other structures is partialled out, and these evolutionary correlations reflect known anatomical and functional links, such as between amygdala and piriform cortex (Barton, in press). Finlay and Darlington themselves made the point that even very small percentages of the total interspecies variance in brain size may represent functionally significant amounts of neural tissue, and they agree that scope for adaptive specialisations exists in spite of the apparent uniformity of brain proportions. Unfortunately, these important comments have been rather lost in some subsequent discussions (e.g. Quartz and Sjenowski, 1997). Finlay and Darlington are right that, in statistical terms, the residual variance unexplained by overall size is generally quite small. The question is, does any of that residual variance represent adaptive specialisation?

Brain specialisation

A number of comparative studies of mammals provide evidence for mosaic evolution of the brain, in which specific parts have evolved independently of the whole. Importantly, the correlations found between lifestyle and the size of specific brain regions accord with the functions of these structures. In primates, for example, activity period is correlated with the size of primary sensory structures (Barton *et al.*, 1995): relative to the size of the rest of the brain, diurnal primates have larger visual cortices, whereas nocturnal primates have larger olfactory bulbs (Fig. 7.3). Clearly, selection has emphasised the sensory systems most appropriate to the ambient light levels associated with each niche. Diet also shows correlations with sensory brain structures in primates; among diurnal species, frugivores have a

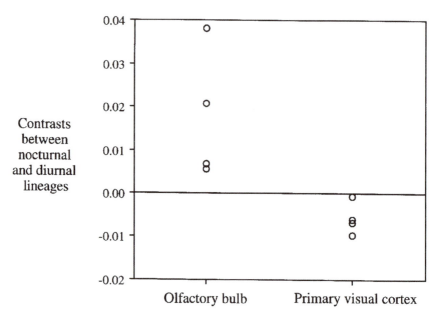

Fig. 7.3   Relative sizes of the olfactory bulb and primary visual cortex in nocturnal versus diurnal primates. The relative size of each structure was calculated as the residuals from the regression of structure size on the size of the rest of the brain (exluding all main sensory structures). The graph plots four independent contrasts in the relative size of each structure. Positive contrasts indicate larger structure size in nocturnal lineages; negative contrasts indicate larger structures in diurnal lineages. The compensatory differences in the size of the olfactory and visual structures may explain why overall brain size does not differ between nocturnal and diurnal lineages. Graph based on analysis in Barton *et al.* (1995).

larger primary visual cortex than do folivores, whereas among nocturnal species, frugivores have relatively enlarged olfactory bulbs and piriform lobes (Barton *et al.*, 1995). The implication that different sensory systems have been favoured for locating fruit depending on activity period has recently found support in an experimental study by Bolen and Green (1997), showing that owl monkeys (*Aotus nancymai*) are more efficient at locating fruit by olfactory cues than are diurnal capuchin monkeys (*Cebus apella*). On the other hand, owl monkeys have lower retinal cone densities than their diurnal relatives, poorer colour vision, and lack a distinct fovea. Also in primates, relative neocortex size is correlated with social group size (Dunbar, 1992; Sawaguchi, 1992; Barton, 1996; Keverne, *et al.*, 1996), and this has been taken as evidence for the 'social intelligence' hypothesis (see Byrne and Whiten, 1988).

Among mammals other than primates, olfactory structures are relatively small in aquatic taxa, within both insectivores (Fig. 7.4) and carnivores

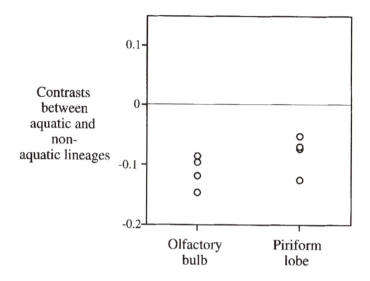

Fig. 7.4   Relative sizes of olfactory brain structures in aquatic versus non-aquatic insectivores. The procedure was the same as in Figure 7.3. Negative contrasts indicate the evolution of smaller relative structure size with aquatic habits. Graph based on analysis in Barton *et al.* (1995).

(Gittleman, 1991), reflecting the relationship between mammalian olfaction and the respiratory system. Predatory behaviour in mammals is correlated with a relatively enlarged tectospinal tract (Barton and Dean, 1993), in accord with the role of this pathway in regulating predatory-type orienting, tracking and attacking movements (Dean, Redgrave and Westby, 1989). In rodents, sex differences in ranging patterns are correlated with differences in the size of a structure implicated in spatial memory, the hippocampus (e.g. Jacobs *et al.*, 1990). Studies of birds also indicate differential evolution of hippocampus (e.g. Krebs, 1990), and other specific brain structures (Healy and Guilford, 1990; Devoogd *et al.*, 1993).

Are such specialisations associated with differences in overall brain size? If the amount that specific brain areas have evolved independently of overall brain size is very small, or if increases in specific areas are compensated by decreases in the size of other areas, then overall brain size may be unaffected. For example, it is known that food-storing and non-storing birds do not differ in overall brain size despite marked differences in relative hippocampus size (Sherry *et al.*, 1989). This is probably because the hippocampus is a small proportion of the total brain size, so that its potential effect on brain size is swamped by other factors. In primates, trade-offs between different brain areas have been noted (Barton *et al.*,

1995; Keverne *et al.*, 1996). For example, the larger visual structures and neocortices of diurnal lineages are compensated by smaller olfactory structures (see Fig. 7.3), resulting in no net difference in brain size. Thus, adaptive neural differences do not necessarily mean differences in brain size. There is, however, evidence that other cases of visual specialisation in primates have had an impact on overall brain size.

Visual specialisation and brain size in primates

As noted by Martin (1996, p. 155)

> no convincing case has been made for the proposal that any particular feature of behaviour . . . has exerted a specific selection pressure favouring an increase in brain size.

The purpose of this section is to make such a case.

Fossil endocasts of early primates indicate both large brain size compared with contemporaneous mammals, and relative expansion of visual cortical areas (Jerison, 1973; Allman, 1987).

> The peculiar, bulbous appearance of the temporal lobe in primates, which reflects the large territory occupied by higher-order visual areas, can be seen in early euprimate endocasts.   (Preuss, 1993, p. 34)

Large relative brain size in fossil primates was thus probably based at least partly on visual specialisation.

Allman (1987; Allman and McGuinness, 1988) draws attention to several unusual features of the primate visual system, including: (i) frontally directed eyes and a high degree of binocular overlap, facilitating stereopsis; (ii) high visual acuity, particularly in diurnal haplorhines, associated with a retinal fovea containing a high density of photoreceptors; (iii) a unique arrangement of the retino-tectal projections (see Allman, 1987); (iv) a lateral geniculate nucleus with up to six distinct layers, including both the two magnocellular layers common to all mammals and two to four parvocellular layers not found in other mammals. These visual specialisations are associated with a complex arrangement of highly interconnected and numerous cortical visual areas: in macaques there are 305 known pathways connecting 32 cortical visual areas, and visual areas make up about half of the entire cortex (van Essen, Anderson and Felleman, 1992). Thus, the expanded primate neocortex (see Fig. 7.2) is a largely visual organ.

Perhaps, then, there is a connection between large brains and the large amount of brain tissue devoted to vision in primates (see also Gibson, 1990). Indeed, Figure 7.5 shows that the size relative to the rest of the brain of both the primary visual cortex (data on secondary and tertiary areas are not available) and the lateral geniculate nucleus (a visual relay in the thalamus that projects to visual cortex) are positively correlated with encephalisation. That is, primates with large brains for their body size have relatively expanded visual structures. Furthermore, there is clear evidence that a specific subsystem within the geniculo-cortical system, the parvocellular subsystem, is selectively expanded in large-brained species. The parvocellular subsystem is implicated in the perception of fine detail and colour, whereas the magnocellular subsystem is primarily involved in movement detection and the analysis of dynamic form (Livingstone and Hubel, 1988; Zeki and Shipp, 1988; Allman and McGuinness, 1988). Crucially, encephalisation is correlated with the relative number of neurons in the parvocellular, but not magnocellular, layers of the lateral geniculate nucleus (Fig. 7.6). Also, like brain size and neocortex size, relative expansion of parvocellular lateral geniculate nucleus layers is correlated with both frugivory and social group size (Table 7.2).

Colour vision and brain size

These findings help with the interpretation of ecological correlates of brain size. As noted above, the correlation between brain size and frugivory has previously been taken as evidence that brain size reflects selection for spatial memory. The link the author has shown with parvocellular specialisation suggests an alternative explanation: colour vision. The parvocellular system processes colour (among other things), whereas the magnocellular system does not (Livingstone and Hubel, 1988; Zeki and Shipp, 1988). Recent work using magnetic resonance imaging of the human brain has shown that large areas of primary and secondary visual cortex carry out colour processing (Engel, Zhang and Wandell, 1997). Evolutionary enhancements of colour vision would therefore be likely to have a significant impact on neocortex size. Furthermore, a link between the evolution of colour vision and frugivory has long been suspected:

> The primary necessity which led to the development of the sense of colour was probably the need of distinguishing objects much alike in form and size, but differing in important properties, such as ripe and unripe, or eatable and poisonous fruits . . .
>
> Wallace (1891)

Contemporary researchers into colour vision in non-human primates have also argued that it is an adaptation for locating and selecting palatable fruit (Mollon, 1989; Jacobs, 1993; Osorio and Vorobyev, 1996).

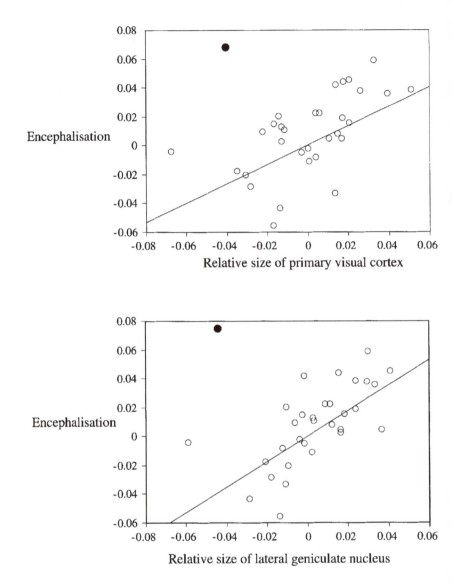

Fig. 7.5   Correlated evolution of encephalisation (brain size relative to body weight) and relative size of geniculo-cortical visual structures. The graphs show that large brain size is associated with relative expansion of visual areas. Encephalisation values are the residuals of independent contrasts in brain volume regressed on contrasts in body weight. Similarly, relative visual area values are the residuals of contrasts in visual area volume regressed on contrasts in volume of the rest of the brain. Because large areas of the neocortex (at least 50% in monkeys) are composed of visual areas, and because the full extent of these areas has not been defined or quantified, neocortex size was subtracted, along

Jacobs (1993; 1995) has reviewed the presence and type of colour vision in primate species. Dichromatic colour vision, present in many New World monkeys and almost certainly in diurnal lemurs, enables animals to distinguish between fruits of different colour, while trichromatic vision (present in some New World monkeys and all Old World monkeys) aids in the detection of fruits against a background of green leaves (Osorio and Vorobyev, 1996). Further research looking directly at the relationship between colour vision abilities and brain size, and at differences in the size of V4 – the cortical area specialised for colour processing (Zeki, 1993) – is needed to clarify the relationship between frugivory and visual specialisation.

Socio-visual cognition and brain size

Two considerations suggest that colour vision is unlikely to be the only aspect of parvocellular function implicated in brain size variation. First, the parvocellular system analyses much more than just colour. It supports high-acuity vision in general. Second, while an obvious link can be made between frugivory and colour vision, a link between sociality and colour vision is less obvious (but see Jacobs, 1995). So why should parvocellular expansion, and hence neocortical expansion, have accompanied increases in group size? The answer probably lies in the rôle of high-acuity vision in the processing of social information. The following argument is made in more detail elsewhere (Barton and Dunbar, 1997; Barton, in press).

Although social cognition is usually thought of as a 'higher' cognitive function, far removed from 'basic' processing of sensory stimuli, it is becoming clear that the distinction between higher cognitive and basic sensory functions is arbitrary and unhelpful (e.g. Jackendoff, 1992; Zeki, 1993; Crick, 1994). It may be most useful to think of social cognition as a

Caption for Fig. 7.5 (*cont.*)

with lateral geniculate nucleus size, in calculating the size of the rest of the brain. The regression lines were fitted excluding the outlier indicated by the filled circle. This outlier is based on contrasts between the subfamilies Daubentoniidae and Indriidae. Significantly, the phylogenetic position of *Daubentonia* is highly contentious (Yoder, 1994), exacerbating the problem of amplified error variance at low taxonomic levels (Purvis and Rambaut, 1995). In addition, it has been suggested that the brain of Daubentonia is 'too large' for its body size, because it is a phyletic dwarf (Stephan, Baron and Frahm, 1988). Even with this datum included, however, the regressions are significant (visual cortex; $r^2 = 0.14$, $t = 2.21$, df $= 31$, $= 0.03$; lateral geniculate nucleus: $r^2 = 0.18$, $t = 2.6$, df $= 31$, $p = 0.014$). Removing it markedly improves the regressions ($r^2 = 0.34$, $p = 0.0005$, and $r^2 = 0.46$, $p < 0.0001$ for visual cortex and lateral geniculate nucleus, respectively).

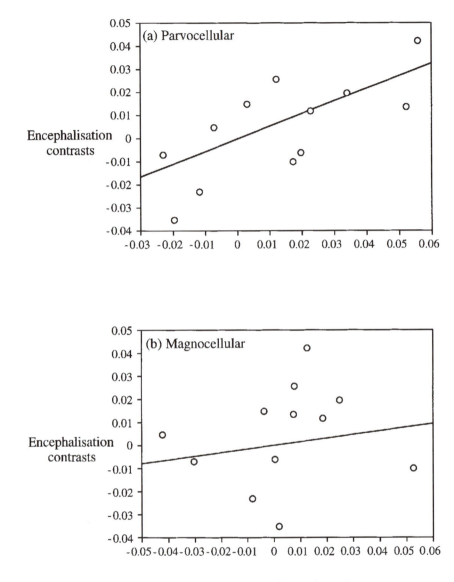

Fig. 7.6   Correlated evolution of encephalisation (brain size relative to body weight) with (a) parvocellular, but not (b) magnocellular, layers of the visual lateral geniculate nucleus. The graphs show that large brain size is associated with relative expansion of the parvocellular system, involved in the analysis of fine detail and colour. Values were calculated in the same way as for Figure 7.3, but lateral geniculate nucleus data used here are the number of neurons in the

Table 7.2. *Ecological correlates of parvocellular and magnocellular layers of the lateral geniculate nucleus in diurnal primates*

Variable	Standardised coefficient	*t* value	*p* value
Parvocellular volume			
Rest of brain volume	0.63	9.48	0.003
Percentage fruit in diet	0.21	2.62	0.050
Social group size	0.36	5.86	0.010
Magnocellular volume			
Rest of brain	0.87	6.45	0.008
Percentage fruit in diet	0.04	0.29	0.789
Social group size	0.13	0.79	0.490

The results of multiple regression analyses of independent contrasts are presented. The analyses were restricted to diurnal primates because (i) parvocellular functions require photic conditions, so would not be expected to diversify in night-active lineages; (ii) the parvocellular layers are significantly larger in diurnal than in nocturnal primates (independent contrasts analysis: $t = 9.7$, df $= 2$, $p = 0.002$), so activity timing could otherwise confound the analysis. Although sample sizes are small, the same ecological factors that explain encephalisation (see Table 7.1) also explain relative expansion of the parvocellular layers, suggesting that visual specialisation for frugivory and sociality was a factor in the evolution of brain size in primates. For overall regressions; parvocellular volume, adjusted $r^2 = 0.98$, $F = 50.2$, df $= 3,3$, $p = 0.005$; magnocellular volume, adjusted $r^2 = 0.96$, $F = 45.3$, df $= 3,3$, $p = 0.005$. Very similar results are obtained if analysis is based on the number of neurons, instead of the volume, of the separate layers.

large array of sensory–cognitive operations occurring in parallel, and vision, particularly vision mediated by the parvocellular system, may be a crucial component. In what way?

The parvocellular system mediates a range of visual processes in the neocortex, particularly those that involve the perception of fine details (Livingstone and Hubel, 1988; Zeki and Shipp, 1988). This kind of processing is critically involved in facial recognition, and perception of gaze direction and facial expression. Several areas of neocortex whose main visual inputs are parvocellular, such as inferotemporal cortex, are specialised for processing social information of this type (e.g. Brothers, 1990; Brothers and Ring, 1992; Perrett *et al.*, 1992). These kinds of complex visual cues must be processed and integrated to achieve what Brothers

Caption for Fig. 7.6 (*cont.*)

separate layers, rathar than volume. Results for volume are, however, much the same. The outlier in Figure 7.5 is excluded. The regression is significant for parvocellular layers ($r^2 = 0.52$, df $= 11$, $p = 0.005$), but not for magnocellular layers ($r^2 = 0.03$, df $= 11$, $p = 0.56$). Data for spearate magnocellular and parvocellular layers are from Shulz (1967).

(1990, p. 28) calls 'the accurate perception of the dispositions and intentions of other individuals'.

Allman (1987, p. 639) states that 'as complex systems of social organisation evolved in haplorhine primates, social communication was increasingly mediated by the visual channel'.

Of course, magnocellular-mediated analysis of motion, as well as auditory processing of vocalisations, must play some role in social information processing. Perhaps, however, the critical evolutionary developments for upgrading such processing were increases in the ability to analyse the fine details of social stimuli, such as facial expressions, and to integrate these with memory and with emotional responses mediated by the limbic system, where cells responsive to social stimuli have also been found (Brothers, 1990). Processing such stimuli may be unusually costly in computational and neural terms (Barton, in press).

Recent evidence suggests that neocortical adaptations for social information processing in primates extend beyond modifications of primary visual mechanisms. Joffe and Dunbar (1997) find that group size and neocortex size remain correlated when the size of the primary area of visual cortex is partialled out. This does not undermine the idea that visual specialisation underlies neocortical evolution in primates, because the neocortex contains extensive higher visual areas in addition to primary visual cortex. What it does suggest is that, in addition to evolutionary changes at lower processing levels (lateral geniculate nucleus and primary visual cortex), there have also been modifications higher up the processing hierarchy, in integrative processing areas (including higher visual areas). This may reflect the addition of new cortical visual areas with specialised functions noted by Allman and McGuinness (1988). It is very likely that the evolution of new higher, integrative areas goes hand in hand with enhancements of the supporting, lower-level architecture.

Conclusion

While a connection between brain size and neural specialisation has sometimes been denied or played down, the evidence presented above suggests that such a connection exists. This is true in two senses. First, the difference in brain size between primates and another mammalian order, the insectivores, is associated with visual specialisation; the size of visual structures in primates is larger relative to the size of the rest of the brain (see Barton, in press). Second, variation in brain size within the primate order is associated with relative expansion of visual areas, specifically the parvocellular visual pathway. Size is not everything, however, and comparative evidence is emerging that size differences are often associated with differences in the connectional architecture of particular neural systems (Preuss, 1993; Young *et al.*, 1994). For example, the enlarged temporal cortex of primates has unusually strong connections with prefrontal cortex (Preuss, 1993), perhaps suggesting an enhanced ability to hold in working memory and manipulate visual information about objects and conspecifics.

Brain size thus cannot be meaningfully intepreted without reference to specific neural adaptations, which will differ from case to case. The fact that brain size is also correlated with life span (or age at first reproduction), even taking account of the correlations with ecology, may give solace to those who view brain size as a life history variable (e.g. Shea, 1987; Parker, 1990). Such results must be treated with caution because of the 'Economos problem'. The possibility remains, however, that both specific ecological adaptations, affecting individual neural systems, and general life history strategies, affecting overall brain size, have been significant factors in brain evolution. This may correspond broadly to the distinction made by Finlay and Darlington (1995) between brain evolution that is 'hard' (change in specific neural systems) and that which is 'easy' (co-ordinated changes in size throughout the brain). Co-ordinated changes in separate brain structures can, however, also arise as a result of neural specialisations. Primates combine visual acuity with hand–eye co-ordination and manual dexterity (Cartmill, 1974; Allman and McGuinness, 1988), functions that bring into play the extensive connections between neocortex and cerebellum. Also, the cerebellum is, like the neocortex, significantly larger in primates than in insectivores (Barton, in press). This suggests that the relevant functional units, or modules of the brain, on which selection has acted cannot be simply equated with single structures or regions, but constitute networks distributed across major brain regions. We should expect to find that arrays of linked nuclei in different brain regions have evolved together, and this provides a functional reason why such regions covary in size.

Acknowledgements

I thank Andy Purvis and Paul Harvey for advice about comparative methods and extensive comments on related work, and Phyllis Lee, Caroline Ross, Ann MacLarnon and Rob Deaner for comments on an early draft.

Note

1 In these and subsequent analyses, I use volumes of separate brain parts measured by Stephan, Frahm and Baron (1981) for primates and insectivores, and from Stephan, Pirlot and Schneider (1974) and Stephan and Pirlot (1970) for bats.

References

Aiello, L.C. and Wheeler, P. 1995. The expensive tissue hypothesis: the brain and digestive system in human and primate evolution. *Current Anthropology* **36**, 199–221.

Allman, J. (1987). Primates, evolution of the brain. In *The Oxford Companion to the Mind*, ed. R.L. Gregory, pp. 633–9. Oxford: Oxford University Press.

Allman, J. and McGuinness, E. (1988). Visual cortex in primates. In *Comparative Primate Biology*, Vol. 4, ed. H.D. Steklis and J. Erwin, pp. 279–326. New York: Alan R. Liss Inc.

Allman, J., McLaughlin, T. and Hakeem, A. (1993). Brain weight and life-span in primate species. *Proceedings of the National Acadamy of Sciences USA* **90**, 118–22.

Armstrong, E. (1983). Relative brain size and metabolism in mammals. *Science* **220**, 1302–4.

Barton, R.A. (1996). Neocortex size and behavioural ecology in primates. *Proceedings of the Royal Society (Biological Sciences)* **263**, 173–7.

Barton, R.A. (in press) Social and ecological factors in primate brain evolution. In *Group Movement: Patterns, Processes and Cognitive Mechanisms*, ed. S. Boiski and P. Garber. Chicago: Chicago University Press.

Barton, R.A. and Dean, P. (1993). Comparative evidence indicating neural specialisation for predatory behaviour in mammals. *Proceedings of the Royal Society (Biological Sciences)* **254**, 63–8.

Barton, R.A. and Dunbar, R.I.M.D. (1997). Evolution of the social brain. In *Machiavellian Intelligence*, 2nd edn, ed. A. Whiten and R.W. Byrne, pp. 240–63. Cambridge: Cambridge University Press.

Barton, R.A. , Purvis, A. and Harvey, P.H. (1995). Evolutionary radiation of visual and olfactory brain systems in primates, bats and insectivores. *Philosophical Transactions of the Royal Society (Biological Sciences)* **348**, 381–92.

Bolen, R.H. and Green, S.M. (1997). Use of olfactory cues in foraging by owl monkeys (*Aotus nancymai*) and capuchin monkeys (*Cebus apella*). *Journal of Comparative Psychology* **111**, 152–8.

Brothers, L. (1990). The social brain: a project for integrating primate behavior and neurophysiology in a new domain. *Concepts in Neuroscience* **1**, 27–51.

Brothers, L. and Ring, B. (1992) A neuroethological framework for the representation of minds. *Journal of Cognitive Neuroscience* **4**, 107–18.

Byrne, R.W. and Whiten, A. (1988). *Machiavellian Intelligence*. Oxford: Oxford University Press.

Cartmill, M. (1974). Rethinking primate origins. *Science* **184**, 436–43.

Charnov, E.L. (1991). Evolution of life history variation among female mammals. *Proceedings of the National Academy of Sciences, USA* **88**, 1134–7.

Clutton-Brock, T.H. and Harvey, P.H. (1980). Primates, brains and ecology, *Journal of Zoology* **207**, 151–69.

Crick, F.H.C. (1994). *The Astonishing Hypothesis*. London: Simon and Schuster.

Dean, P., Redgrave, P. and Westby, G.W.M. (1989). Event or emergency? Two response systems in the mammalian superior colliculus. *Trends in Neuroscience* **12**, 137–47.

Devoogd, T.J., Krebs, J.R., Healy, S.D. and Purvis, A. (1993). Relations between song repertoire size and the volume of brain nuclei related to song: comparative evolutionary analyses amongst oscine birds. *Proceedings of the Royal Society, London, Series B* **254**, 75–82.

Dunbar, R.I.M. (1992). Neocortex size as a constraint on group size in primates. *Journal of Human Evolution* **20**, 469–93.

Economos, A.C. (1980). Brain–life span conjecture: a re-evaluation of the evidence. *Gerontology* **26**, 82–9.

Eisenberg, J.F. and Wilson, D.E. (1978). Relative brain size and feeding strategies in the chiroptera. *American Naturalist* **32**, 740–51.

Elgar, M.A. and Harvey, P.H. (1987). Basal metabolic rates in mammals: allometry, phylogeny and ecology. *Functional Ecology* **1**, 25–36.

Engel, S., Zhang, X. and Wandell, B. (1997). Colour tuning in human visual cortex measured with functional magnetic resonance imaging. *Nature* **388**, 68–71.

Finlay, B.L. and Darlington, R.B. (1995). Linked regularities in the development and evolution of mammalian brains. *Science* **268**, 1578–84.

Foley, R.A. and Lee, P.C. (1992). Ecology and energetics of encephalization in hominid evolution. In *Foraging Strategies and Natural Diet of Monkeys, Apes and Humans*, ed. A. Whitten and E.M. Widdowson, pp. 63–72. Oxford: Clarendon Press.

Gibson, K.R. (1990). New perspectives on instincts and intelligence: brain size and the emergence of hierarchical mental constructional skills. In *'Language' and Intelligence in Monkeys and Apes*, ed. S.T. Parker and K.R. Gibson, pp. 97–128. Cambridge: Cambridge University Press.

Gittleman, J.L. (1991). Carnivore olfactory bulb size – allometry, phylogeny and ecology. *Journal of Zoology* **225**, 253–72.

Grant, J.W.A., Chapman, C.A. and Richardson, K.S. (1992). Defended versus undefended home range size of carnivores, ungulates and primates. *Behavioural Ecology and Sociobiology* **31**, 149–61.

Harvey, P.H. and Krebs, J.R. (1990). Comparing brains. *Science* **249**, 140–6.

Harvey, P.H., Martin, R.D. and Clutton-Brock, T.H. (1987). Life histories in comparative perspective. In *Primate Societies*, ed. B.B. Smuts, D.L. Cheney, R.M. Seyfarth, R.W. Wrangham and T.T. Struhsaker, T.T., pp. 181–96. Chicago: University of Chicago Press.

Harvey, P.H. and Pagel, M.D. (1991). *The Comparative Method in Evolutionary Biology*. Oxford: Oxford University Press.

Harvey, P.H. and Read, A.F. (1988). How and why do mammalian life histories vary? In *Evolution of Life Histories of Mammals: Theory and Pattern*, ed. M.S. Boyce, pp. 213–31. New Haven: Yale University Press.

Healy, S. and Guilford, T. (1990). Olfactory bulb size and nocturnality in birds. *Evolution* **44**, 339–46.

Hofman, M.A. (1984). On the presumed coevolution of brain size and longevity in hominids. *Journal of Human Evolution* **13**, 371–6.

Jackendoff, R. (1992). *Consciousness and the Computational Mind*. Cambridge, Mass.: MIT Press.

Jacobs, G.H. (1993). The distribution and nature of colour vision among the mammals. *Biological Reviews of the Cambridge Philosophical Society* **68**, 413–71.

Jacobs, G.H. (1995). Variations in primate colour vision: mechanisms and utility. *Evolutionary Anthropology* **3**, 196–205.

Jacobs, L.F., Gaulin, S.C., Sherry, D.F. and Hoffman, G.E. (1990). Evolution of spatial cognition: sex-specific patterns of spatial behavior predict hippocampal size. *Proceedings of the National Academy of Sciences, USA* **87**, 6349–52.

Jerison, H.J. (1973). *Evolution of the Brain and Intelligence*. New York: Academic Press.

Joffe, T.H. (1997). Social pressures have selected for an extended juvenile period in primates. *Journal of Human Evolution* **32**, 593–605.

Joffe, T.H. and Dunbar, R.I.M. (1997). Visual and socio-cognitive information processing in primate brain evolution. *Proceedings of the Royal Society, Series B* **264**, 1303–7.

Kappeler, P.M. and Heymann, E.W. (1996). Nonconvergence in the evolution of primate life history and socio-ecology. *Biological Journal of the Linnean Society* **59**, 297–326.

Keverne, E.B., Martel, F.L. and Nevison, C.M. (1996). Primate brain evolution – genetic and functional considerations. *Proceedings of the Royal Society, Series B* **263**, 689–96.

Krebs, J.R. (1990). Food-storing birds: adaptive specialization in brain and behaviour? *Philosophical Transactions of the Royal Society* **329**, 153–60.

Lande, R. (1979). Quantitative genetic analysis of multivariate evolution, applied to brain: body size allometry. *Evolution* **33**, 402–16.

Livingstone, M.S. and Hubel, D.H. (1988). Segregation of form, color, movement and depth: anatomy, physiology and perception. *Science* **240**, 740–9.

Mace, G. and Eisenberg, J.F. (1982). Competition, niche specialisation and the evolution of brain size in the genus *Peromyscus*. *Biological Journal of the Linnean Society* **17**, 243–57.

McNab, B.K. (1995). Energy-expenditure and conservation in frugivorous and mixed-diet carnivorans. *Journal of Mammalogy* **76**, 206–22.

Martin, R.D. (1981). Relative brain size and basal metabolic rate in terrestrial vertebrates. *Nature* **293**, 57–60.

Martin, R.D. (1990). *Primate Origins and Evolution: a Phylogenetic Reconstruction*. London: Chapman and Hall.

Martin, R.D. (1996). Scaling of the mammalian brain: the maternal energy hypothesis. *News in Physiological Sciences* **11**, 149–56.

Milton, K. (1988). Foraging behaviour and the evolution of primate intelligence. In *Machiavellian Intelligence*, ed. R.W. Byrne and A. Whiten, pp. 285–306. Oxford: Clarendon Press.

Mollon, J.D. (1989). Tho she kneeld in that place where they grew, the uses and origins of primate color-vision. *Journal of Experimental Biology* **146**, 21–38.

Osorio, D. and Vorobyev, M. (1996). Colour vision as an adpatation to frugivory in primates. *Proceedings of the Royal Society, London, Series B* **263**, 593–9.

Pagel, M.D. and Harvey, P.H. (1988a). The taxon-level problem in the evolution of mammalian brain size: facts and artifacts. *American Naturalist* **132**(3), 344–59.

Pagel, M.D. and Harvey, P.H. (1988b). How mammals produce large-brained offspring. *Evolution* **42**, 948–57.

Pagel, M.D. and Harvey, P.H. (1989). Taxonomic differences in the scaling of brain on body weight among mammals. *Science* **244**, 1589–93.

Parker. S.T. (1990). Why big brains are so rare: energy costs of intelligence and brain size in anthropoid primates. In *'Language' and Intelligence in Monkeys and Apes*, S.T. Parker and K.R. Gibson, pp. 129–54. Cambridge: Cambridge University Press.

Perrett, D.I., Hietanen, J.K., Oram, M.W. and Benson, P.J. (1992). Organization and function of cells responsive to faces in the temporal cortex. *Philosophical Transactions of the Royal Society, London, Series B* **335**, 23–30.

Preuss, T.M. (1993). The role of the neurosciences in primate evolutionary biology. In *Primates and their Relatives in Phylogenetic Perspective*, ed. R.S.D.E. Macphee, pp. 333–62. New York: Plenum Press.

Purvis, A. (1995). A composite estimate of primate phylogeny. *Philosphical Transactions of the Royal Society, London, Series B* **348**, 405–21.

Purvis, A. and Harvey, P.H. (1995). Mammal life history evolution: a comparative test of Charnov's model. *Journal of Zoology, London* **237**, 259–83.

Purvis, A. and Rambaut, A. (1995). Comparative analysis by independent contrasts (CAIC): an Apple Macintosh application for analysing comparative data. *Computer Application in the Biosciences* **11**, 247–51.

Quartz, S. and Sjenowski, T. (1997). The neural basis of cognitive development: a constructivist manifesto. *Behavioural and Brain Sciences*, **20**, 537–80.

Rosenberger, A.L. and Strier, K.B. (1989). Adaptive radiation of ateline primates. *Journal of human evolution* **18**, 717–50.

Ross, C. (1988). The intrinsic rate of natural increase and reproductive effort in primates. *Journal of zoology* **214**, 199–219.

Ross, C. (1992). Basal metabolic rate, body weight and diet in primates: an evaluation of the evidence. *Folia Primatologica* **58**, 7–23.

Sacher, G.A. (1959). Relationship of lifespan to brain weight and body weight inmammals. In *CIBA Foundation Symposium on the Lifespan of Animals*, ed. G.E.W. Wolstenholme and M. O'Connor, pp. 115–33. Boston: Little Brown.

Sawaguchi, T. (1992). The size of the neocortex in relation to ecology and social structure inmonkeys and apes *Folia Primatologica* **58**, 131–45.

Schmid, J. and Ganzhorn, J. (1996). Resting metabolic rates of *lepilemur ruficaudatus. American Journal of Primatology* **38**, 169–74.

Shea, B.T. (1987). Reproductive strategies, body size and encephalization in primate evolution. *International Journal of Primatology* **8**, 139–56.

Sherry, D.F., Vaccarino, A.L., Buckenham, K. and Herz, R.S. (1989). The hippocampal complex of food-storing birds. *Brain, Behavior and Evolution* **34**, 308–17.

Shulz, H-D. (1967). Metrische untersuchungen an den schichten des corpus geniculatum laterale tag- und nachtaktiver primaten. Doctoral dissertation,

Johann Wolfgang Goethe-Universität, Frankfurt.

Smuts, B.B., Cheney, D.L., Seyfarth, R.M., Wrangham, R.W. and Struhsaker, T.T. (1987). *Primate Societies*. Chicago: University of Chicago Press.

Stephan, H., Baron, G. and Frahm, H. (1988). Comparative size of brains and brain components. In Steklis, H.D. and Erwin, J. (eds) *Comparative Primate Biology*, Vol. 4, ed. H.D. Steklis and J. Erwin, pp. 1–37. New York: Alan R. Liss.

Stephan, H., Frahm, H.D. and Baron, G. (1981). New and revised data on volumes of brain structures in insectivores and primates. *Folia Primatologica* **35**, 1–29.

Stephan, H. and Pirlot, P. (1970). Volumetric comparisons of brain structures in bats. *Zeitschrift Zoologiste Systematik Evolutionforschung* **8**, 200–36.

Stephan, H., Pirlot, P. and Schneider, R. (1974). Volumetric analysis of pteropid brains. *Acta Anatomica* **87**, 161–92.

van Essen, D.C., Anderson, C.H. and Felleman, D.J. (1992). Information processing in the primate visual system: an integrated systems perspective. *Science* **255**, 419–23.

Wallace, A.R. (1891). *Natural Selection and Tropical Nature*. London: Macmillan.

Yoder, A.D. (1994). Relative position of the Cheirogaleidae in strepsirhine phylogeny: a comparison of morphological and molecular methods and results. *American Journal of Physical Anthropology* **94**, 25–46.

Young, M.P., Scannell, J.W., Burns, G.A.P.C. and Blakemore, C. (1994). Analysis of connectivity: neural systems in the cerebral cortex. *Reviews in the Neurosciences* **5**, 227–49.

Zeki, S. (1993). *A Vision of the Brain*. Oxford: Blackwell Scientific Publications.

Zeki, S.M. and Shipp, S. (1988). The functional logic of cortical connections *Nature* **335**, 311–17.

Appendix 7.1
Main primate database, listed alphabetically by species name

Species	Adult brain weight (g)	Mean body weight (g)	Body weight (BMR) (g)	BMR	BMR	Age first breeding (female) (months)	Maximum life span (years)	Per cent fruit	Group size	Activity period
Allenopithecus nigroviridis	62.5	5495	—	—	—	—	—	—	—	D
Alouatta caraya	56.7	6637	—	—	—	—	—	—	8.1	D
Alouatta palliata	55.1	6577	4670	2000	—	43.2	20	32	12.9	D
Alouatta seniculus	57.9	6577	—	—	—	54	25	42	7.2	D
Aotus trivirgatus	18.2	733	925	465	—	28.8	20	45.5	4.5	N
Arctocebus calabarensis	7.7	313	206	131	—	—	9.5	14	—	N
Ateles belzebuth	106.6	8222	—	—	—	—	—	83	14.5	D
Ateles fusciceps	114.7	9036	—	—	—	58.8	24	—	—	D
Ateles geoffroyi	110.9	7568	—	—	—	60	27.3	79.8	42	D
Ateles paniscus	109.9	8810	—	—	—	60	33	82.9	20	D
Avahi laniger	10	1285	—	—	—	—	—	0	2.5	N
Brachyteles arachnoides	120.1	9506	—	—	—	—	—	23	21	D
Cacajao calvus	73.3	3828	—	—	—	—	—	—	17.4	D
Callicebus moloch	19	900	—	—	—	36	10	53.7	3.2	D
Callicebus torquatus	22.4	1074	—	—	—	48	—	70	3.7	D
Callimico goeldii	10.8	590	—	—	—	15.6	9.3	—	7.6	D
Callithrix jacchus	7.9	288	—	—	—	18	12	22	9	D
Cebuella pygmaea	4.2	140	117	110.6	—	22.8	11.7	—	6	D
Cebus albifrons	82	2489	—	—	—	48	44	24.6	25	D
Cebus apella	71	2742	—	—	—	66	44	52	15	D
Cebus capucinus	79.2	3006	—	—	—	48	46.9	67.5	18.8	D
Cercocebus albigena	99.1	7362	—	—	—	49.2	32.7	64	15.4	D
Cercocebus galeritus	114.7	7834	—	—	—	—	19	80.1	22.4	D
Cercocebus torquatus	109.6	8933	—	—	—	56.4	27	79	26.9	D
Cercopithecus aethiops	59.8	4178	—	—	—	42	31	71	21.4	D
Cercopithecus ascanius	66.5	3606	—	—	—	—	—	50.6	30	D

Cercopithecus campbelli	65.8	3243	—	—	—	—	—	10.5	D
Cercopithecus cephus	63.6	3381	—	—	—	—	79	11.5	D
Cercopithecus diana	77.3	3811	—	—	64.8	34.8	41.4	26.9	D
Cercopithecus erythrogaster	65.2	4295	—	—	—	—	—	—	D
Cercopithecus lhoesti	76	6607	—	—	—	—	—	17.4	D
Cercopithecus mitis	75	5821	8500	3391.5	51.6	20	54.5	18.6	D
Cercopithecus mona	66	3451	—	—	48	—	—	4	D
Cercopithecus neglectus	70.8	5559	—	—	48	22	77	4	D
Cercopithecus nictitans	78.6	5408	—	—	—	—	67.1	18.2	D
Cercopithecus pogonias	71.1	3767	—	—	—	—	82.9	15.1	D
Cheirogaleus major	5.9	469	—	—	—	8.8	—	1.5	N
Cheirogaleus medius	2.9	207	300	195	—	9	—	1	N
Chiropotes satanas	53	2642	—	—	—	—	91	19.1	D
Colobus angolensis	73.5	9840	—	—	—	—	35	—	D
Colobus badius	73.8	7998	—	—	—	—	26	34	D
Colobus guereza	82.3	9863	10450	2978	56.4	22.3	14	6.9	D
Colobus polykomos	76.7	9397	—	—	102	26	36	—	D
Colobus satanas	80.2	10740	—	—	—	—	57	15.5	D
Daubentonia madagascariensis	45.2	2742	—	—	28.8	23.3	75	1	N
Erythrocebus patas	106.6	8690	—	—	36	21.6	75	28	D
Euoticus elegantulus	5.8	290	262	215.6	—	—	5	—	N
Galago alleni	6.1	248	—	—	—	—	73	6	N
Galago senegalensis	4.8	178	275	198	10.8	16	0	6	N
Galagoides demidoff	2.7	65	68	63.2	12	14	29	10	N
Gorilla gorilla	505.9	114 551	—	—	118.2	50	3	11	D
Hapalemur griseus	14.7	916	—	—	28.6	12.1	—	3.6	N*
Hylobates agilis	131.7	5715	—	—	—	—	61	4.4	D
Hylobates hoolock	108.5	6699	—	—	—	—	67	3.2	D
Hylobates klossi	91.1	5794	—	—	—	—	72	3.8	D
Hylobates lar	107.7	5559	—	—	111.6	31.5	60	3.4	D
Hylobates moloch	113.7	5821	—	—	—	—	61	—	D
Hylobates pileatus	114.2	—	—	—	—	—	79.4	3.7	D
Hylobates syndactylus	121.7	10839	—	—	108	35	47	4	D

Species	Adult brain weight (g)	Mean body weight (g)	Body weight (BMR) (g)	BMR	Age first breeding (female) (months)	Maximum life span (years)	Per cent fruit	Group size	Activity period
Indri indri	34.5	8375	—	—	—	—	43	4	D
Lagothrix lagothricha	96.4	6109	—	—	60	25.9	79	33.1	D
Lemur catta	25.6	2466	—	—	24	27.1	54	16	D
Leontopithecus rosalia	12.9	558	—	—	28.8	14.2	—	—	D
Lepilemur mustelinus	9.5	630	693	225			6	2	N
Loris tardigradus	6.7	270	128	284	18	12	15	2	N
Macaca arctoides	104.1	8590	—	—	42	30	—	—	D
Macaca fascicularis	69.2	4977	—	—	46.8	37.1	66.9	24.5	D
Macaca fuscata	109.1	10447	—	—	66	33	38	36.3	D
Macaca mulatta	95.1	5902	6225	2239	45	29	63	33	D
Macaca nemestrina	106	7762	—	—	47.3	26.3	75	26.9	D
Macaca nigra	94.9	8492	—	—	57.6	—	—	—	D
Macaca radiata	76.8	5000	—	—	—	—	—	34.7	D
Macaca silenus	85	5902	—	—	—	—	—	—	D
Macaca sinica	69.9	4656	—	—	—	—	97	18.6	D
Macaca sylvanus	93.2	9750	—	—	57.6	22	33	24	D
Mandrillus leucophaeus	152.7	14223	—	—	60	29	—	—	D
Mandrillus sphinx	159.4	16444	—	—	60.5	29.1	—	251.2	D
Microcebus murinus	1.8	67	—	—	12	15.5	51	1	N
Miopithecus talapoin	37.7	1250	—	—	52.8	27.7	54	91.2	D
Nasalis larvatus	94.2	15066	—	—	—	—	43	12	D
Nycticebus coucang	10	659	1160	272.6	—	14.5	60	—	N
Otolemur crassicaudatus	11.8	1167	—	—	16.5	15	16.5	6	N
Pan troglodytes	410.3	37844	34150	9000	122.4	53	66	28	D
Papio cynocephalus	169.1	17140	—	—	73	—	62	41	D

Species									
Papio hamadryas	142.5	13 996	—	—	—	—	88	66.5	D
Papio papio	165.3	17 579	—	—	—	—	—	—	D
Papio ursinus	214.4	21 777	—	—	—	—	73	34.7	D
Perodicticus potto	14.3	953	964	326.6	24	22.3	81	2	N
Petterus fulvus	25.2	2377	2330	746	27.6	30.8	54	9.3	D*
Petterus macaco	25.6	2472	—	—	—	27.1	—	10	D*
Petterus mongoz	21.8	1698	—	—	—	—	—	2.3	N*
Petterus rubriventer	27.2	2009	—	—	—	—	—	—	D*
Phaner furcifer	7.3	420	—	—	—	—	85	—	N
Pithecia monachus	38.1	2328	—	—	—	—	—	—	D
Pithecia pithecia	31.7	1710	—	—	25.2	13.8	92	2.7	D
Pongo pygmaeus	413.3	55 208	—	—	128.4	57.3	64	2	D
Presbytis aygula	80.3	6730	—	—	—	—	28	7.6	D
Presbytis cristata	64	6412	—	—	—	—	—	—	D
Presbytis entellus	135.2	12 647	—	—	51	25	52	19.1	D
Presbytis geei	81.3	8356	—	—	—	—	—	9.5	D
Presbytis johnii	84.6	13 397	—	—	—	—	21	9	D
Presbytis melalophos	80	6531	—	—	—	—	61	16.5	D
Presbytis obscura	67.6	7261	—	—	—	—	60.5	10	D
Presbytis rubicunda	92.7	6761	—	—	—	—	60	—	D
Presbytis senex	64.9	6067	—	—	—	—	40	7.9	D
Procolobus verus	57.8	3890	—	—	—	—	24	—	D
Propithecus diadema	37	5821	—	—	—	—	—	—	D
Propithecus verreauxi	27.5	3664	3350	670	33.6	18.2	41	7	D
Pygathrix nemaeus	108.5	9550	—	—	—	—	—	—	D
Pygathrix roxellanae	121.7	12 246	—	—	—	—	—	—	D
Saguinus fuscicollis	9.3	395	—	—	—	—	60.4	6.5	D
Saguinus midas	10.5	543	—	—	24	13.3	69	4.7	D
Saguinus nigricollis	8.9	465	—	—	—	—	—	6.8	D
Saguinus oedipus	10	417	—	—	22.8	13.5	—	6	D
Saimiri oerstedii	25.7	815	—	—	—	—	—	25.1	D
Saimiri sciureus	24.4	752	825	677	30	21	28	34.7	D
Tarsius spectrum	3.8	200	173	149	16.8	12	0	—	N

Species	Adult brain weight (g)	Mean body weight (g)	Body weight (BMR) (g)	BMR	Age first breeding (female) (months)	Maximum life span (years)	Per cent fruit	Group size	Activity period
				BMR					
Tarsius syrichta	4	122	113	76.7	—	—	0	—	N
Theropithecus gelada	131.9	15 560	—	—	48	19.3	26	10	D
Varecia variegata	34.2	3524	—	—	20.4	13	—	2.8	

Brain weights (in grams) are from Harvey, Martin and Clutton-Brock (1987). Mean male/female body weights (in grams) are calculated from Harvey *et al.* (1987), Kappeler and Heymann (1996), and recent primary sources (available on request). Basal metabolic rates (BMR) and associated body weights are from Ross (1992), plus Schmid and Ganzhorn (1996). Age at first breeding (months) and maximum recorded life span (years) are from Ross (1988). Per cent fruit in the diet (mean per cent of feeding time) is collated from chapters in Smuts *et al.* (1987), Rosenberger and Strier (1989), and recent primary sources. Group size is collated from chapters in Smuts *et al.* (1987) and Kappeler and Heymann (1996), supplemented from Grant, Chapman and Richardson (1992). Activity period is from Smuts *et al.* (1987); asterisks indicate species properly classified as cathemeral (active by day and/or night), following Kappeler and Heymann (1996), but classified as nocturnal or diurnal according to their primary activity period, for the purposes of dichotomous CAIC analysis. Gestation lengths are not included in the table as they were taken from Ross's compilation (see Chapter 4).

Appendix 7.2
Data on separate brain parts used in the analyses

Species	Volume of brain parts (mm³)						Number of LGN neurons	
	Brain	Brain (neocortex + LGN)	V1	LGN	Magnocellular LGN	Parvocellular LGN	Magnocelluar	Parvocellular
Alouatta spp.	49 009	17 219	2374	79	—	—	—	—
Aotus trivirgatus	16 191	6194	1144	33	2	2.6	57 544	164 437
Ateles geoffroyi	100 925	29 992	4742	151	10	24	237 137	824 138
Avahi laniger	9795	4955	537	28	2.2	3	47 424	146 893
Callicebus moloch	17 944	6730	1503	54	—	—	—	—
Callithrix jacchus	7241	2844	692	25	0.9	3.7	51 168	248 313
Cebuella pygmaea	4305	1754	—	—	—	—	—	—
Cebus capucinus	66 988	20 417	4688	137	4.8	16.2	94 406	709 578
Cercocebus albigena	97 499	28 576	6823	182	—	—	—	—
Cercopithecus ascanius	63 533	18 197	5152	147	5.8	18.9	109 648	820 352
Cercopithecus mitis	70 632	20 606	5272	150	—	—	—	—
Cheirogaleus major	6373	3404	478	21	—	—	—	—
Cheirogaleus medius	2961	1726	217	11	—	—	—	—
Colobus badius	73 790	22 751	3981	128	6.2	20.3	121 060	719 449
Daubentonia madagascariensis	42 611	20 464	1355	58	4.5	5.9	64 565	165 577
Galago senegalensis	4512	2355	334	16	—	—	—	—
Galagoides demidoff	3206	1629	237	11	—	—	—	—
Gorilla gorilla	469 894	128 529	15 171	384	—	—	—	—
Hylobates lar	97 499	31 623	—	—	—	—	—	—
Indri indri	36 285	16 106	1738	94	—	—	—	—
Lagothrix lagothricha	95 499	29 444	6295	164	—	—	—	—
Lepilemur mustelinus	7175	3882	357	16	1.3	1.8	39 719	98 628
Loris tardigradus	6269	2716	587	25	2.6	3.7	72 778	183 654
Macaca mulatta	87 902	24 210	6592	158	—	—	—	—

Microcebus murinus	1680	931	149	7	0.8	1	31 261	79 799
Miopithecus talapoin	37 757	11 220	3048	109	—	—	—	—
Nycticebus coucang	11 755	5521	822	37	—	—	—	—
Otolemur crassicaudatus	9668	4920	571	22	—	—	—	—
Pan troglodytes	381 944	89 743	14 689	356	—	—	—	—
Papio cynocephalus	190 985	50 350	—	—	—	—	—	—
Perodicticus potto	13 213	6501	552	24	—	—	—	—
Petterus fulvus	22 101	9863	1528	52	2.9	7.8	68 077	342 768
Pithecia monachus	32 885	11 776	2153	76	—	—	—	—
Propithecus verreauxi	25 194	11 940	1560	64	3.5	6.9	79 799	297 852
Saguinus oedipus	9535	3606	1002	34	—	—	—	—
Saimiri sciureus	22 594	6998	2328	63	—	—	—	—
Tarsius spp.	3393	1611	370	21	1.9	2.6	110 408	199 986
Varecia variegata	29 713	14 388	2113	69	—	—	—	—

The data are from Stephan *et al.* (1981), except for the volumes and number of neurons in separate lateral geniculate layers (LGN), which are from Shulz (1967).

8 Sex and social evolution in primates

CAREL P. VAN SCHAIK, MARIA A. VAN NOORDWIJK AND
CHARLES L. NUNN

Introduction

Life history and male infanticide risk

Infanticide by males unlikely or unable to have fathered a female's current
dependent offspring is an adaptive male reproductive strategy if the mother
can soon be fertilised again and the infanticidal male is in a position to be
the likely sire of her next infant (Hrdy, 1979; Hrdy, Janson and van Schaik,
1995). It is remarkably common among primates (Hausfater and Hrdy,
1984; Struhsaker and Leland, 1987; Hiraiwa-Hasegawa, 1988). Its importance as a source of infant mortality probably varies widely, but it is
estimated to be responsible for 31–64% of all infant mortality in some
well-studied species (hanuman langurs: Sommer, 1994; Borries, 1997;
mountain gorillas: Watts, 1989; red howlers: Crockett and Sekulic, 1984).
Several detailed studies have demonstrated that it is an adaptive behaviour
for males (*op. cit.*; Hrdy, 1979).

Male infanticide is potentially common in primates because many species have prolonged lactational amenorrhoea. In species without lactational amenorrhoea, early resumption of mating activity ('postpartum
oestrus') means that killing infants will not advance the female's next birth.
The incidences of postpartum mating and lactational amenorrhoea, in
turn, are determined by the relative length of gestation and lactation.
Where lactation is longer than gestation, we find lactational amenorrhoea,
in both primates and other mammals (van Schaik, in press).

Among primates, there is a strong relationship between the mode of
infant care, the incidence of postpartum mating and the incidence of male
infanticide. Primates have two radically different modes of infant care:
absentee care, in which the offspring are left in a nest or parked somewhere
(galagos and lorises, several lemurs, some tarsiers), and permanent care, in
which the offspring are carried around by the mother (all monkeys and
apes, one tarsier and several lemurs: see van Schaik and Kappeler, 1997).
Among species in which females carry their infants, there is an important
difference between those in which mothers basically rear their offspring

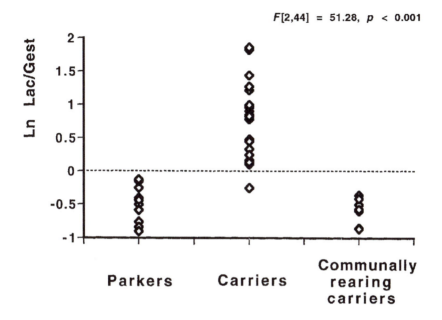

Fig. 8.1 The relative length of lactation (lactation/gestation) as a function of mode of infant care in primates (Reproductive data from Harvey, Martin and Clutton-Brock, 1987; infant care from Smuts *et al.*, 1987; van Schaik and Kappeler, 1993).

alone and those in which there is communal rearing, i.e. the mother consistently receives help in rearing the offspring, especially by carrying and sometimes by provisioning. In these communal breeders, postnatal development can be speeded up due to the input by others. Among primates, they are represented by callitrichids and the pair-living cebids (cf. Lee, 1996).

Figure 8.1 shows that species with these three types of care vary in the relative length of lactation: the carrying primates with mother-only rearing have relatively very long lactation periods because their infants develop slowly (cf. Charnov, 1993). As predicted, none of these species shows postpartum mating and conception, whereas they are very common among species with absentee infant care and among species with communal infant rearing (82% of 11 species and 63% of 8, respectively; data on postpartum mating from Appendix 8.1). The frequent incidence of postpartum mating among the communally rearing infant carriers is clearly a derived condition.

The pattern in postpartum mating implies that infanticide is only expected to be common among species that carry their young but do not rear

them communally, i.e. some lemuroids, some ceboids, and all cercopithecoids and hominoids. The data in Figure 8.1 are consistent with this
expectation. Male infanticide is only reported for the non-communal infant
carriers, for which it is reported in 49% of 61 well-studied species (Appendix 8.1). Although under-reporting may explain its absence in the parkers
(0% of 14), which are nocturnal and solitary, it cannot account for the
absence among the communal infant carriers (0% of 8).

Female counterstrategies

In non-communally breeding infant-carrying primates, the risk of male
infanticide should have selected for female counterstrategies. Female primates in these taxa have evolved three major ways of reducing the risk of
infanticide: (i) to gain protection for the infant by associating with the likely
sire; (ii) to cut losses through abortion or premature weaning; and (iii) to
prevent infanticidal attacks by employing sexual strategies.

   There is strong evidence that likely fathers play a particularly important
role in protecting primate infants (reviewed in van Schaik, 1996; in press).
Male infanticide is therefore expected when this protection is removed. In
nature, infanticide is almost always associated with take-overs by outsiders
or, less commonly, with dominance upheavals inside groups (Hrdy et al.,
1995; Borries, 1997; Steenbeek, 1996). In captivity, infanticide can be
provoked reliably by experimentally replacing the adult (or dominant)
male (Angst and Thommen, 1977; Kyes et al., 1995). This pattern suggests
that females have evolved association with males as a counterstrategy
against male infanticide. As expected, permanent male–female association
in primates is absent among species with absentee infant care, whereas it is
virtually universal among the infant-carrying species (van Schaik and
Kappeler, 1997). It is retained among the communally rearing ceboids.
However, while effective, it is not a perfect strategy: males can die or be
ousted.

   Where protection fails, females may respond to acute infanticide risk by
terminating their investment in vulnerable offspring. Thus, a pregnant
female could resorb or abort a fetus. This has been reported for patas
(Rowell and Hartwell, 1978), hanuman langurs (Sommer, 1994), gelada
baboons (Mori and Dunbar, 1985), captive hamadryas baboons (Colmenares and Gomendio, 1988), and for a yellow baboon group with an
unusually aggressive immigrant male (Pereira, 1983).[1] When the female
already has an infant, and infanticide risk suddenly arises, she could wean
her infant sooner than she would otherwise have done, even if it would
reduce the infant's survival, as noted for wild gelada (Dunbar, 1980),

hamadryas baboons (Sigg *et al.*, 1982), and captive vervets (Fairbanks and McGuire, 1987). Resorption, abortion and infant abandonment are probably rare among primates.

A less costly alternative counterstrategy would be to prevent infanticide by manipulating the male's assessment of paternity or his actual chances of it through sexual behaviour.[2] This chapter explores the possibility that the variation in primate sexual behaviour and physiology is an adaptation to reduce the risk of male infanticide (see also Hrdy, 1981; Hrdy and Whitten, 1987). The hypothesis examined in this chapter is that female sexuality in species vulnerable to male infanticide has been molded by the dual need for paternity concentration and confusion: concentration in order to elicit infant protection from the likely father, confusion in order to prevent infanticide from non-likely fathers (see also Nunn, in press). The right balance of confusion and concentration is achieved by (i) prolonged mating periods, often accompanied by actively pursued polyandrous mating, (ii) mating during non-fertile periods, such as pregnancy, (iii) making ovulation unpredictable, thus blurring the correlation with attractivity, and (iv) attracting males of all ranks by producing gradually changing, exaggerated sexual swellings.

Sexual counterstrategies against male infanticide

Paternity concentration and confusion

Whenever the threat of male infanticide occurs, a female would benefit from having a protector male, most likely the infant's sire. This protection is best obtained by concentrating paternity into a single male. Such concentration is normally achieved by default because, regardless of morphological advertising, dominant males mate preferentially with females around the time of likely ovulation and often guard them so as to monopolise mating (Kaufman, 1965; Hausfater, 1975; Glander, 1980; Harcourt *et al.*, 1980; van Noordwijk, 1985).

On the other hand, female primates can also reduce the risk of infanticide by mating with at least some of the other males present before birth and thus confusing paternity. Paternity confusion is a feasible option because mammalian males cannot recognise infants as kin (Elwood and Kennedy, 1994). Thus, male decisions about whether to defend an infant, to ignore it, or, alternatively, to attempt to kill it, should be based on assessments of the likelihood of paternity, weighted against likely paternity of the future infant. These assessments are necessarily based entirely on mating history. If a male has exclusive mating access to a female and if matings only take place during the cycle in which fertilisation occurs, the male rule can be simple: if he mated with the female, his paternity is ensured. Field studies on primates with unimale groups suggest that when a newly immigrated male mates with a pregnant female, the probability that he will subsequently commit infanticide is reduced (e.g. blue monkey: Fairgrieve, 1995; hanuman langur: Sommer, 1994), although it is not clear how long before birth this mating must take place. Experimental work on rodents confirms these rules (Perrigo and vom Saal, 1994). Where multiple males mate with the female before she gives birth, more complex rules are needed; these have not been studied in detail.

Females have two fundamentally different tactics at their disposal to confuse paternity. First, they mate polyandrously during regular ovarian cycles. This tactic should reduce the risk of infanticide by males that were already in the group but did not mate (enough) with the female and that rose dramatically in rank after conception (only high-ranking males would be likely to benefit from committing infanticide). Second, they pursue matings with one or more males when these males appear when fertilisation is impossible (e.g. during pregnancy).[3] This tactic should reduce the risk from males who immigrate into the group after the female's conception cycle and become dominant, thus gaining good prospects of siring the female's next infant. In addition, post-conception matings could provide additional paternity confusion for males already present during the conception cycle.

Paternity is most easily concentrated when a female mates briefly, and exclusively or predominantly with a single male, whereas paternity confusion requires mating with multiple males, and therefore in most cases extended periods of mating. A review of primate mating behaviour (van Noordwijk, unpublished) shows that males can use a variety of anatomical, chemical and behavioural cues to assess the female's oestrogen activity, and thus likelihood of ovulation. However, the feasibility of deceptive receptivity during pregnancy underscores that ovulation itself is truly concealed to males at all times. The cues used by males can also be produced during pregnancy (when ovulation is effectively blocked), although females will be less attractive because of the inevitable endocrine differences (for which they may compensate by being more proceptive, e.g. patas: Loy, 1981). Indeed, males mating with females during pregnancy tend to be lower-ranking or subadult males (e.g. gorilla: Watts, 1991; sooty mangabey: Gust, 1994). We conclude that the mechanisms of male paternity assessment are so crude that there is considerable room for female manipulation of it.

Natural selection may have achieved the balance between paternity confusion and concentration by making ovulation unpredictable. Martin (1992) and Nunn (in press) review quantitative studies of several species and note that ovulation can take place over a range of at least 6 to 13 days. Thus, in species with swellings, ovulation is most likely during peak swelling, but not exclusively so (e.g. Wildt *et al.*, 1977; Whitten and Russell, 1996). In species without morphological advertising, this unpredictability is harder to demonstrate. However, in humans, ovulation can take place during a surprisingly broad time window after last menstruation (Martin, 1992). This implies that the length of the follicular phase is highly variable. Indeed, across studies the standard deviation in the length of the follicular phase is consistently far greater than that of the luteal phase (see Hayssen, van Tienhoven and van Tienhoven, 1993). Thus, although male primates in polyandrous mating situations have endocrine and behavioural indications that ovulation is likely, its exact timing, and thus the best timing for fertile matings, is hard to predict.

Unfortunately, there are no data to test the assumption that the degree of unpredictability depends on the need for confusion. However, the concept of unpredictable ovulation also suggests that there should be a good but not perfect correlation between male rank and paternity, with a small proportion of infants being sired by low-ranking males. A recent review of paternity studies in primates supports this contention (Paul, 1998). We assume that the female can manipulate male paternity assessments and make them non-zero for all males who have mated with her before she gave birth. The males cannot improve much on their rules; they can use the correlation between the strength of the female's signal of likelihood of ovulation and the temporal proximity of mating to birth to obtain approximate clues, but the probabilistic nature of ovulation makes it impossible for natural selection to design a male decision rule that does much better.

Predictions

We can now develop predictions for species in which infanticide risk is acute and paternity confusion is required. We can use the variation within primates in vulnerability to male infanticide (see Figure 8.1) to explore broad differences in sexual behaviour between the infant parkers and non-communal infant carriers (see below). Specifically, we expect that non-communal carriers should differ from parkers in the following ways. First, females should actively pursue promiscuity. Second, females should show situation-dependent receptivity in situations when new males appear that are potentially infanticidal and when not enough males mated during

Table 8.1. *The range of social conditions in which females need to concentrate paternity by mating exclusively or predominantly with a single dominant male around the most likely time of conception, confuse it by mating with all potentially infanticidal males over a minimum frequency and confuse it by mating (deceptively) during times of non-fertility*

Case	Number of potentially infanticidal males:		Female sexual response:		
			Concentrate paternity	Confuse paternity	
	in group or at periphery	entering post-conception	in dominant	within regular cycles	situation dependent
1	1	−	+	−	−
2	1	+	+	−	+
3	More	+	+	+	+
4	More	−	−	+	− +

the regular cycles. Third, females should have longer mating periods within a cycle, and perhaps ovarian cycles with longer follicular phases, because mating over longer periods (blue monkey: Rowell, 1994) or more frequently (rhesus macaque: Wilson, Gordon and Chikazawa, 1982; barbary macaque: Small, 1990) implies doing so with more males. Fourth, where needed, females should use additional signals that alter their attractivity to manipulate male behaviour in ways that reduce infanticide.

Among the non-communal infant carriers, the extent to which these predictions hold should depend on details of the social systems, in particular the number of males in the group and the presence of male immigration. These predictions are examined again in a later section in light of this social variation, both among and within species. To place these predictions in the broader social context, consider Table 8.1, which catalogues the impact of male number in the group and male immigration. The first case, in which one male is resident but there is little risk of infanticide by intruding males, is indistinguishable from that in which there is no risk of male infanticide. This situation should be rare among primates facing male infanticide risk. The second case is when females live in pairs or unimale groups subject to take-overs by male invaders. In this case, no confusion during ovarian cycles is needed, so mating periods are probably short, but situation-dependent receptivity is expected. A combination of concentration and confusion is called for in the third case, in which females live in multi-male groups with male immigration (or unimale groups with influxes during the mating season and unpredictably occurring take-over attempts). Here, we expect active promiscuity by females, longer mating periods, frequent situation-dependent receptivity, and a high incidence of attractivity-enhancing signals. The extent to which these traits are developed should covary not only with infanticide risk but also with the degree to which females have control over whom they mate with. Finally, in the rare situation in which groups contain multiple males but immigrant males never enter, confusion in the regular cycles is required but neither concentration nor situation-dependent receptivity is (although females may, as in case 3 of Table 8.1, continue to show sexual activity during pregnancy to increase confusion).

Sex and infant care

Table 8.2 gives a summary of the range of variation within each of the major primate radiations and vulnerability to male infanticide (proxied by mode of infant care; cf. Fig. 8.1). Too little is known about the sexual behaviour of infant parkers in the wild to test their tendency towards polyandrous mating, but we know of no reports in captivity. On the other hand, females in polygynandrous groups of infant carriers actively pursue polyandrous matings. First, females in species as different as ring-tailed lemurs (Pereira, 1991), brown capuchins (Janson, 1984) and woolly spider monkeys (Milton, 1985) actively invite multiple males to mate. Second, female guenons or macaques surreptitiously invite a male to mate with them while their dominant consort partner is temporarily distracted (Wilson et al., 1982; van Noordwijk, 1985; Cords, 1988; Small, 1993). Even pair-living primates are known to engage in extra-pair copulations with males from neighbouring groups (dusky titi: Mason, 1966; gibbons: Palombit, 1994; Reichard, 1995).

Matings during pregnancy are not reported for infant parkers (Table 8.2). Indeed, in many infant-parking strepsirhines, females are not receptive outside the main mating period because their vaginal opening is sealed by an impenetrable membrane (Van Horn and Eaton, 1979; Hrdy and Whitten, 1987). However, pregnancy matings are common among infant carriers. They may reduce the risk of male infanticide, at least in part because those males are attracted that did not mate much during the conception period. However, its high incidence in communally-rearing ceboids suggests that such matings may also have additional functions.

Infant parkers have short periods of receptivity. Indeed, strepsirhines have retained many aspects of classic oestrus so conspicuously absent

Table 8.2. *Reproductive behaviour and mode of infant care in primates, largely based on Appendix 8.1*

	Parkers (%)	Communal carriers (%)	Non-communal carriers (%)	Lem	Ceb	Cer	Hom
Active polyandrous mating	?	+	+	+	+	+	+
Situation-dependent mating (e.g. pregnancy)	0 (6)	100 (6)	60 (48)	+	+	+	+
Mating period (days)							
median	2	10	6	2	3	9	13
range	1–3 (11)	6–28 (6)	1–33 (44)				
Some sex skin	100 (13)	0 (7)	34 (64)	+	+	+	+
Exaggerated sex skin	0	0	39	–	–	+	+
Mating calls	20 (10)	0 (8)	49 (47)	–	+	+	+

Numbers in parentheses are numbers of species with relevant information.
Lem, lemuroids; Ceb, ceboids; Cer, cercopithecoids; Hom, hominoids.

among anthropoids (Keverne, 1987). The range of mating periods is quite broad among carriers (and dependent on the mating system, see below) but few are as short as among the parkers. Finally, exaggerated sex skins are absent and mating calls uncommon in parkers, whereas in the carriers both are common, and linked to the social system (see below). These general differences support the idea that the threat of male infanticide has molded primate sexuality.

This variation is not simply due to taxonomy, because the predicted differences also hold within the lemuroids. Among the infant-carrying lemurs, mating periods tend to be longer than among the parkers, active female polyandry is observed in ring-tailed lemurs (Pereira, 1991), and situation-dependent receptivity in Verreaux's sifaka (Brockman and Whitten, 1996). However, among the infant carriers, cercopithecoids and hominoids have much longer mating periods than the lemuroids and ceboids (see Table 8.2). This contrast may be explained by systematic variation in susceptibility to sexual coercion by males, which is especially pronounced among catarrhines (Smuts and Smuts, 1993). Lemur females tend to dominate males (Kappeler, 1993), and their greater behavioural control is reflected in descriptions of mating behaviour (e.g. Richard, 1992). Similarly, ceboid females may have greater control over mating behaviour. First, ceboid females in various species are not subordinate to males (e.g. red-backed squirrel monkey: Boinski, 1987; woolly spider monkey: Strier, 1990), and males have to gang up in order to be able to assess a female's reproductive state (Boinski, 1987; Symington, 1987). Second, females more often take the initiative to mating and maintaining consortships (Robinson and Janson, 1987). Third, male–male aggression over females is reportedly rare, even where males tend to cluster around attractive females (Janson, 1984; Milton, 1985; Symington, 1987; Strier, 1990). Thus, female lemurs and ceboids seem to be able to mate with as many males as they need to in a relatively short period. For instance, in woolly spider monkeys, a female attracts (through a twittering call) up to nine males around her and then mates with up to five of them during a single day (Milton, 1985).

The communally breeding ceboids obviously do not fit. They have long mating periods and ubiquitous post-conception receptivity – in the absence of male infanticide. Because their communal infant care is derived, their ancestors were probably vulnerable to infanticide and thus had the associated sexuality, including prolonged receptivity and unpredictable ovulation. They may have retained these features for other functions. For instance, it is possible that females trade matings for infant care (Price, 1990; but see Tardiff and Bales, 1997), and may thus benefit from prolonged attractivity.

Sex among species vulnerable to male infanticide

Active polyandrous mating

In some primate species, the female tendency towards polyandry varies with social conditions. In long-tailed and bonnet macaques and brown capuchins, the females' tendency towards polyandrous mating is stronger during times of instability of the male dominance relations (Samuels, Silk and Rodman, 1984; van Noordwijk, 1985; C. Janson in Manson, 1995). In species with multi-male influxes, females end up mating with many of the invading males, and often make this easier for the invaders by approaching them (Cords, 1988).

One non-communal ceboid, the red-backed squirrel monkey, living in large groups, may be an exception to the rule that females actively pursue polyandrous matings. During the very brief period of receptivity (two days on average), a female only mates with one or two males (Boinski, 1987). However, pregnant females or females with young infants freely migrate between groups, suggesting there is no risk of male infanticide even though lactation lasts longer than gestation. This exception may well prove the rule, because this species is a highly seasonal breeder and all females have been observed to give birth each year. Thus, males derive no advantage from committing infanticide.

In sum, the evidence for the active pursuit of polyandrous matings by females in the high-infanticide risk category is overwhelming, and variation in this tendency, while still under-reported, is also consistent with the general hypothesis. Thus, while there may be other, unidentified benefits to these matings, this aspect of sexual behaviour supports the general hypothesis.

Situation-dependent receptivity

Sexual activity during pregnancy in hanuman langurs was interpreted by Hrdy (1979) as serving to reduce the risk of infanticide by a new male. However, Sommer (1994) found that sex during pregnancy is not conditional on the immigration of a male, i.e. it always takes place, and also that it is not always effective in averting infanticide. Thus, Sommer rejects Hrdy's functional interpretation. Nevertheless, Sommer's data do suggest that pregnant females show proceptivity and mate more often and later into pregnancy when a new male has taken over the group (Fig. 8 in Sommer, 1994). Thus, while receptive periods during pregnancy cannot be made facultative, their flexible deployment during exposure to a new resident male is clearly consistent with this hypothesised function.

Some degree of facultative sexual activity and attractivity (swelling) during pregnancy is suggested by the contrasting findings of Caldecott (1986) and Oi (1996) on pigtailed macaques. In Caldecott's population, adult males had strictly enforced dominance hierarchies, and the dominant male tended to guard a female during her peri-ovulatory period, mating almost exclusively. Post-conception sexual activity was common. In Oi's population, by contrast, mating was far less exclusive, and no sexual activity during pregnancy was observed. Although many more comparisons are needed, this contrast suggests that females only engage in sexual activity during pregnancy when paternity confusion has not been attained before or around conception.

In some cases, receptivity seems to be truly facultative. Okayasu (1992), studying Japanese macaques, noted immediate proceptivity and rapid changes of the sex skin upon the appearance of new males in the group. Experimental introduction of males into captive groups also reliably produces swellings in species with such anatomical attractivity indicators, in both pregnant females and females with older infants (e.g. talapoins: Rowell, 1977). Male immigration in the wild has the same effect (e.g. baboons: Stein, 1984).

Variation in the length of the mating period

Interspecific variation

Although variation in the length of the mating period has a strong taxonomic component (see Table 8.2), one can still predict that mating periods should be longer in cases in which infanticide risk is high if longer mating periods would allow females to mate with all potentially infanticidal males. Because the actual number of mates per female in each ovarian cycle is often very difficult to determine, the authors have classified them into three categories (see Appendix 8.1): (i) only a single male mates in most cases; (ii) mating is predominantly with one male but others occasionally mate too; (iii) multiple males routinely mate in an average cycle. This relationship is examined by radiation.

Within the taxa subject to a high risk of male infanticide, the predicted relationship between mating period and number of mates is indeed found (Fig. 8.2). This relationship is significant in the whole data set ($r_s = +0.55$, $n = 35$, $p < 0.01$) and remains positive within each of the three radiations. Figure 8.2 also shows the expected difference between non-communal ceboids and catarrhines, in that the former have shorter mating periods

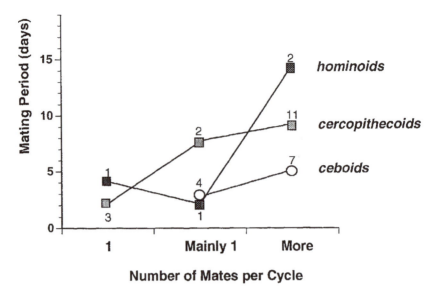

Fig. 8.2   Length of mating period in the cycle and the number of male partners a female has. Data from Appendix 8.1.

across the board. An analysis controlling for phylogenetic inertia[4] (including only the taxa with high vulnerability to infanticide) yields five contrasts in the predicted direction, one of zero, and two in the opposite direction, a non-significant result ($t_7 = 1.11$, n.s.). However, one of these negative contrasts is between the *Papio–Theropithecus* clade and the *Cercocebus–Mandrillus* clade; it is caused by the reconstruction of the ancestor of the *Papio-Theropithecus* clade as having predominantly one-male mating (probably an incorrect reconstruction). The second negative contrast is between lion-tailed macaques and the other macaques. Lion-tails were scored as having predominantly one-male mating, but it is likely that groups of this species other than the one used here have multi-male mating, as in other macaques. The other contrasts are all plausible. Thus, phylogenetic analysis does not contradict the conclusion that longer mating periods are associated with more mates.

Intraspecific variation

Several observational studies in the wild are consistent with an increased duration of receptivity, or mating period, in situations with increased potential for infanticide. The first example concerns hanuman langurs living in unimale or multi-male groups. C. Borries (personal communication) noted that the mating period lasts about four days in the population of Jodhpur in which unimale groups predominate, whereas it lasts about eight days in Ramnagar, where multi-male groups are the norm. Hrdy (1977) had also noted that mating in this species at Abu lasted much longer than the usual five to seven days after new males had entered the group. The second example comes from a comparison of adjacent groups of Japanese macaques (*Macaca fuscata yakui*) that had comparable adult sex ratios (Takahata *et al.*, 1994). The average mating period lasted 5.3 days in one group, but 27.8 in another group. The first group was stable and had only three non-troop males mating with the group's females, whereas the second group was invaded by 19 non-troop males, one of whom eventually took over the top-dominance position. The third example concerns wild redtail monkeys, which live in unimale groups that are occasionally subject to large influxes of extra-group males (Cords, 1984). During a normal mating season, a female's mating period lasts up to one week (mean 3.6 days), but when the group experiences an influx, it can even last up to one month (mean 4.7 days), and the intervals between the receptive periods become much shorter than expected, indicating situation-dependent receptivity as well.

Attractivity-enhancing signals: sex skins

In many species, females signal impending ovulation through vulval swelling or reddening. In several catarrhines, however, such indices of the likelihood of ovulation have become exaggerated, in the form of striking red patches or pink swellings, involving not just the vulva but also adjacent 'perineal' areas (Rowell, 1972; Dixson, 1983; Pagel, 1994; Nunn, in press), which raise their attractivity to potential mates (Bielert and Girolami, 1986). Since Darwin (1876) first drew attention to them, no subject has generated as much interest among students of primate sexuality as these 'sex skins', which have evolved at least three times (Dixson, 1983). However, to this day no satisfactory explanation exists for this enigmatic aspect of female reproductive physiology. A variety of hypotheses has been suggested to account for the evolution of these exaggerated swellings (reviewed by Nunn, in press). However, none of these ideas is entirely satisfactory (Pagel, 1994; Nunn, in press).

Distribution of sex skins

Any hypothesis for the function of exaggerated swellings should explain the following patterns: (i) sex skins (exaggerated signals) are limited to catarrhines; (ii) adolescent females often have stronger forms of sex skins than adults (Anderson and Bielert, 1994); (iii) sex skins are concentrated in species that live in multi-male groups or situations in which groups routinely have close contact with other males (Clutton-Brock and Harvey, 1976); (iv) seasonal breeders often lack the sex skins, even though the social system seems appropriate; (v) species with sex skins have longer mating periods and ovarian cycles. The last three patterns are discussed here.

The association between actually or potentially polyandrous mating and sex skins (Appendix 8.1; Fig. 8.3) is supported by intraspecific variation in red colobus (*Colobus badius*) the Tana River subspecies (*C.b. rufomitratus*) of which lives predominantly in unimale groups and has barely noticeable swellings as compared to the other red colobus subspecies in which females have conspicuous swellings (Struhsaker, 1975; M. Kinnaird, personal communication). Thus, the pattern is overwhelming: sex skins are only found in those catarrhine species that routinely or potentially mate polyandrously.

However, the opposite does not hold: several taxa with multi-male groups (vervets, several macaques, some populations of hanuman langur) do not have sex skins. Neither do most guenons and patas monkeys that live in unimale groups subject to influxes of non-group males during some of their mating seasons (Harding and Olson, 1986; Cords, 1987). Cursory inspection suggests that seasonally breeding primates lack sex skins (Appendix 8.1). Some species are highly seasonal; they have over 67% of their births (often over 90%) concentrated in a single three-month period. These species, when kept at temperate latitudes, will mostly retain highly seasonal breeding despite superabundant food, indicating photoperiodic regulation of female cycling. Here we will call them seasonal breeders. Other species are moderately seasonal (33–67%) or nonseasonal (< 33% of births in three months) in the wild. When held in captivity at temperate latitudes, both lose their seasonality, so they are combined here.

As shown in Figure 8.3, most non-seasonal breeders among polyandrous species have sex skins (91% of 23 species), whereas only 1 of the 11 seasonal ones (9%) do (including three species of temperate snub-nosed langurs – *Pygathrix*, subgenus *Rhinopithecus* – not in Appendix 1, which live in multilevel societies and are highly seasonal breeders: Rowe, 1996). This difference is highly significant ($\chi^2 = 22.02$, $p < 0.001$). The exceptions are all macaques, but the pattern remains significant if only macaques are considered ($\chi^2 = 6.35$, $p = 0.01$).[5] A phylogenetically controlled analysis, using Maddison's (1990) concentrated changes test, suggests that losses of sex skins in multi-male or multilevel species are more likely in taxa in which seasonal breeding has evolved ($p = 0.012$; assuming three gains and losses

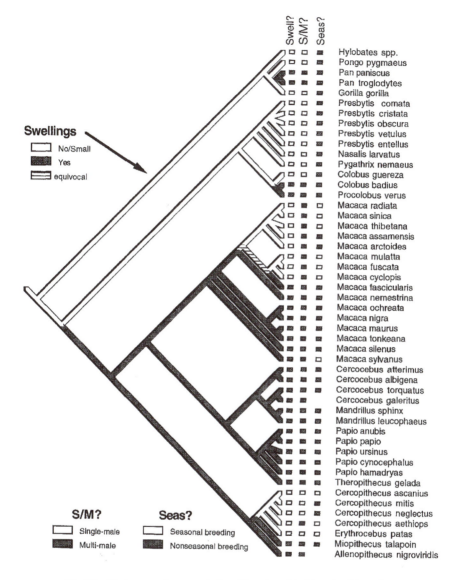

Fig. 8.3   Phylogeny of exaggerated swellings (sex skins) in catarrhine primates (including only known species). Added to the character states at the tips are the social system (predominantly unimale or multi-male) and the degree of seasonal breeding (for criteria, see text). Note the concentration of sex skins in species that live in non-seasonally breeding, multi-male groups.

of sex skins and a seasonal ancestor for the *sinica* and *arctoides* groups of *Macaca*).

Finally, polygynandrous species with sex skins have longer mating periods than their counterparts without swellings (Nunn, in press). They also have longer ovarian cycles (Mann-Whitney U: $n_1 = 8$, $n_2 = 16$, $U = 12.5$, $p < 0.01$), and this effect is retained if all catarrhine species without sex skins are included in the comparison, regardless of their social system ($n_1 = 14$, $n_2 = 16$, $U = 22.5$, $p < 0.001$). A phylogenetic analysis of cycle length in relation to the presence of exaggerated swellings yielded five contrasts, and all were in the predicted direction ($t_4 = 5.42$, $p < 0.01$).[6] The effect of seasonality is less strong, with non-seasonal breeders having longer cycles (contrasts in taxa vulnerable to infanticide; five of seven in predicted direction; $t_6 = 1.60$, n.s.). This suggests that the effect of swellings on cycle length (as proxy for follicular phase) is stronger than that of seasonality.

Male infanticide and sex skins

The hypothesis presented in this chapter would suggest that sex skins are needed if a female cannot achieve the right balance of paternity bias and confusion when she advertises her peri-ovulatory period without exaggerated signals. The need for sex skins is most acute when a female will be monopolised for too long by dominant males, so that she is unable to mate with subordinate males. Dominant males are less likely to intervene if a female mates with a subordinate if she is not at peak swelling. This suggests that females need this graded signal when they do not have enough behavioural freedom to choose their mates freely. Thus, we call this the graded signal hypothesis for exaggerated swellings (for details, see Nunn, in press).

This idea may go far in explaining the additional patterns in sex skins just documented (see also Nunn, in press). First, the opportunity to confuse paternity in regular ovarian cycles arises only in situations with the potential for polyandrous matings, which explains the absence of sex skins in unimale breeding groups. Second, where a female has sufficient behavioural freedom to select her mates, there is no need to attract dominant males during particular periods. This explains the absence of sex skins in infant-carrying lemurs and non-communal ceboids, in which, as noted earlier, females are much less subject to harassment by males. Third, in non-seasonal breeders females are more likely to be monopolised by a dominant male throughout their attractive period as females will overlap less often (cf. Thierry *et al.*, 1996). Graded swellings may help to break the dominant male's monopoly. In contrast, seasonal breeders are more likely to have multiple females that are sexually active and attractive to males, creating more behavioural freedom for females, and that are therefore more likely to achieve the right balance of matings with dominant and subordinate males. Fourth, the idea may account for adolescent exaggeration: once swellings have evolved as generally honest signals of ovulation probability, females may use this signal to deceive males (as also occurs in pregnancy swellings). Finally, the gradual build-up of the signal and the accompanying matings by males of varying rank obviously require more time than situations with 'normal' signals, thus explaining the longer mating periods and ovarian cycles of species with sex skins.

Attractivity-enhancing signals: mating vocalisations

Some female primates emit calls that are only given when they are ready to mate (indicating receptivity and proceptivity); these calls are usually referred to as 'oestrous' calls, perhaps better referred to as proceptivity calls. In many species, females also (or only) give vocalisations during or right after copulation. The distinction between these two contexts is not always hard and fast. For instance, in tonkean macaques, females start giving proceptivity calls but these calls turn into copulation calls when mating commences (Masataka and Thierry, 1993). Nonetheless they are considered separately here.

Proceptivity calls carry far and may, above all, attract distant potential mates. Several nocturnal strepsirrhines give proceptivity vocalisations, though none of the lorisids does, perhaps to avoid attracting predators (Appendix 8.1; see Table 8.2). Dispersed diurnal species could use sex calls for the same purpose. Woolly spider monkey females give calls that attract males from all over the range (Milton, 1985). However, their occurrence in many gregarious species is not readily explained in this way.

Copulation calls are also loud and can usually be heard from tens of metres away (soft calls emitted by some species in which individuals are usually widely dispersed, e.g. gibbons or slow lorises, are not counted because they probably serve a function in within-pair communication). Mating calls may serve to reduce infanticide risk just like sex skins by producing the right balance of paternity bias and confusion (cf. O'Connell and Cowlishaw, 1994). First, like sex skins, they indicate the likelihood of ovulation: mating calls of female baboons become longer towards ovulation (O'Connell and Cowlishaw, 1994), and they may change in structure in other species (Hauser, 1996), suggesting that they can be considered graded signals and are basically honest. Second, although females use them in most matings (e.g. van Noordwijk, 1985; O'Connell and Cowlishaw, 1994), they can suppress calls when mating with low-ranking males (Tibetan macaques: Zhao, 1993; chimpanzees: Hauser, 1990). Third, in species in which males cannot monopolise females and sperm competition therefore occurs, the calls announce that the female is mating again, thus enticing the other potential mates to mate again. These patterns are consistent with a function to reduce infanticide risk; there is little support for other suggested functions.

If mating calls function just like sex skins, their taxonomic distributions should tend to be similar. Although the distribution of female mating calls is not as completely known as that of sex skins, the correspondence is indeed quite good. All species with sex skins also have female mating calls ($n = 17$), whereas only 36% of the 14 species without sex skins have them (G-test with Williams' correction ($G_{adj.}$) = 17.97, $p < 0.001$).

Discussion

It is argued in this chapter that the risk of infanticide by males is ultimately responsible for many derived features of the sexuality of female primates. The basic mammalian pattern of brief receptivity generally concentrates paternity in a single male (cf. Brockman, 1994). While this will favour paternal behaviour, such as infant protection, it increases the risk of attacks by non-sires. Primate cycles allow longer receptive periods that serve to confuse paternity and so reduce infanticide risk. The various aspects of female sexuality can be regarded as tools at the female's disposal to produce the appropriate balance between paternity concentration and confusion, i.e. mate mostly with dominant males when ovulation is most likely, but also mate with subordinate males when ovulation is less likely. Longer cycles (especially follicular phases), longer mating periods and polyandrous mating, unpredictable timing of ovulation, and sexual activity outside periods of regular ovarian cycles obviously all serve to bring about paternity confusion. The most obvious function of the signalling of ovulation probability through sex skins and mating calls is that they attract dominant males when the signal is strong, allowing others to mate when the signal is weak, thus again helping to bring about the right timing and relative proportion of matings by dominant and subordinate males.

How extensive these traits are should also covary with the degree to which females have control over whom they mate with. This is indeed what we have found. The extent to which a female can choose her mates freely in multi-male groups depends on the potential for one dominant male to monopolise her; this, in turn, largely depends on the degree to which breeding is seasonal and thus the number of females expected to be sexually active simultaneously (given a certain group size) and on intersexual dominance relations. Thus, both highly seasonal species and the non-communal lemurs and ceboids, whose females tend to have more behavioural freedom, lack sexual swellings and also tend to have shorter mating periods, because they can achieve the right balance of paternity bias and confusion without these additional features.

There is not enough space here to discuss the evidence concerning primate reproductive physiology and behaviour in the detail it deserves. Neither can alternative ideas for the various phenomena be explored, at least in part because carefully formulated alternatives are scarce. However, we do not wish to convey the impression that all variation in primate sexuality, let alone mammalian sexuality, is related to infanticide risk. Other ecological and social factors (e.g. Wrangham, 1993), as well as phylogenetic constraints, must all play important roles; some of these factors may explain much more variation in mammalian orders other than primates. Nonetheless, a remarkable amount of variation in the primate order is consistent with the selective impact of infanticide risk, in interaction with other factors and within immutable phylogenetic constraints.

Many of our conclusions remain tentative in the absence of solid information on key species or social effects. Systematic evaluations of the effect of social context on the length of the constituent parts of ovarian cycles are sorely needed. While more extensive information on interspecific comparisons in taxa with variation in the number of males will be quite useful, more convincing tests will come from intraspecific comparisons of cycle length and mating periods of different populations of species such as red colobus or hanuman langur that vary in the number of males per group. The most convincing demonstration of social effects on the supposedly entirely spontaneous ovarian cycles would come from variation in cycle lengths due to experimental manipulation of the number of males (and females) in female groups.

The framework presented here may also help to resolve the debate over concealed ovulation. What is usually called concealed ovulation (see Wallen, 1995) is perhaps best considered as highly unpredictable ovulation. The hypothesis explored here suggests that it should be found where the threat of male infanticide comes only from inside the group, and there is therefore no need to bias paternity towards the most protective male. This situation should be found where male immigration must be exceedingly rare (as in bonobos, red colobus, and chimpanzees in which males are sufficiently cohesive), or where females have control over the identity of immigrating males (e.g. vervets: Smuts, 1987).

Acknowledgements

We thank Maurits van der Graaf for his efforts in locating technical references, Diane Brockman and Sarah Hrdy for comments on a draft, and many members – but especially Carola Borries and Keith Hodges – of the reproductive biology and behavioural ecology groups at the German Primate Center and of the Department of Anthropology, all in Goettingen, for references and discussion. During the conception of this chapter, CvS was supported by a Forschungspreis of the Alexander von Humboldt Foundation. CLN was supported by an NSF graduate student fellowship and dissertation improvement grant.

Notes

1 In general, we expect that the Bruce effect – pregnancy block and early resorption of embryos when females are exposed to a new male (Labov *et al.*, 1985) – should be more common among taxa unable to confuse paternity during regular cycles or during pregnancy (pregnancy matings are absent in baboons: see Appendix 8.1).

2 Discussions of sexual behaviour are facilitated by distinguishing three separate components of sexual behaviour (first proposed by Beach, 1976), because there is abundant evidence that these components are controlled by different endocrine mechanisms. *Proceptivity* is the tendency of females to seek matings; *receptivity* is the extent to which the females co-operate in male-initiated copulation; and *attractivity* is the extent to which males are attracted to females (or the stimulus value of the female to the male without female behaviour factored in). There is also a place for a term that describes overall female attractivity based on all signals and behaviours; for lack of a generally used term, we refer to this as *perceived quality*.

3 Although mating periods during pregnancy may be cyclical (e.g. Hadidian and Bernstein, 1979), they should not be regarded as an accidental by-product of continuing cyclicity of hormone levels during pregnancy. For instance, some species with perineal swellings do and others do not continue these swellings during pregnancy (e.g. mangabeys do but the closely related baboons do not).

4 The results are reported of both non-phylogenetic and phylogenetic tests (Harvey and Pagel, 1991). Phylogenetic analyses are based on Purvis' (1995) composite estimate of primate phylogeny. Comparisons of two continuous or continuous and discrete variables are derived using CAIC (Purvis and Rambaut, 1995); those involving two discrete characters were derived using the concentrated changes test of Maddison (1990), as implemented in MacClade (Maddison and Maddison, 1992).

5 Some hanuman langurs are also an exception. Some populations contain mainly multi-male groups, but females are not reported to have sex skins, although reproduction is not always highly seasonal.

6 A preliminary analysis of data on the length of follicular phase supports this conculsion (Hodges *et al.*, in preparation).

References

Anderson, C.M. and Bielert, C.F. (1994). Adolescent exaggeration in female Catarrhine primates. *Primates* **35**, 283–300.

Angst, W. and Thommen, D. (1977). New data and a discussion of infant killing in old world monkeys and apes. *Folia Primatologica* **27**, 198–229.

Beach, F.A. (1976). Sexual attractivity, proceptivity and receptivity in female mammals. *Hormones and Behavior* **7**, 105–38.

Bielert, C. and Girolami, L. (1986). Experimental assessments of behavioral and anatomical components of female chacma baboon (*Papio ursinus*) sexual attractiveness. *Psychoneuroendocrinology* **11**, 75–90.

Boinski, S. (1987). Mating patterns in squirrel monkeys (*Saimiri oerstedi*). *Behavioral Ecology and Sociobiology* **21**, 13–21.

Borries, C. (1997). Infanticide in seasonally breeding multi-male groups of Hanuman langurs (*Presbytis entellus*) in Ramnagar (South Nepal). *Behavioral Ecology and Sociobiology* **41**, 139–50.

Brockman, D.K. (1994). Reproduction and mating system of Verrreaux's sifaka, *Propithecus verreauxi*, at Beza Mahafaly, Madagascar. PhD thesis, Yale University.

Brockman, D.K. and Whitten, P.L. (1996). Reproduction in free-ranging *Propithecus verrauxi*: estrus and the relationship between multiple partner matings and fertilization. *American Journal of Physical Anthropology* **100**, 57–69.

Caldecott, J.O. (1986). *An Ecological and Behavioral Study of the Pig-tailed Macaque*. Basel: Karger.

Charnov, E.L. (1993). *Life History Invariants*. Oxford: Oxford University Press.

Clutton-Brock, T.H. and Harvey, P.H. (1976). Evolutionary rules and primate societies. In *Growing Points in Ethology*, ed. P.P.G. Bateson and R.A. Hinde, pp. 195–237. Cambridge: Cambridge University Press.

Colmenares, F. and Gomendio, M. (1988). Changes in female reproductive condition following male take-overs in a colony of hamadryas and hybrid baboons. *Folia Primatologica* **50**, 157–74.

Cords, M. (1984). Mating patterns and social structure in redtail monkeys (*Cercopithecus ascanius*). *Zeitschrift für Tierpsychologie* **64**, 313–29.

Cords, M. (1987). Forest guenons and patas monkeys: male–male competition in one-male groups. In *Primate Societies*, ed. B.B. Smuts, D.L. Cheney, R.M. Seyfarth, R.W. Wrangham and T.T. Struhsaker, pp. 98–111. Chicago: University of Chicago Press.

Cords, M. (1988). Mating systems of forest guenons: a preliminary review. In *A Primate Radiation: Evolutionary Biology of the African Guenons*, ed. A. Gauthier-Hion, F. Bourliere, J.P. Gauthier and J. Kingdon, pp. 323–39. Cambridge: Cambridge University Press.

Crockett, C.M. and Sekulic, R. (1984). Infanticide in red howler monkeys (*Alouatta seniculus*). In *Infanticide; Comparative and Evolutionary Perspectives*, ed. G. Hausfater and S.B. Hrdy, pp. 173–91. New York: Aldine.

Darwin, C. (1876). Sexual selection in relation to monkeys. *Nature* **15**, 18–19.

Dixson, A.F. (1983). Observations on the evolution and behavioral significance of 'sexual skin' in female primates. *Advances in the Study of Behavior* **13**, 63–106.

Dunbar, R.I.M. (1980). Demographic and life history viaibles of a population of gelada baboons (*Theropithecus gelada*). *Journal of Animal Ecology* **49**, 485–506.

Elwood, R.W. and Kennedy, H.F. (1994). Selective allocation of parental and infanticidal responses in rodents: a review of mechanisms. In *Infanticide and Parental Care*, ed. S. Parmigiani and F.S. vom Saal, pp. 397–425. Chur, Switzerland: Harwood Academic Publishers.

Fairbanks, L.A. and McGuire, M.T. (1987). Mother–infant relationships in vervet monkeys: response to new adult males. *International Journal of Primatology* **8**, 351–66.

Fairgrieve, C. (1995). Infanticide and infant eating in the blue monkey (*Cercopithecus mitis stuhlmanni*) in the Budongo forest reserve, Uganda. *Folia Primatologica* **64**, 69–72.

Glander, K.E. (1980). Reproduction and population growth in free-ranging mantled howling monkeys. *American Journal of Physical Anthropology* **53**, 25–36.

Gust, D.A. (1994). Alpha-male sooty mangabeys differentiate between females' fertile and their postconception maximal swellings. *International Journal of Primatology* **15**, 289–302.

Hadidian, J. and Bernstein, I.S. (1979). Female reproductive cycles and birth data from an Old World monkey colony. *Primates* **20**, 429–42.

Harcourt, A.H., Fossey, D, Stewart, K.J. and Watts, D.P. (1980). Reproduction in wild gorillas and some comparisons with chimpanzees. *Journal of Reproduction and Fertility Supplement* **28**, 59–70.

Harding, R.S.O. and Olson, D.K. (1986). Patterns of mating among male patas monkeys (*Erythrocebus patas*) in Kenya. *American Journal of Primatology* **11**, 343–58.

Harvey, P.H., Martin, R.D. and Clutton-Brock, T.H. (1987). Life histories in comparative perspective. In *Primate Societies*, ed. B. B. Smuts, D. L. Cheney, R. W. Seyfarth, R. W. Wrangham and T. T. Struhsaher, pp. 179–96. Chicago: Chicago University Press.

Harvey, P.H. and Pagel, M.D. (1991). *The Comparative Method in Evolutionary*

Biology. Oxford: Oxford University Press.

Hauser, M.D. (1990). Do chimpanzee copulatory calls incite male–male competition? *Animal Behaviour* **39**, 596–7.

Hauser, M.D. (1996). *The Evolution of Communication*. Cambridge, Mass.: MIT Press.

Hausfater, G. (1975). *Dominance and Reproduction in Baboons: a Quantitative Analysis*. Basel: Karger.

Hausfater, G. and Hrdy, S.B. (1984). *Infanticide: Comparative and Evolutionary Perspectives*. New York: Aldine de Gruyter.

Hayssen, V., van Tienhoven, A. and van Tienhoven, A. (1993). *Asdell's Patterns of Mammalian Reproduction: a Compendium of Species-specific Data*. Ithaca, New York: Cornell University Press.

Hiraiwa-Hasegawa, M. (1988). Adaptive significance of infanticide in primates. *Trends in Ecology and Evolution* **3**, 102–5.

Hrdy, S.B. (1977). *The Langurs of Abu: Female and Male Strategies of Reproduction*. Cambridge, Mass.: Harvard University Press.

Hrdy, S.B. (1979). Infanticide among animals: a review, classification, and examination of the implications for the reproductive strategies of females. *Ethology and Sociobiology* **1**, 13–40.

Hrdy, S.B. (1981). *The Woman that Never Evolved*. Cambridge, Mass.: Harvard University Press.

Hrdy, S.B., Janson, C.H. and van Schaik, C.P. (1995). Infanticide: let's not throw out the baby with the bath water. *Evolutionary Anthropology* **3**, 151–4.

Hrdy, S.B. and Whitten, P.L. (1987). Patterning of sexual activity. In *Primate Societies*, ed. B.B. Smuts, D.L. Cheney, R.W. Seyfarth, R.W. Wrangham and T.T. Struhsaker, pp. 370–84. Chicago: University of Chicago Press.

Janson, C.H. (1984). Female choice and mating system of the brown capuchin monkey *Cebus apella* (Primates: Cebidae). *Zeitschrift für Tierpsychologie* **65**, 177–200.

Kappeler, P.M. (1993). Female dominance in primates and other mammals. In *Perspectives in Ethology*, Vol. 10, ed. P.P.G. Bateson, N. Thompson and P. Klopfer, pp. 143–58. New York: Plenum Press.

Kaufman, J.H. (1965). A three year study of mating behaviour in a free ranging band of monkeys. *Ecology* **46**, 500–12.

Keverne, E.B. (1987). Processing of environmental stimuli and primate reproduction. *Journal of Zoology, London* **213**, 395–408.

Kyes, R.C., Rumawas, R.E., Sulistiawati, E. and Budiarsa, N. (1995). Infanticide in a captive group of pig-tailed macaques (*Macaca nemestrina*). *American Journal of Primatology* **36**, 135–6.

Labov, J.B., Huck, U.W., Elwood, R.W. and Brooks, R.J. (1985). Current problems in the study of infanticidal behavior of rodents. *Quarterly Review of Biology* **60**, 1–20.

Lee, P.C. (1996). The meanings of weaning: growth, lactation, and life history. *Evolutionary Anthropology* **5**, 87–96.

Loy, J. (1981). The reproductive and heterosexual behaviour of adult patas monkeys in captivity. *Animal Behaviour* **29**, 714–26.

Maddison, W. (1990). A method for testing the correlated evolution of two binary characters: are gains or losses concentrated on certain branches of a phylogenetic tree? *Evolution* **44**, 539–57.

Maddison, W.P. and Maddison, D.R. (1992). *MacClade. Analysis of Phylogeny and Character Evolution.* Sunderland, Mass.: Sinauer Associates.

Manson, J.H. (1995). Female mate choice in primates. *Evolutionary Anthropology* **4**, 192–5.

Martin, R. D. (1992). Female cycles in relation to paternity in primate societies. In *Paternity in Primates: Genetic Tests and Theories*, ed. R.D. Martin, A.F. Dixson and E.J. Wickings, pp. 238–74. Basel: Karger.

Masataka, N. and Thierry, B. (1993). Vocal communication of Tonkean macaques in confined environments. *Primates* **34**, 169–80.

Mason, W.A. (1966). Social organization of the South American monkey, *Callicebus moloch*: a preliminary report. *Tulane Studies in Zoology* **13**, 23–8.

Milton, K. (1985). Mating patterns of woolly spider monkeys, *Brachyteles arachnoides*: implications for female choice. *Behavioral Ecology and Sociobiology* **17**, 53–9.

Mori, U. and Dunbar, R.I.M. (1985). Changes in the reproductive condition of female gelada baboons following the takeover of one-male units. *Zeitschrift für Tierpsychologie* **67**, 215–24.

Nunn, C.L. (in press). The evolution of exaggerated sexual swellings in primates: a critical review and consolidation of existing hypotheses. *Animal Behaviour*.

O'Connell, S. and Cowlishaw, G. (1994). Infanticide avoidance, sperm competition and mate choice: the function of copulation calls in female baboons. *Animal Behaviour* **48**, 687–94.

Oi, T., ed. (1996). Sexual behaviour and mating system of the wild pig-tailed macaque in West Sumatra. In *Evolution and Ecology of Macaque Societies*, ed. J.E. Fa and D.G. Lindburg, pp. 342–68. Cambridge: Cambridge University Press.

Okayasu, N. (1992). Prolonged estrus in female Japanese macaques (*Macaca fuscata yakui*) and the social influence on estrus: with special reference to male intertroop movement. In *Topics in Primatology*. Vol. 2: *Behavior, Ecology, and Conservation*, ed. N. Itoigawa, Y. Sugiyama and G.P. Sackett, pp. 163–78. Tokyo: Tokyo University Press.

Pagel, M. (1994). The evolution of conspicuous oestrous advertisement in Old World monkeys. *Animal Behaviour* **47**, 1333–41.

Palombit, R.A. (1994). Extra-pair copulations in a monogamous ape. *Animal Behaviour* **47**, 721–3.

Paul, A. (1998). *Von Affen und Menschen: Verhaltensbiologie der Primaten.* Darmstadt: Wissenschaftliche Buchgesellschaft.

Pereira, M.E. (1983). Abortion following the immigration of an adult male baboon (*Papio cynocephalus*). *American Journal of Primatology* **4**, 93–8.

Pereira, M.E. (1991). Asynchrony within estrous synchrony among ringtailed lemurs (Primates: Lemuridae). *Physiology and Behavior* **49**, 47–52.

Perrigo, G. and vom Saal, F.S. (1994). Behavioral cycles and the neural timing of infanticide and parental behavior in male house mice. In *Infanticide and Parental Care*, ed. S. Parmigiani and F.S. vom Saal, pp. 365–96. Chur, Switzerland: Harwood Academic Publishers.

Price, E.C. (1990). Infant carrying as a courtship strategy of breeding male cotton-top tamarins. *Animal Behaviour* **40**, 785–6.

Purvis, A. (1995). A composite estimate of primate phylogeny. *Philosophical Transactions of the Royal Society, London, Series B* **348**, 405–21.

Purvis, A. and Rambaut, A. (1995). Comparative analysis by independent contrasts (CAIC): an Apple Macintosh application for analysing comparative data. *Computer Applications in the Biosciences* **11**, 247–51.

Reichard, U. (1995). Extra-pair copulation in a monogamous gibbon *(Hylobates lar)*. *Ethology* **100**, 99–112.

Richard, A.F. (1992). Aggressive competition between males, female-controlled polygyny and sexual monomorphism in a Malagasy primate, *Propithecus verreauxi. Journal Human Evolution* **22**, 395–406.

Robinson, J.G. and Janson, C.H. (1987). Capuchins, squirrel monkeys, and atelines: socioecological convergence with Old World primates. In *Primate Societies*, ed. B.B. Smuts, D.L. Cheney, R.W. Seyfarth, R.W. Wrangham and T.T. Struhsaker, pp. 69–82. Chicago: University of Chicago Press.

Rowe, N. (1996). *The Pictorial Guide to the Living Primates*. East Hampton, New York: Pogonias Press.

Rowell, T.E. (1972). Female reproductive cycles and social behaviour in primates. *Advances in the Study of Behaviour* **4**, 69–105.

Rowell, T.E. (1977). Reproductive cycles of the talapoin monkey (*Miopithecus talapoin*). *Folia Primatologica* **28**, 188–202.

Rowell, T.E. (1994). Choosy or promiscuous – it depends on the time scale. In *Current Primatology*, Vol. II, ed. J.J. Roeder, B. Thierry, J.R. Anderson and N. Herrenschmidt, pp. 11–24. Strasbourg: Université Louis Pasteur.

Rowell, T.E. and Hartwell, K.M. (1978). The interaction of behaviour and reproductive cycles in patas monkeys. *Behavioural Biology* **24**, 141–67.

Samuels, A., Silk, J.B. and Rodman, P.S. (1984). Changes in the dominance rank and reproductive behaviour of male bonnet macaques (*Macaca radiata*). *Animal Behaviour* **32**, 994–1003.

Sigg, H., Stolba, A., Abegglen, J.-J. and Dasser, V. (1982). Life history of hamadryas baboons: physical development, infant mortality, reproductive parameters and family relationships. *Primates* **23**, 473–87.

Small, M.F. (1990). Promiscuity in barbary macaques (*Macaca sylvanus*). *American Journal of Primatology* **20**, 267–82.

Small, M. (1993). *Female Choices*. Ithaca, New York: Cornell University Press.

Smuts, B.B., Cheney, D.L., Seyfarth, R.W., Wrangham, R.W. and Struhsaker, T.T. eds. (1987). *Primate Societies*. Chicago: University of Chicago Press.

Smuts, B.B. (1987). Gender, aggression, and influence. In *Primate Societies*, ed. B.B. Smuts, D.L. Cheney, R.W. Seyfarth, R.W. Wrangham and T.T. Struhsaker, pp. 400–12. Chicago: University of Chicago Press.

Smuts, B.B. and Smuts, R.W. (1993). Male aggression and sexual coercion of females in nonhuman primates and other mammals: evidence and theoretical implications. *Advances in the Study of Behavior* **22**, 1–63.

Sommer, V. (1994). Infanticide among the langurs of Jodhpur: testing the sexual selection hypothesis with a long-term record. In *Infanticide and Parental Care*, ed. S. Parmigiani and F.S. vom Saal, pp. 155–98. London: Harwood Academic Publishers.

Steenbeek, R. (1996). What a maleless group can tell us about the constraints on female transfer in Thomas's langurs (*Presbytis thomasi*). *Folia Primatologica*

67, 169–81.

Stein, D.M. (1984). *The Sociobiology of Infant and Adult Male Baboons.* Norwood, NJ: Ablex Publishing Company.

Strier, K.B. (1990). New world primates, new frontiers: insights from the woolly spider monkey, or muriqui (*Brachyteles arachnoides*). *International Journal of Primatology* **11**, 7–19.

Struhsaker, T.T. (1975). *The Red Colobus Monkey.* Chicago: University of Chicago Press.

Struhsaker, T.T. and Leland, L. (1987). Colobines: infanticide by adult males. In *Primate Societies*, ed. B.B. Smuts, D.L. Cheney, R.W. Seyfarth, R.W. Wrangham and T.T. Struhsaker, pp. 83–97. Chicago: University of Chicago Press.

Symington, M.M. (1987). *Ecological and Social Correlates of Party Size in the Black Spider Monkey, Ateles paniscus chamek.* Princeton: Princeton University.

Takahata, Y., Sprague, D.S., Suzuki, S. and Okayasu, N. (1994). Female competition, co-existence, and the mating structure of wild Japanese macaques on Yakushima island, Japan. In *Animal Societies: Individuals, Interactions and Organisation*, ed. P.J. Jarman and A. Rossiter, pp. 163–79. Kyoto: Kyoto University Press.

Tardiff, S.D. and Bales, K. (1997). Is infant-carrying a courtship strategy in callitrichid primates? *Animal Behaviour* **53**, 1001–7.

Thierry, B., Heistermann, M., Aujard, F. and Hodges, J.K. (1996). Long-term data on basic reproductive parameters and evaluation of endocrine, morphological, and behavioral measures for monitoring reproductive status in a group of semifree-ranging Tonkean macaques (*Macaca tonkeana*). *American Journal of Primatology* **39**, 47–62.

Van Horn, R.N. and Eaton, G.G. (1979). Reproductive physiology and behavior in prosimians. In *The study of Prosimian Behavior*, ed. G.A. Doyle and R.D. Martin, pp. 79–122. New York: Academic Press.

van Noordwijk, M.A. (1985). Sexual behaviour of Sumatran long-tailed macaques (*Macaca fascicularis*). *Zeitschrift für Tierpsychologie* **70**, 277–96.

van Schaik, C.P. (1996). Social evolution in primates: the role of ecological factors and male behaviour. *Proceedings of the British Academy* **88**, 9–31.

van Schaik, C. P. (in press). Social counterstrategies against male infanticide in primates and other mammals. In *Primate Males*, ed. P.M. Kappeler. Cambridge: Cambridge University Press.

van Schaik, C.P. and Kappeler, P.M. (1993). Life history, activity period and lemur social systems. In *Lemur Social Systems and their Ecological Basis*, ed. P.M. Kappeler and J.U. Ganzhorn, pp. 241–60. New York: Plenum Press.

van Schaik, C.P. and Kappeler, P.M. (1997). Infanticide risk and the evolution of permanent male–female association in nonhuman primates: a new hypothesis and comparative test. *Proceedings of the Royal Society London, Series B* **264**, 1687–94.

Wallen, K. (1995). The evolution of female sexual desire. In *Sexual Nature, Sexual Culture*, ed. P.R. Abramson and S.D. Pinkerton, pp. 57–79. Chicago: University of Chicago Press.

Watts, D.P. (1989). Infanticide in mountain gorillas: new cases and a reconsideration of the evidence. *Ethology* **81**, 1–18.

Watts, D.P. (1991). Mountain gorilla reproduction and sexual behavior. *American*

Journal of Primatology **24**, 211–26.

Whitten, P.L. and Russell, E. (1996). Information content of sexual swellings and fecal steroids in sooty mangabeys (*Cercocebus torquatus atys*). *American Journal of Primatology* **40**, 67–82.

Wildt, D.E., Doyle, L.L., Stone, S.C. and Harrison, R.M. (1977). Correlation of perineal swelling with serum ovarian hormone levels, vaginal cytology, and ovarian follicular development during the baboon reproductive cycle. *Primates* **18**, 261–70.

Wilson, M.E., Gordon, T.P. and Chikazawa, D. (1982). Female mating relationships in rhesus monkeys. *American Journal of Primatology* **21**, 21–7.

Wrangham, R.W. (1993). The evolution of sexuality in chimpanzees and bonobos. *Human Nature* **4**, 47–79.

Zhao, Q.-K. (1993). Sexual behavior of Tibetan macaques at at Mt Emei, China. *Primates* **34**, 431–44.

Appendix 8.1
Primate sex

Super-family	Species	Name	Cycle length (d)	Size of sex skin	Female vocalisations proc./cop.	Median mating period (d)	Mating during pregnancy	Post-partum 'oestrus'	Number of mates per cycle
Lemuroidea	Daubentonia madagascariensis	Aye-aye	50	1	E	3	?	0	1 + m
	Cheirogaleus major	Eastern/Greater dwarf lemur	30	1	?	3	?	?	
	Cheirogaleus medius	Fat-tailed dwarf lemur	20	1	C	3	0	1	M
	Microcebus murinus	Gray mouse lemur	50	1	?	1	?	1	
	Mirza coquereli	Coquerel's mouse lemur	22	1	E	1	0	1	
	Propithecus verreauxi	Verreaux's sifaka	31	1	E	2	1	0	M
	Lemur catta	Ring-tailed lemur	39	1	O	1	0	0	M
	Lemur coronatus	Crowned lemur	34	?	?	1	?	0	
	Lemur fulvus	Brown lemur	28	1	?	2	?	0	
	Lemur macaco	Black lemur	33	1	0	?	?	0	
	Varecia variegata	Ruffed or variegated lemur	42	?	0	2	?	0	1 + m
Lorisoidea	Galago crassicaudatus	Thick-tailed bushbaby	44	1	0	?	0	1	
	Galago senegalensis	Lesser bushbaby	35	1	0	2	0	1	
	Arctocebus calabarensis	Angwantibo	38	1	?	1	?	1	
	Loris tardigradus	Slender loris	34	1	0	2	0	1	
	Nycticebus coucang	Slow loris	42	1	?	1	?	1	
	Perodicticus potto	Potto	38	1	?	2	?	1	
Tarsoidea	Tarsius bancanus	Western or Bornean tarsier	24	1	C	1	0	1	1 + m
	Tarsius syrichta	Philippine tarsier	24	1	E	?	?	?	
Ceboidea	Callimico goeldii	Goeldi's marmoset	24	0	0	7	1	1	1
	Callithrix jacchus	Common marmoset	28	0	0	28	1	1	1
	Cebuella pygmaea	Pygmy marmoset	16	?	0	6	?	1	1 + m
	Leontopithecus rosalia	Lion tamarin	18	0	0	18	1	1	1 + m
	Saguinus fuscicollis	Saddleback tamarin	17	0	0	10	1	0	1 + m
	Saguinus oedipus	Cottontop tamarin	23	0	0	> 10	1	1	1 + m
	Alouatta caraya	Black howler	20	0	?	3	?	0	1 + m
	Alouatta palliata	Mantled howler	16	1	0	3	1	0	M
	Alouatta seniculus	Red howler	20	1	?	3	1	0	1 + m
	Ateles paniscus	Black spider monkey	27	0	0	6	0	0	M
	Brachyteles arachnoides	Woolly spider monkey (Muriqui)	21	0	E	2	?	0	M

Super-family	Species	Name	Cycle length (d)	Size of sex skin	Female vocalisations proc./cop.	Median mating period (d)	Mating during pregnancy	Post-partum 'oestrus'	Number of mates per cycle
	Lagothrix lagotricha	Woolly monkey	28	0	?	8	0	0	M
	Cebus albifrons	White-fronted capuchin	18	0	0	5	0	0	M
	Cebus apella	Brown capuchin	21	0	E + C	5	0	0	M
	Cebus capucinus	White-throated capuchin	?	0	?	?	1	1	M
	Cebus olivaceus	Wedge-capped capuchin	?	?	0	?	?	0	M
	Saimiri oerstedi	Red-backed squirrel monkey	7	1	E	2	0	0	1 + m
	Saimiri sciureus	Common squirrel monkey	9	0	E	2	0	0	1 + m
	Aotus trivirgatus	Night or Owl monkey	16	0	0	?	1	0	1
	Cacajao calvus	Bald uakari	28	0	0	4	0	0	M
	Callicebus moloch	Dusky titi	19	0	0	?	?	0	1 + extra-p
	Chiropotes spp.	Bearded saki	?	1	C	?	0	0	?
	Pithecia spp.	Saki	16.5	0	0	?	?	0	1 + m
Cercopithecoidea	*Allenopithecus nigroviridis*	Allen's swamp monkey	?	2	?	?	?	0	?
	Cercocebus albigena	Gray-cheeked mangabey	30	2	C	4	0	0	M
	Cercocebus aterrimus	Black mangabey	31	2	?	?	?	0	?
	Cercocebus galeritus	Agile mangabey	33	2	C	?	1	1	M
	Cercocebus torquatus	White-collared mangabey	32	2	C	?	1	0	M
	Cercopithecus aethiops	Vervet monkey	33	0	0	33	2	0	?
	Cercopithecus ascanius	Redtail monkey	?	0	0	3 [> 3 influx]	1	0	1 [M if infl.]
	Cercopithecus mitis	Blue or Sykes's monkey	32	0	0	2 [> 2 influx]	1	0	1 [M if infl.]
	Cercopithecus neglectus	De Brazza's monkey	?	0	?	?	?	0	?
	Erythrocebus patas	Patas monkey	31	1	0	1 [>> 1 influx]	1	0	1 [M if infl.]
	Macaca arctoides	Stump-tailed macaque	30	0	0	30	1	0	M
	Macaca assamensis	Assamese macaque	32	1	?	?	?	0	?
	Macaca cyclopis	Taiwan or Formosan macaque	29	1	C	?	2	0	M
	Macaca fascicularis	Long-tailed macaque	31	2	C	15	2	0	M
	Macaca fuscata	Japanese macaque	27	1	C	11	1	0	M
	Macaca maurus	Moor macaque	35	2	?	15	0	0	M
	Macaca mulatta	Rhesus macaque	28	1	0	9	0	0	M
	Macaca nemestrina	Pig-tailed macaque	36	2	E + C	13	1	0	M
	Macaca nigra	Celebes or Crested black macaque	36	2	?	9	0	0	M
	Macaca radiata	Bonnet macaque	28	0	C	5	1	0	M

	Species	Common name	Cycle length	Size of sex skin	Female vocalisations	Median mating period	Mating during pregnancy	Breeding seasonality	Number of mates per cycle
	Macaca silenus	Lion-tailed macaque	34	2	E + C	18	0	?	1 + m
	Macaca sinica	Toque macaque	29	0	?	14	0	?	?
	Macaca sylvanus	Barbary macaque	31	2	C	14	0	1	M
	Macaca thibetana	Tibetan macaque	?	1	E + C	?	0	?	M
	Macaca tonkeana	Tonkean macaque	35	2	?	10	0	1	M
	Mandrillus leucophaeus	Drill	34	2	?	?	0	1	?
	Mandrillus sphinx	Mandrill	34	2	C	?	0	0	?
	Miopithecus talapoin	Talapoin	34	2	C	11	0	0	?
	Papio anubis	Olive or Anubis baboon	38	2	C	6	0	0	M
	Papio cynocephalus	Yellow baboon	33	2	C	9	0	0	M
	Papio hamadryas	Hamadryas baboon	30	2	C	5	0	0	1 + m
	Papio ursinus	Chacma baboon	36	2	C	9	0	0	M
	Theropithecus gelada	Gelada baboon	34	2	E + C	9	0	0	1 + m
	Colobus badius	Red colobus	?	2	C	5	0	1	?
	Colobus guereza	Black and white colobus	?	0	?	?	0	?	?
	Colobus verus	Olive colobus	?	2	?	?	0	?	?
	Nasalis larvatus	Proboscis monkey	?	1	?	?	0	?	1
	Presbytis cristata	Silver leaf monkey	?	0	?	?	0	?	1 + m/M
	Presbytis entellus	Gray or hanuman langur	24	0	0	6	0	1	1
	Presbytis obscura	Dusky leaf monkey	21	1	?	?	0	0	1
	Presbytis senex	Purple-faced langur	?	0	?	?	0	1	?
	Pygathrix nemaeus	Douc langur	29	1	0	?	0	2	?
Hominoidea	*Hylobates* spp.	Gibbon	30	0	0	4	0	1	1 + extra-p
	Gorilla gorilla	Gorilla	29	1	C	2	0	1	1 + m
	Pan paniscus	Bonobo or Pygmy chimpanzee	42	2	C	15	0	2	M
	Pan troglodytes	Chimpanzee	37	2	C	14	0	2	M
	Pongo pygmaeus	Orangutan	30	0	0	13	0	1	1 + m

Cycle length rounded to nearest whole number, sometimes mean of multiple studies.

Size of sex skin: 0 = no visible changes near ovulation; 1 = slight swelling or coloration of vulva or adjacent region in adults; 2 = exaggerated swelling of large area, often including vulva and anus.

Female vocalisations: proc. = proceptive (or 'oestrus') calls = E; cop. = copulation calls = C; 0 = no calls known; ? = unknown. Additional information from compilation by Hauser (1996).

Median mating period: median of sample of cycles, or mid-range if only range is given (preference for wild, pair-tests excluded).

Mating during pregnancy: 0 = none; 1 = few; 2 = regularly.

Number of mates per cycle: 1 = only 1 mate; 1 + m = most copulations with 1 mate, but also with other males in group; M = routinely with most or all males in group; 1 + extra-p = most with pair-bonded mate, but also some with extra-pair males; M if infl. = polyandrous during invasion of males.

Breeding seasonality: 0 = < 33% of births in peak 3-month period; 1 = between 33% and 67%; 2 = > 67% in peak 3-month period.

Infanticide: y = reported to occur; additional information from compilation by C.P. van Schaik (unpublished).

Species	Number of adult males in unit	Breeding seasonality	Commun. breeding	Male infanticide	Sources[a]
Daubentonia madagascariensis	sol	2			Sterling (1993); Winn (1994)
Cheirogaleus major	sol	2			Van Horn & Eaton (1979)
Cheirogaleus medius	sol	2			Foerg (1982a)
Microcebus murinus	sol	2			Van Horn & Eaton (1979)
Mirza coquereli	sol	2			Stanger et al. (1995)
Propithecus verreauxi	1.7	3		y	Brockman & Whitten (1996)
Lemur catta	5.2	2		y	Evans & Goy (1968)
Lemur coronatus	2	2			Kappeler (1987)
Lemur fulvus	3	2		y	Izard et al. (1993)
Lemur macaco	3.4	2		y	Pollock (1986)
Varecia variegata	2.5	2		y	Foerg (1982b); Morland (1993)
Galago crassicaudatus	sol	2			Van Horn & Eaton (1979)
Galago sensgalensis	sol	1			Van Horn & Eaton (1979)
Arctocebus calabarensis	sol	0			Van Horn & Eaton (1979); Hrdy & Whitten (1987)
Loris tardigradus	sol	1			Izard & Rasmussen (1985)
Nycticebus coucang	sol	0			Izard et al. (1988)
Perodicticus potto	sol	1			Van Horn & Eaton (1979)
Perodicticus potto	sol	1			Van Horn & Eaton (1979)
Tarsius bancanus	sol	1			Wright et al.(1986); Crompton & Andau (1987); Roberts (1994)
Tarsius syrichta	sol	0			Van Horn & Eaton (1979)

Species	C	y			Reference
Callimico goeldii	C		1	1	Heltne et al. (1981); Carrol et al. (1989); Ziegler et al. (1989)
Callithrix jacchus	C		2.3	0	Hearn (1983); Kohlkute (1984); Digby & Ferrari (1994); L.J. Digby (personal communication)
Cebuella pygmaea	C		1.5	0	Soini (1987); Converse et al. (1995)
Leontopithecus rosalia	C		?	1	Stribley et al. (1990); Baker et al. (1993)
Saguinus fuscicollis	C		1.8	1	Epple & Katz (1983); Terborgh & Goldizen (1985)
Saguinus oedipus	C		2.7	1	Brand & Martin (1983)
Alouatta caraya		y	2.3	0	Kinzey (1997)
Alouatta palliata		y	2.9	0	Glander (1980)
Alouatta seniculus		y	1.7	0	Kinzey (1997)
Ateles paniscus			4	1	van Roosmalen (1985); Symington (1987)
Brachyteles arachnoides			7	1	Milton (1985); Strier & Ziegler (1997)
Lagothrix lagotricha			7.3	1	Kinzey (1997)
Cebus albifrons			3.8	1	C.H. Janson (personal communication)
Cebus apella		y	3.6	2	Janson (1984)
Cebus capucinus		y	?	?	Manson et al. (1997)
Cebus olivaceus		y	1.2	2	Valderrama et al. (1990)
Saimiri oerstedi			10	2	Boinski (1987)
Saimiri sciureus			4.3	2	Mitchell (1990); Kinzey (1997)
Aotus trivirgatus			1	1	Dixson (1994)
Cacajao calvus	C		?	2	Fontaine & DuMond (1977); Kinzey (1997)
Callicebus moloch	C		1	2	Kinzey (1997)
Chiropotes spp.			8	2	Kinzey (1997)
Pithecia spp.			2	2	Kinzey (1997)

Species	Number of adult males in unit	Breeding seasonality	Commun. breeding	Male infanticide	Sources[a]
Allenopithecus nigroviridis	M	?			Hrdy & Whitten (1987)
Cercocebus albigena	4.3	0			Wallis (1983)
Cercocebus aterrimus	3	1			Butler (1974); Horn (1987)
Cercocebus galeritus	2	?		y	Field & Walker (1995); Gust (1994); Kinnaird (1990)
Cercocebus torquatus	?	1		y	Butler (1974); Gautier & Gautier-Hion (1976); Gust (1994)
Cercopithecus aethiops	4.1	2		y	Andelman (1987)
Cercopithecus ascanius	1	1–2		y	Cords (1984)
Cercopithecus mitis	1	1		y	Tsingalia & Rowell (1984); Else et al. (1985)
Cercopithecus neglectus	1	1			Hrdy & Whitten (1987)
Erythrocebus patas	1.5	2			Loy (1981); Sly et al. (1983)
Macaca arctoides	?	1			Dukelow et al. (1979); Nieuwenhuijsen et al. (1986)
Macaca assamensis	3	1			Hayssen et al. (1993)
Macaca cyclopis	1.8	2			Peng et al. (1973)
Macaca fascicularis	4.7	1		y	Dukelow (1977); van Noordwijk (1985)
Macaca fuscata	3.4	2			Takahata (1980); Takahata et al. (1994); Tokuda (1961)
Macaca maurus	?	0–1			Matsumura & Wanatabe (1994)
Macaca mulatta	4.5	2		y	Loy (1970), (1971); MacDonald (1971); Gordon (1981)
Macaca nemestrina	2.4	0–1		y	Caldecott (1986); Oi (1996)

Species				Reference
Macaca nigra	?	0–1		Dixson (1977); Hadidian & Bernstein (1979); Thierry et al. (1996)
Macaca radiata	7.3	2		Srinath (1980); Samuels et al. (1984)
Macaca silenus	1.6	1	y	Kumar & Kurup (1985); Kumar (1987)
Macaca sinica	2.7	2		Roonwal & Mohnot (1977); Mitani et al. (1996)
Macaca sylvanus	4.7	2	y	Kuester & Paul (1984); Ménard et al. (1985)
Macaca thibetana	5.5	2		Zhao & Deng (1988); Zhao (1993)
Macaca tonkeana	?	0–1		Thierry et al. (1996)
Mandrillus leucophaeus	?	1	y	Butler (1974)
Mandrillus sphinx	1	1		Hadidian & Bernstein (1979); Hrdy & Whitten (1987)
Miopithecus talapoin	13	2?		Dixson et al. (1973); Rowell (1977)
Papio anubis	7.2	0	y	Bercovitch (1991)
Papio cynocephalus	7	0	y	Hausfater (1975)
Papio hamadryas	1	1	y	Kummer (1968); Loy (1987)
Papio ursinus	6.7	0	y	Saayman (1970); Anderson (1992)
Theropithecus gelada	1.5	1	y	Dunbar & Dunbar (1974); Dunbar (1978); Aich et al. (1990)
Colobus badius	4.4	1	y	Struhsaker & Leland (1987); Starin (1988)
Colobus guereza	1.5	0	y	Struhsaker & Leland (1987)
Colobus verus	2	1		Struhsaker & Leland (1987)
Nasalis larvatus	1	1		Yeager (1990); Gorzitse (1996)
Presbytis cristata	1	1	y	Hrdy & Whitten (1987)
Presbytis entellus	1/2.6	1	y	Hrdy (1977); Sommer et al. (1992)
Presbytis obscura	2.5	1		Badham (1967)
Presbytis senex	1	1	y	Rudran (1973)
Pygathrix nemaeus	?	?	y	Lippold (1977)

Species	Number of adult males in unit	Breeding seasonality	Commun. breeding	Male infanticide	Sources[a]
Hylobates spp.	1	0			Hrdy & Whitten (1987); Martin (1992); Reichard (1995)
Gorilla gorilla	1.8	0		y	Harcourt et al. (1980); Watts (1991)
Pan paniscus	6.7	0			Furuichi (1987), (1992)
Pan troglodytes	6.7	0		y	Wallis (1985); Nadler et al. (1985); Tutin (1980)
Pongo pygmaeus	1	0			Graham (1981); Schürmann & van Hooff (1986); van Schaik (unpublished)

[a]All references for the data in the appendix are available from the authors on request.
Sol = solitary.

9 Mating systems, intrasexual competition and sexual dimorphism in primates

J. MICHAEL PLAVCAN

Many primates are sexually dimorphic. Collectively, sexual dimorphism refers to any character – behavioural, morphological or physiological – that differs between the sexes. Some primates show conspicuous differences in pelage and skin colour, and numerous musculoskeletal differences have been documented for a wide variety of species. However, most comparative studies of primate sexual dimorphism focus on body size and canine tooth size.

Strepsirhines and haplorhines differ fundamentally in dimorphism. Most male haplorhine primates are larger than females, and possess larger canine teeth. In contrast, most strepsirhine species are either monomorphic or only slightly dimorphic (Fig. 9.1). Consequently, most comparative studies of primate sexual dimorphism focus on haplorhines.

Dimorphism, mating systems and intrasexual competition

Sexual dimorphism in haplorhine primates is widely viewed as a product of sexual selection. Sexual selection comprises two broad mechanisms: mate choice and mate competition (Andersson, 1994). Mate choice in primates is usually referred to as 'female choice'. There is clearly evidence that female choice plays a role in the evolution of sexual dimorphism in primates (Boinski, 1987; Richard, 1992). However, because female choice is difficult to quantify, and data are not available for most species, the exact role that it plays in explaining interspecific variation in dimorphism is currently unknown (Small, 1989).

Primate sexual dimorphism is commonly viewed as a product of male mate competition. Because males are effectively limited in their reproductive success by the number of females that they can inseminate, large differences in male reproductive success will occur if some males can

241

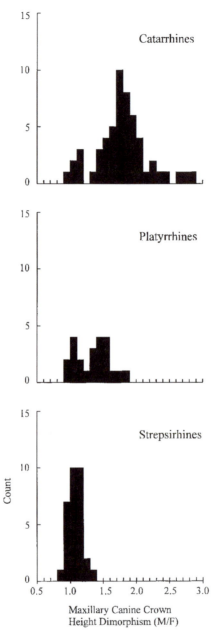

Fig. 9.1   Frequency histograms of maxillary canine crown height dimorphism in catarrhines, platyrrhines and strepsirhines. Data are listed in Appendix 9.1. Dimorphism is calculated simply as the ratio of male divided by female body mean values.

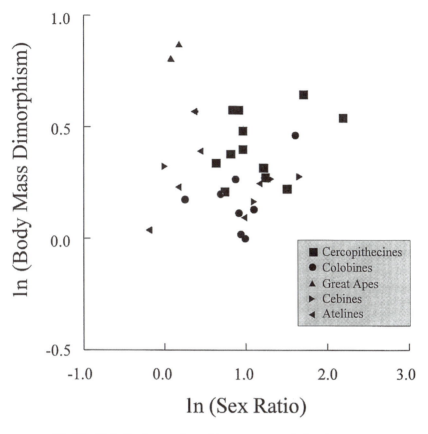

Fig. 9.2   Plot of body mass dimorphism versus sex ratio in polygynous
haplorhines. All data are ln-transformed.

exclude others from access to females. Males should therefore compete for
access to females. Any character, such as large body size or large canine
teeth, that helps males to win fights for access to females should be under
strong selection, resulting in dimorphism.

As predicted by sexual selection theory, polygynous haplorhines are
more dimorphic in their canine teeth and body mass than monogamous or
polyandrous species (Clutton-Brock, Harvey and Rudder, 1977; Harvey,
Kavanagh and Clutton-Brock, 1978b). However, it has long been noted
that polygynous species show a tremendous range of dimorphism that is
not associated with variation in mating system or sex ratio (Fig. 9.2). A
number of studies have debated the causes for this variation in the magnitude of sexual dimorphism (Coelho, 1974; Gautier-Hion, 1975; Clutton-

Brock *et al.*, 1977; Harvey *et al.*, 1978b; Phillips-Conroy and Jolly, 1981; Leutenegger and Cheverud, 1982, 1985; Gaulin and Sailer, 1984; Cheverud, Dow and Leutenegger, 1985; Lucas, Corlett and Luke, 1986; Kay *et al.*, 1988; Ely and Kurland, 1989; Kappeler, 1990, 1991, 1993; Plavcan and van Schaik, 1992, 1997b; Ford, 1994; Martin, Wilner and Dettler, 1994; Mitani, Gros-Louis and Richards, 1996). It is now widely acknowledged that factors such as phylogeny, allometric effects, predation pressure, correlated response, substrate and locomotor constraints, diet and energetic constraints influence the expression of dimorphism. However, there is no consensus on the relative roles that these factors play in the evolution of dimorphism. For example, Ford (1994) suggests that comparative data demonstrate that selection related to female energetic requirements plays little role in the evolution of size dimorphism, while Martin *et al.* (1994) propose just the opposite. Following from these debates, some studies suggest that sexual selection may play at best a secondary role in the evolution of dimorphism (Cheverud *et al.*, 1985; Lucas *et al.*, 1986; Martin *et al.*, 1994).

Testing the sexual selection hypothesis is not straightforward. Whereas sexual dimorphism is relatively easy to measure, sexual selection is not. Ideally, the sexual selection hypothesis should be evaluated by comparing the reproductive consequences for males that win and loose fights, and by testing the correlation between body mass, canine size, and the ability to win and loose fights (Clutton-Brock, 1985). Such data are not available for comparative analyses. Consequently, all broadly based comparative evaluations of the sexual selection hypothesis must use a surrogate measure of sexual selection – the most common being mating system (e.g. Gaulin and Sailer, 1984; Cheverud *et al.*, 1985). Studies usually classify species as monogamous or polygynous, or as monogamous, polyandrous, single-male/multi-female, or multi-male/multi-female. Unfortunately, these classifications of mating system are poor estimates of the strength of sexual selection associated with male–male competition.

A close look at patterns of male–male competition within several polygynous species shows why this is so. For example, *Pan*, *Brachyteles*, *Macaca* and *Papio* all have polygynous, multi-male/multi-female mating systems, and so should be dimorphic. However *Pan* and *Brachyteles* show relatively low degrees of dimorphism by comparison to *Papio* and *Macaca* (Plavcan and van Schaik, 1992, 1997b). *Pan* and *Brachyteles* are 'male bonded' whereas *Papio* and *Macaca* show agonistic male dominance hierarchies. These two mating systems fundamentally differ in the way that males interact with one another. The males of *Pan* and *Brachyteles* do not transfer out of their natal troops, resulting in groups composed of related males that engage in promiscuous mating with females, even though male dominance hierarchies can be discerned (Strier, 1986, 1992; Morin, 1993). Males of *Papio* and *Macaca*, on the other hand, transfer regularly, resulting in troops with unrelated males that compete intensely for access to mates (Walters and Seyfarth, 1987). Consequently, sexual selection stemming from male–male competition in *Pan* and *Brachyteles* should be less intense than that in *Papio* and *Macaca*.

Recent analyses of dimorphism attempt a more refined approach to estimating sexual selection, either through classifications of intrasexual competition (Kay *et al.*, 1988; Plavcan and van Schaik, 1992; Plavcan, van Schaik and Kappeler, 1995) or through estimates of the operational sex ratio (Mitani *et al.*, 1996). These studies provide evidence that variation in the strength of sexual selection produces variation in the magnitude of dimorphism (Kay *et al.*, 1988; Plavcan and van Schaik, 1992, 1997b; Greenfield, 1992b; Ford, 1994; Plavcan *et al.*, 1995; Mitani *et al.*, 1996). They also provide some insight into the relation between sexual selection, agnostic competition and mating systems.

Building on Kay *et al.* (1988), Plavcan and van Schaik (1992) derive 'competition levels' from dichotomous categorical estimates of the 'intensity' and 'potential frequency' of male–male competition. Males are classified as high intensity when they are intolerant of one another, escalated fighting is common, and they form easily detected agonistic dominance hierarchies. Males are ranked as low intensity when dominance hierarchies are difficult to detect, escalated fighting is rare, and they are relatively tolerant of one another. The frequency category is (with a few exceptions) a demographic measure. High-frequency competition occurs when there is typically more than one male in a breeding group, and low-frequency competiton occurs when there is typically only a single male or when competition is limited to a short breeding season.

The competition classifications were ranked into four competition levels, 1 through 4. Canine dimorphism and body mass dimorphism are strongly associated with competition levels (Fig. 9.3), even when monogamous and polyandrous species are excluded from the analysis. The relationship holds when controlling for phylogeny, allometry, diet and substrate (Plavcan and van Schaik, 1992, 1997b).

The competition levels are broadly correlated with mating system, but there are important differences. Polygynous haplorhines are ranked into competition levels 2 through 4. The most important distinction is the classification of eight multi-male/multi-female species into competition level 2. These species are polygynous, but male–male competition is not particularly intense by comparison to other polygynous species. In several

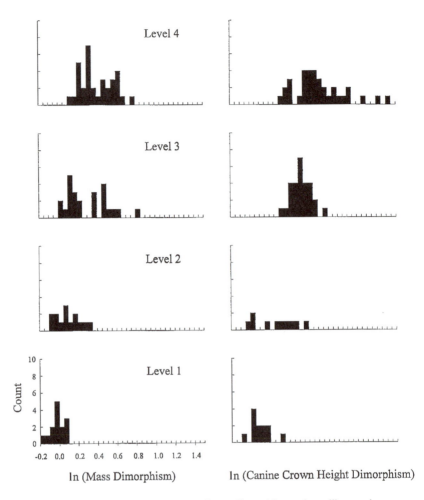

Fig. 9.3   Frequency histograms of mass dimorphism and maxillary canine crown height dimorphism in competition levels for haplorhine primates. Dimorphism is the natural logarithm of the ratio between male and female variables.

of these species, mating is reported as promiscuous, implying that sexual selection is not intense. These species show significantly less canine and mass dimorphism than other polygynous haplorhines (Table 9.1).

With only a few exceptions, the remaining polygynous species comprise competition levels 3 and 4, which correspond to single-male/multi-female and multi-male/multi-female species respectively. Predicted levels of dimorphism for these competition levels are the opposite to those predicted by Harvey *et al.* (1978b) for the multi-male/single-male contrast. Harvey *et*

Table 9.1. *Results of analyses of variance testing for significant differences in dimorphism between competition level 2 species (excluding polyandrous species) and other polygynous haplorhine primates*

Measure	n	F	p
Body mass dimorphism	72	9.580	0.003
Canine dimorphism			
Mandibular			
Mesiodistal	68	5.943	0.017
Buccolingual	68	7.941	0.006
Crown height	68	4.000	0.050
Maxillary			
Mesiodistal	68	8.968	0.004
Buccolingual	68	6.654	0.012
Crown height	68	3.626	0.061

Analyses are for species values. results are repeated within lower taxonomic groups and using phylogenetic contrasts (Plavcan and van Schaik, 1992, 1997b). Dimorphism is quantified as the natural logarithm of the male mean value divided by the female mean value for each species.

al. (1978b) predict that differential male reproductive success should be greater for single-male species than for multi-male species, because in the former a single male gains exclusive access to mates. Plavcan and van Schaik (1992) predicted that the higher frequency of competition in multi-male groups produces stronger selection for the development of weaponry. Dimorphism tends to be greater in competition level 4 species.

This frequency effect may reflect selection for the development of weaponry for display. Males are frequently wounded and sometimes killed in fights (Plavcan and van Schaik, 1992). Fighting between two males can potentially benefit a third male if both contestants are wounded or exhausted. Therefore, selection may favour the development of weaponry for display by reducing the risk of determining every contest through physical contact. Alternatively, the number of males in a group might not inversely reflect differential reproductive success. If male tenure in single-male troops is short enough, differential male reproductive success might not be greater than among males of some multi-male groups, where a single male might consistently gain access to mates for an extended period of time.

Monogamous species such as the hylobatids and *Callicebus* comprise competition level 1. In both genera, a lack of differential male reproductive success associated with monogamy should result in monomorphism. However, these two species show very different patterns of male–male competition. In *Callicebus*, both sexes are relatively peaceful, with infrequent overt aggression (Robinson, Wright and Kinzey, 1987). Logically, these species belong in competition level 1 (low intensity, low frequency). However, both male and female hylobatids occasionally show intense, sometimes lethal intrasexual competition in territorial defense (Mitani, 1985). Considered alone, male hylobatids belong in competition level 3 (high intensity, low frequency). Hylobatids should show low dimorphism because males and females show similar types of agonistic competition (Plavcan and van Schaik, 1992, 1997b), and because this competition does not generate grossly different patterns of male and female reproductive success (Emlen and Oring, 1977; Andersson, 1994). Notably, male and female hylobatids both possess canines as tall as competition level 3 species, whereas male and female *Callicebus* have very small canines.

The competition levels clearly support the sexual selection hypothesis and underscore problems with the use of mating system as a measure of the strength of sexual selection. Even so, the competition levels are a broad measure of male–male competition. Mitani *et al.* (1996) provide a more restricted but more refined analysis using operational sex ratios.

Theoretically, male–male competition should covary with the ratio of adult males to females in groups. Excluding monogamous and polyandrous species, there is no correlation between sex ratio and mass dimorphism in primates (Clutton-Brock *et al.*, 1977; Kappeler, 1990, 1991; Martin *et al.*, 1994). However, sex ratio is a poor measure of male–male competition for a variety of reasons (Clutton-Brock *et al.*, 1977). The operational sex ratio (OSR) is a better measure of male–male competition (Emlen and Oring, 1977; Mitani *et al.*, 1996). The OSR is the ratio of males to available breeding females in a group. Consider two groups, each with 12 adult males and 12 adult females. Both are characterised by the same sex ratio (1.0). However, in one group, all the females become receptive at the same time once a year. In this group, a male cannot exclude other males from access to more than one female, effectively resulting in no differential male reproductive success, and no sexual selection. In the other group, one female becomes receptive each month. In this group, a single male can potentially exclude all other males from access to all females, since he only needs to defend one receptive female at a time. The latter group has a very high OSR by comparison to the former, so sexual selection should be strong.

Mitani *et al.* (1996) quantify the OSR for 18 haplorhine primates. Their formula weights adult sex ratios with data on breeding seasonality, interbirth intervals, duration of mating periods, and the number of cycles before conception. Using the phylogenetic contrast method (Felsenstein, 1985; Pagel, 1992), they demonstrate a significant correlation between allomet-

Table 9.2. *Correlations between socionomic sex ratio (SR), operational sex ratio (OSR) and body mass dimorphism in haplorhines*

Analysis	n	r	p
Phylogenetic contrasts			
Mitani body mass data			
OSR	15	0.881	< 0.001
SR	15	0.641	0.007
Appendix 9.1 body mass data			
OSR	15	0.395	0.130
SR	15	0.311	0.242
Species values			
Mitani body mass data			
OSR	18	0.585	0.011
SR	18	0.470	0.049
Appendix 9.1 body mass data			
OSR	18	0.504	0.033
SR	18	0.453	0.059

All data were ln-transformed. Separate analyses were carried out using body mass data from Mitani *et al.* (1996), Plavcan and van Schaik (1997b) and Smith and Jungers (1997). Results are presented for phylogenetic contrasts and analysis of species values. Phylogenetic contrasts were calculated by hand using the phylogeny presented in Mitani *et al.* (1996), but differing from this latter study in that the *Alouatta* trichotomy was left unresolved, branch lengths were initially set at 1, and dimorphism was estimated as the natural log of the ratio of male and female mass. Contrasts of both dimorphism and OSR estimates were first regressed against female body mass contrasts. Regressions of contrasts were forced through the origin.

rically adjusted mass dimorphism and OSR. Using similar methods, there is also a significant correlation between the simple sex ratio and mass dimorphism in the sample used by Mitani *et al.* (Plavcan and van Schaik, 1997b). Additionally, using data from Plavcan and van Schaik (1997b), there is no correlation between the OSR and mass dimorphism using the phylogenetic contrast method, although there is a significant correlation using species values (Table 9.2). This suggests that a larger sample is necessary to test whether the results of Mitani *et al.* represent a sampling artifact. Unfortunately, reliable data for calculating the OSR are not available for most primates. This limits the utility of the OSR for investigating the joint relation between sexual selection and other factors in a broad comparative analysis.

Assuming that the findings of Mitani *et al.* (1996) are correct, they corroborate the sexual selection hypothesis, and imply that at least some of the variation not accounted for by mating system and competition levels

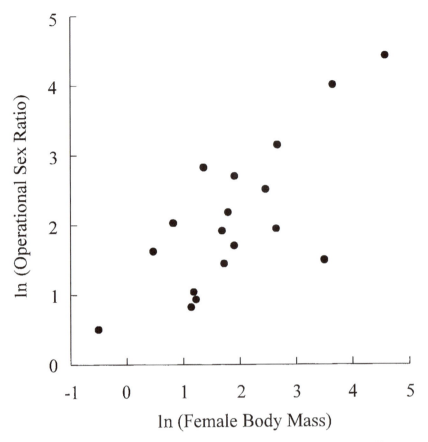

Fig. 9.4   Plot of the operational sex ratio and female body mass in haplorhine primates. Operational sex ratios were taken from Mitani *et al.* (1996). Female body weight is from Appendix 9.1.

reflects the imprecision of the categorical estimates of male–male competition.

Dimorphism increases with size in primates, but the underlying cause of this correlation is widely debated (Clutton-Brock *et al.*, 1977; Leutenegger and Cheverud, 1982, 1985; Gaulin and Sailer, 1984; Cheverud *et al.*, 1985; Pickford, 1986; Plavcan and Kay, 1988; Kappeler, 1990, 1992; Godfrey *et al.*, 1993; Plavcan and van Schaik, 1992, 1997b). Leutenegger and Cheverud (1982, 1985) offer a quantitative genetic model that explains the correlation as a result of lower male heritability coupled with an evolutionary increase in overall size. Kappeler (1990, 1991) suggests that larger size is not as important in winning fights in small species as speed

Table 9.3. *Correlations between the operational sex ratio and canine dimorphism in haplorhines*

Analysis	n	r	p
Phylogenetic contrasts			
Mandibular			
Mesiodistal	14	0.541	0.037
Bucculingual	14	0.361	0.186
Crown height	14	0.539	0.038
Maxillary			
Mesiodistal	14	0.375	0.168
Bucculingual	14	0.205	0.463
Crown height	14	0.475	0.074
Species values			
Mandibular			
Mesiodistal	16	0.449	0.081
Bucculingual	16	0.313	0.238
Crown height	16	0.275	0.303
Maxillary			
Mesiodistal	16	0.168	0.535
Bucculingual	16	0.186	0.490
Crown height	16	0.447	0.083

Phylogenetic contrasts were calculated using the phylogeny of Mitani *et al.* (1996) following the procedure detailed in Table 9.2. Sample size is 14 because the *Alouatta* trichotomy was not resolved, and because canine data were not available for *Cebus olivaceus*. All data were ln-transformed.

and agility. Plavcan and van Schaik (1992) suggest that the correlation reflects an interaction between size and male–male competition, but specify no mechanism. Most recently, Mitani *et al.* (1996) demonstrated a significant correlation between body size and the OSR (Fig. 9.4), probably reflecting the correlation between increasing interbirth intervals and size. This implies that the correlation between dimorphism and size may not be a direct allometric phenomenon, but rather an indirect result of the relation between size and mate competition.

The relation between canine dimorphism and the OSR is not clear (Table 9.3). The OSR is modestly significantly correlated with mandibular canine mesiodistal and crown height dimorphism using the phylogenetic contrast method, but not with maxillary canine crown height dimorphism. The values of the correlations change depending on whether or not the phylogenetic contrasts are allometrically adjusted. Using species values, there is no correlation between any measure of canine dimorphism and the OSR. Maxillary canine crown height is thought to be the target of selection for weaponry (Greenfield, 1992a, 1992b; Plavcan *et al.*, 1995) and bears a stronger relation to competition levels than any other measure of dimorphism (Plavcan and van Schaik, 1997a).

The sample size of the canine dimorphism phylogenetic contrasts is only 14, and these results may reflect sampling error. Assuming that these results are meaningful, they suggest that the relation between mass dimorphism and the OSR is spurious, that the OSR and competition levels reflect different selective pressures, or that canine and size dimorphisms arise from different mechanisms. The latter two are more likely hypotheses. As discussed above, intensity and frequency effects reflect selection for different functions – fighting and display. The OSR is intended to reflect a continuum of the intensity of male mate competition. Furthermore, unlike canine size, body size is a critical trait in an animal's adaptation to its environment, and therefore probably reflects a larger spectrum of influences. Understanding why OSR and competition levels produce different results for dimorphism of different characters is likely greatly to enhance our understanding of the evolution of dimorphism.

Sexual dimorphism as a function of female trait variation

Recent studies emphasise that dimorphism arises from different selection pressures on both male and female traits (Harvey, Kavanagh and Clutton-Brock, 1978a; Greenfield, 1992a, 1992b; Leigh, 1992; Martin *et al.*, 1994; Leigh and Shea, 1995; Plavcan *et al.*, 1995). Consequently, sexual dimorphism should not be viewed as a singular product of sexual selection, but rather as a compromise between selective pressures acting separately on male and female traits.

Recent studies have begun to unravel how dimorphism varies as a function of male and female trait variation. Leigh (1992) demonstrates that variation in female developmental pathways can alter the magnitude of adult size dimorphism. Plavcan *et al.* (1995) demonstrate that selection for the development of canine teeth as weapons operates in females as well as in males.

Shea (1986) provided detailed predictions about how male and female growth patterns can vary to produce sexual size dimorphism. In short, both males and females can change the timing of growth cessation, the rate of growth, or both factors. Males can become larger than females either by growing faster or by delaying maturation and thus growing for a longer period of time. Conversely, females can become smaller than males either by ceasing growth earlier or by slowing the rate of development. Leigh (1992) verified such variation in the ontogenetic basis of mass dimorphism.

Interestingly, Leigh's data suggest that the magnitude of adult dimorphism is not constrained by ontogeny. Furthermore, even species within a single genus show substantial variation in growth patterns, suggesting considerable plasticity in the mechanisms leading to size dimorphism.

Following from the work of Jarman (1983) and Janson and van Schaik (1993), Leigh (1995) suggests that juvenile life history parameters and social system strongly influence male growth trajectories. To summarise, in single-male 'harem'-style mating systems, males usually leave groups in adolescence, leading to rapid increase in the risks that the juveniles face as they mature. These risks are directly correlated with size: the larger the individual, the fewer risks it faces during development. Selection therefore favours a rapid size increase to minimise the risks encountered as a subadult. Conversely, in those species for which risks do not change markedly during development (exemplified by multi-male, multi-female troops), males can increase body size simply by delaying maturation and thus prolonging growth. This period of extended development can also allow males to work their way slowly into new groups and, through social interactions and experience, work their way up the dominance hierarchy (following Jarman, 1983).

Leigh and Shea (1995) present evidence that variation in female ontogeny alters the magnitude of adult size dimorphism in hominoids. Specifically, they demonstrate early cessation of female growth in *Gorilla gorilla* and *Pan paniscus* relative to *Pan troglodytes*. They suggest that variation in female growth reflects variation in foraging adaptations. Species such as gorillas and bonobos that rely more heavily on foliage face lower risks and less intrasexual competition for resources. Consequently, females can mature earlier at smaller sizes in order to increase reproductive output without facing any disadvantage in resource competition (see also Martin *et al.*, 1994).

Early studies of female canine size in haplorhines came to conflicting conclusions about the relation between mating system, competition and female canine size (Harvey *et al.*, 1978a; Smith, 1981; Lucas *et al.*, 1986). Greenfield (1992a) suggested that large female canines offer no selective advantage, and that female canine size represents the conflict between correlated response to male canine size and selection to incorporate the canines into an incisal functional field.

Large female canines also appear to be selected as weapons, affecting the magnitude of canine dimorphism. Plavcan *et al.* (1995) classify separately male and female competition frequency and intensity. They find that in both sexes, males and females classified as showing 'high-intensity' competition posses relatively larger canines than those classified as showing 'low-intensity' competition. For males, high-frequency species have relatively larger canines than low-frequency species, as predicted, but for females the effect is reversed. This 'reversed' frequency effect could reflect greater costs of winning and loosing fights for access to restricted resources where single females defend territories.

There is some evidence that when agonistic competition occurs regularly between coalitions of individuals, selection for the development of weaponry is relatively less intense than when individuals fight without partners (Plavcan *et al.*, 1995). Females of many cercopithecine monkeys are female bonded, forming matrilineal groups (Wrangham, 1980). Females maintain their position within an agonistic dominance hierarchy with the assistance of kin, with the outcome of contests frequently determined by the enlistment of partners in coalitions (Gouzoules and Gouzoules, 1987; Walters and Seyfarth, 1987). A similar pattern is seen in male baboons (Noë, 1990). Lower ranking males are known to form temporary alliances in order to defeat a higher ranking male. Inevitably, when the higher ranking male is defeated, the coalition breaks down. Herein is the primary difference between males and females. Whereas male coalitions are effective in overcoming the individual fighting advantage that a particular male might have over others, the alliances are temporary. Ultimately, a male gains the alpha position through individual fighting skills. Females, on the other hand, participate in alliances throughout their lives.

The males of those cercopithecines characterised by female coalitionary behaviour do not posses relatively larger canines than other cercopithecines, even though canine dimorphism is greater in these species. This suggests that the extreme canine dimorphism of many cercopithecines partly reflects a reduction in female canine size, rather than an extreme hypertrophy of the male canines. As such, patterns of female social behaviour can directly affect dimorphism.

While selection apparently favours the development of large female canines for fighting, most female haplorhines do not have canines as large as those of males, probably reflecting the different objectives of male and female fighting (Plavcan *et al.*, 1995). Generally, males fight for access to mates, and females fight for access to resources (Trivers, 1972; Wrangham, 1980). The consequences of winning or losing fights are therefore probably greater for males than for females, resulting in stronger selection for the development of hypertrophied canines in males. The female intensity classifications roughly parallel the distinction between scramble and contest competition. In this sense, the development of female canine size can at least partly be attributed to variation in resource distribution and exploitation (Plavcan and van Schaik, 1994; Plavcan *et al.*, 1995). The end

Table 9.4. *Correlations between male and female relative maxillary canine crown height in strepsirhines, and haplorhines in which females are classified as showing low-intensity and high-intensity female–female competition*

Comparison	n	r	p
Strepsirhines	31	0.927	< 0.001
Haplorhines			
Low-intensity	21	0.804	< 0.001
High-intensity	27	0.150	0.445

Relative canine crown height was calculated as the least-squares residual from an isometric line passed through a comparison between maxillary canine crown height (dependent variable) and either male or female body mass (independent variable). All data were ln-transformed.

result is that at least some canine dimorphism persists even when both males and females show agonistic competition.

Strepsirhines show a variety of mating systems, but are characterised by little sexual dimorphism (Kappeler, 1990, 1991, 1993). The fundamental phylogenetic difference in dimorphism between haplorhines and strepsirhines has been very difficult to explain. An interesting mechanism that may explain the difference in canine dimorphism is correlated response. Correlated response is simply a mechanism whereby character changes in one sex are accompanied by similar changes in the other sex because of a common genetic control for the trait (Lande, 1980). Greenfield (1992a) suggests that most variation in female canine size reflects correlated response. However, considering jointly the competition classifications and the correlation between relative male and female canine tooth size, it appears that haplorhines and strepsirhines show very different patterns (Plavcan, in press). Strepsirhines show a strong correlation between relative male and female canine tooth size, regardless of a classification of female competition. Among haplorhines, male and female relative canine size is strongly correlated in species in which females do not fight. But in species in which females do fight, there is no correlation between relative male and female canine tooth size (Table 9.4).

These results suggest that canine monomorphism in strepsirhines is the product of correlated response. That is, where males develop large canines, females do too. Conversely, the simple presence of dimorphism in haplorhines demonstrates that the genetic control of canine size differs between the sexes. Selection operates to develop or maintain large female canines in a number of species, generating considerable variation in relative female canine size and in canine sexual dimorphism. However, where there is no selection to develop or maintain large female canines, much interspecific variation in female canine size can be explained as a correlated response to male canine size.

If correlated response operates in strepsirhines, then the lack of canine dimorphism does not necessarily contradict the sexual selection hypothesis. Rather, it reflects a constraint on the independent expression of male and female traits. Notably, strepsirhines also lack size dimorphism. This may also be explained by correlated response, but it is much more difficult to demonstrate the effect for mass dimorphism.

Van Schaik and Kappeler (1994) suggest that oestrous synchrony in many strepsirhines may result in multi-male/multi-female groups that are characterised by monogamous mating. In other words, the OSR would be very low. If so, this may explain the lack of dimorphism in a number of polygynous strepsirhines (Kappeler, 1993; van Schaik and Kappeler, 1994). Notably, though, males and females of many of these strepsirhines possess relatively large canines, suggesting that selection for the development of weaponry may still operate (Plavcan et al., 1995).

Conclusion

Dimorphism in primates is a complex phenomenon. This chapter cannot possibly present a fair or balanced discussion of all the factors that generate variation in dimorphism in primates. For example, half of Darwin's (1871) model – female choice – has not been discussed here, even though there is mounting evidence for the important role that mate choice plays in the evolution of primate dimorphism (Boinski, 1987; Richard, 1992). Similarly, phylogeny clearly plays an important role in understanding variation in dimorphism, and the complex interaction between phylogeny in behaviour and dimorphism alone could easily fill another chapter.

This chapter only attempts a review of the relationship between mating systems, intrasexual competition, and sexual selection in primates. As we come to understand better the relationship between these factors, comparative analyses will be able more precisely to tease apart the interactions of other selective and non-selective factors that influence dimorphism. For example, the analysis of the joint roles of correlated response and intrasexual competition would be difficult, if not impossible, without first deriving a model for evaluating the relationship between canine size and competition in both sexes. Similarly, quantification of the OSR not only provides evidence that sexual selection is important in the evolution of dimorphism, but also suggests a mechanism to explain the relationship between dimorphism and body size. Thus, the results summarised here should not be viewed as an endorsement that sexual selection is the *only* factor generating dimorphism in primates. However, until we understand the relationship between behaviour and selection, it is difficult to evaluate competing hypotheses about the evolution of dimorphism. Future analyses of dimorphism should focus on better understanding the relationship between behaviour, life history traits and sexual selection, and on carefully setting up testable, predictive models for the joint relationship between factors hypothesised to influence dimorphism.

Acknowledgements

The author thanks Phyllis Lee for kindly inviting him to contribute to this volume, and Carel van Schaik, Peter Kappeler, Steve Leigh, Jay Kelley and Karen Strier for many helpful discussions.

References

Andersson, M. (1994). *Sexual Selection*. Princeton: Princeton University Press.
Boinski, S. (1987) Mating patterns in squirrel monkeys (*Saimiri oerstedi*). *Behavioral Ecology and Sociobiology* **21**, 13–21.
Cheverud, J.M., Dow, M.M. and Leutenegger, W. (1985). The quantitative assessment of phylogenetic constraints in comparative analysis: sexual dimorphism in body weight among primates. *Evolution* **38**, 1335–51.
Clutton-Brock, T.H. (1985). Size, sexual dimorphism, and polygyny in primates. In *Size and Scaling in Primate Biology*, ed. W.L. Jungers, pp. 51–60. New York: Plenum Press.
Clutton-Brock, T.M., Harvey, P.H. and Rudder, B. (1977). Sexual dimorphism, socionomic sex ratio and body weight in primates. *Nature* **269**, 191–5.
Coelho A.M. (1974). Socio-bioenergetics and sexual dimorphism in primates. *Primates* **15**, 262–9.
Darwin, C. (1871). *The Descent of Man, and Selection in Relation to Sex*. London: J. Murray.
Ely, J. and Kurland, J.A. (1989). Spatial autocorrelation, phylogenetic constraints, and the causes of sexual dimorphism in primates. *International Journal of Primatology* **10**, 151–71.
Emlen S. and Oring, L. (1977). Ecology, sexual selection and the evolution of mating systems. *Science* **197**, 215–23.
Felsenstein, J. (1985). Phylogenies and the comparative method. *American Naturalist* **125**, 1–15.
Ford, S.M. (1994). Evolution of sexual dimorphism in body weight in platyrrhines. *American Journal of Primatology* **34**, 221–4.
Gaulin, S.J.C. and Sailer, L.D. (1984). Sexual dimorphism in weight among

primates: the relative impact of allometry and sexual selection. *International Journal of Primatology* **5**, 515–35.

Gautier-Hion, A. (1975). Dimorphisme sexuel et organisation sociale chez les cercopithécinés forestiers africains. *Mammalia* **39**, 365–74.

Godfrey, L.R., Lyon, S.K. and Sutherland, M.R. (1993) Sexual dimorphism in large bodied primates: the case of the subfossil lemurs. *American Journal of Physical Anthropology* **90**. 315–34.

Gouzoules, S. and Gouzoules, H. (1987). Kinship. In *Primate Societies*, ed. B.B. Smuts, D.L. Cheney, R.M. Seyfarth, R.W. Wrangham and T.T. Struhsaker, pp. 299–305. Chicago: University of Chicago Press.

Greenfield, L.O. (1992a). Origin of the human canine: a new solution to an old enigma. *Yearbook of Physical Anthropology* **35**, 153–85.

Greenfield, L.O. (1992b). Relative canine size, behavior, and diet in male ceboids. *Journal of Human Evolution* **23**, 469–80.

Harvey, P.H., Kavanagh, M. and Clutton-Brock, T.H. (1978a). Canine tooth size in female primates. *Nature* **276**, 817–18.

Harvey, P.H., Kavanagh, M. and Clutton-Brock, T.H. (1978b). Sexual dimorphism in primate teeth. *Journal of Zoology, London* **186**, 474–85.

Janson, C.H. and van Schaik, C.P. (1993). Ecological risk aversion in juvenile primates: slow and steady wins the race. In *Juvenile Primates*, ed. M.E. Pereira and L.A. Fairbanks, pp. 57–74. New York: Oxford University Press.

Jarman, P.J. (1983). Mating system and sexual dimorphism in large terrestrial mammalian herbivores. *Biology Reviews* **58**, 485–520.

Kappeler, P.M. (1990). The evolution of sexual size dimorphism in prosimian primates. *American Journal of Primatology* **21**, 201–14.

Kappeler, P.M. (1991). Patterns of sexual dimorphism in body weight among prosimian primates. *Folia Primatologica* **57**, 132–46.

Kappeler, P.M. (1992). Female Dominance in Malagasy Primates. PhD dissertation, Duke University, Durham, North Carolina.

Kappeler, P.M. (1993). Sexual selection and lemur social systems. In *Lemur Social Systems and Their Ecological Basis*, ed. P.M. Kappeler and J.U. Ganzhorn, pp. 223–40. New York: Plenum Press.

Kay, R.F., Plavcan, J.M., Glander, K.E. and Wright, P.C. (1988). Sexual selection and canine dimorphism in New World monkeys. *American Journal of Physical Anthropology* **77**, 385–97.

Lande, R. (1980). Sexual dimorphism, sexual selection and adaptation in polygenic characteristics. *Evolution* **34**, 292–307.

Leigh, S.R. (1992). Patterns of variation in the ontogeny of primate body size dimorphism. *Journal of Human Evolution* **23**, 27–50.

Leigh, S.R. (1995). Socioecology and the ontogeny of sexual size dimorphism in anthropoid primates. *American Journal of Physical Anthropology* **97**, 339–56.

Leigh, S.R. and Shea, B.T. (1995). Ontogeny and the evolution of adult body size dimorphism in apes. *American Journal of Primatology* **36**, 37–60.

Leutenegger, W. and Cheverud, J.M. (1982). Correlates of sexual dimorphism in primates: ecological and size variables. *International Journal of Primatology* **3**, 387–402.

Leutenegger, W. and Cheverud, J.M. (1985). Sexual dimorphism in primates: the effects of size. In *Size and Scaling in Primate Biology*, ed. W.L. Jungers, pp.

33–50. New York: Plenum Press.

Lucas, P.W., Corlett, R.T. and Luke, D.A. (1986). Sexual dimorphism in tooth size in anthropoids. *Human Evolution* **1**, 23–39.

Martin, R.D., Wilner, L.A. and Dettler, A. (1994). The evolution of sexual size dimorphism in primates. In *The Differences Between the Sexes*, ed. R.V. Short and E. Balaban, pp. 159–200. Cambridge: Cambridge University Press.

Mitani, J.C. (1985). Territoriality and monogamy among agile gibbons (*Hylobates agilis*). *Behavioral Ecology and Sociobiology* **20**, 265–9.

Mitani, J., Gros-Louis, J. and Richards, A.F. (1996). Sexual dimorphism, the operational sex ratio, and the intensity of male competition in polygynous primates. *American Naturalist* **147**, 966–80.

Morin, P.A. (1993). Reproductive strategies in chimpanzees. *Yearbook of Physical Anthropology* **36**, 179–212.

Noë, R. (1990). A veto game played by baboons: a challenge to use of the prisoner's dilemma as a paradigm for reciprocity and cooperation. *Animal Behaviour* **39**, 78–90.

Pagel, M.D. (1992). A method for the analysis of comparative data. *Journal of Theoretical Biology* **156**, 431–42.

Phillips-Conroy, J.E. and Jolly, C.J. (1981). Sexual dimorphism in two subspecies of Ethiopian baboons (*Papio hamadryas*) and their hybrids. *American Journal of Physical Anthropology* **56**, 115–29.

Pickford, M. (1986). On the origins of body size dimorphism in primates. In *Sexual Dimorphism in Living and Fossil Primates*, ed. M. Pickford and B. Chiarelli, pp. 77–91. Florence: Il Sedicesimo.

Plavcan, J.M. (1990). Sexual Dimorphism in the Dentition of Extant Anthropoid Primates. PhD dissertation, Duke University, Durham, North Carolina.

Plavcan, J.M. (in press). The role of correlated response in the evolution of female canine size in anthropoid primates. *American Journal of Physical Anthropology*.

Plavcan, J.M. and Kay, R.F. (1988). Sexual dimorphism and dental variability in Platyrrhine primates. *International Journal of Primatology* **9**, 169–78.

Plavcan, J.M. and van Schaik, C.P. (1992) Intrasexual competition and canine dimorphism in anthropoid primates. *American Journal of Physical Anthropology* **87**, 461–77.

Plavcan, J.M. and van Schaik, C.P. (1994). Canine dimorphism. *Evolutionary Anthropology* **2**, 208–14.

Plavcan, J.M. and van Schaik, C.P. (1997a). Interpreting hominid behavior on the basis of sexual dimorphism. *Journal of Human Evolution* **32**, 345–74.

Plavcan, J.M. and van Schaik, C.P. (1997b). Intrasexual competition and body weight dimorphism in anthropoid primates. *American Journal of Physical Anthropology* **103**, 37–68.

Plavcan, J.M., van Schaik, C.P. and Kappeler, P.M. (1995). Competition, coalitions, and canine size in primates. *Journal of Human Evolution* **28**, 245–76.

Richard, A.F. (1992). Aggressive competition between males, female-controlled polygyny and sexual monomorphism in a Malagasy primate, *Propithecus verreauxi*. *Journal of Human Evolution* **22**, 395–406.

Robinson, J.G., Wright, P.C., and Kinzey, W.G. (1987). Monogamous cebids and their relatives: intergroup calls and spacing. In *Primate Societies*, ed. B.B.

Smuts, D.L. Cheney, R.M. Seyfarth, R.W. Wrangham and T.T. Struhsaker, pp. 44–53. Chicago: University of Chicago Press.

Shea, B.T. (1986). Ontogenetic approaches to sexual dimorphism in anthropoids. *Human Evolution* **1**, 97–110.

Small, M.F. (1989). Female choice in nonhuman primates. *Yearbook of Physical Anthropology* **32**, 103–27.

Smith, R.J. (1981). Interspecific scaling of maxillary canine size and shape in female primates: Relationships to social structure and diet. *Journal of Human Evolution* **10**, 165–73.

Smith, R.J. and Jungers, W.L. (1997). Body mass in comparative primatology. *Journal of Human Evolution* **32**, 523–59.

Strier, K.B. (1986). The Behavior and Ecology of the Woolly Spider Monkey, or Miriqui, *Brachyteles arachnoides*, E. Geoffroy 1806. PhD dissertation, Harvard University, Cambridge, Massachusetts.

Strier, K.B. (1992). Ateline adaptations: behavioral strategies and ecological constraints. *American Journal of Physical Anthropology* **88**, 515–24.

Trivers, R.L. (1972). Parental investment and sexual selection. In *Sexual Selection and the Descent of Man 1871–1971*, ed. B. Campbell, pp. 136–79. Chicago: Aldine Publishing Company.

van Schaik, C.P. and Kappeler, P.M. (1994). Life history, activity period and lemur social systems. In *Lemur Social Systems and their Ecological Basis*, ed. P.M. Kappeler and J.U. Ganzhorn, pp. 241–60. New York: Plenum Press.

Walters, J.R. and Seyfarth, R.M. (1987). Conflict and cooperation. In *Primate Societies*, ed. B.B. Smuts, D.L. Cheney, R.M. Seyfarth, R.W. Wrangham and T.T. Struhsaker, pp. 306–17. Chicago: University of Chicago Press.

Wrangham, R.W. (1980). An ecological model for female-bonded groups. *Behaviour* **75**, 26–300.

Appendix 9.1

The following table shows absolute dimensions (in mm) for haplorhine primate canines. Three dimensions were taken for each mandibular and maxillary canine tooth: the mesiodistal length (MD) is the longest dimension at the base of the tooth; the buccolingual breadth (BL) is the widest dimension perpendicular to the mesiodistal length at the base of the tooth; and the crown height (Hgt) is measured from the apex to the cementum–enamel junction on the mesiobuccal face of the tooth. Detailed descriptions of the measurements, as well as sample sizes and descriptions of the sample populations, are presented in Plavcan (1990). Male (M) and female (F) sample sizes are the *minimum* for each canine dimension.

Species	n(F/M)	Mandibular						Maxillary					
		MD		Hgt		BL		MD		Hgt		BL	
		F	M	F	M	F	M	F	M	F	M	F	M
Catarrhines													
Cercocebus agilis agilis	6/9	6.65	8.40	8.17	14.05	3.84	4.80	6.29	9.50	8.65	16.97	4.94	6.55
Cercocebus albigena atterimus	16/19	6.18	7.76	8.46	14.07	3.69	4.65	5.92	8.25	8.82	16.73	4.71	5.69
Cercocebus torquatus atys	19/17	6.07	8.78	8.56	17.14	3.84	5.33	6.46	10.35	9.53	23.53	4.99	7.04
Cercopithecus aethiops hilgerti	16/16	5.07	6.52	8.01	13.12	3.36	4.04	5.17	7.37	9.76	17.66	3.96	5.14
Cercopithecus ascanius whitesidei	20/23	4.53	5.59	7.79	11.37	3.05	3.62	5.23	6.80	10.30	16.13	3.90	4.51
Cercopithecus cephus cephus	15/24	4.51	5.88	7.78	11.92	3.01	3.72	4.85	6.89	9.21	16.32	3.78	4.62
Cercopithecus diana diana	17/8	5.56	7.20	8.66	12.55	3.71	4.50	6.19	8.45	12.33	19.94	4.34	5.43
Cercopithecus lhoesti lhoesti	11/15	5.18	7.43	8.37	14.69	3.23	4.59	5.60	8.39	10.82	19.93	4.16	5.76
Cercopithecus mitis kolbi	8/12	4.80	6.77	8.42	13.55	3.31	4.14	5.22	8.29	9.77	19.24	4.33	5.62
Cercopithecus mona	9/17	4.68	6.54	6.83	12.66	2.91	4.12	5.34	7.82	9.32	17.76	3.79	5.03
Cercopithecus neglectus	15/22	5.19	7.14	9.00	13.67	3.38	4.59	5.68	8.56	11.64	20.02	4.04	5.72
Cercopithecus nictitans nictitans	21/24	5.11	6.99	8.26	12.96	3.43	4.58	5.54	8.50	11.22	18.10	4.19	5.60
Cercopithecus pogonias grayi	18/19	4.76	5.79	7.36	10.75	3.04	3.85	5.27	7.01	10.10	15.73	4.00	4.65
Cercopithecus preussi preussi	7/6	4.76	6.80	8.32	14.21	3.16	4.48	5.25	8.25	10.66	18.91	3.93	5.00
Cercopithecus wolfi wolfi	10/11	4.29	6.14	6.67	12.61	2.79	4.11	4.90	7.41	9.59	18.06	3.71	4.95
Colobus angolensis cottoni	9/20	6.47	8.23	8.58	15.28	4.34	5.49	7.05	9.30	9.57	17.66	5.03	6.66
Colobus badius badius	25/24	6.20	8.00	7.57	13.20	4.18	5.15	6.78	8.30	9.32	17.60	5.11	6.28
Colobus guereza caudatus	23/18	7.37	8.54	11.42	15.84	5.00	5.59	8.19	9.84	13.85	20.31	6.21	6.71
Colobus polykomos polykomos	25/23	7.34	7.92	9.33	15.46	5.13	5.83	8.05	9.20	10.81	19.13	6.18	6.68
Colobus satanas	23/20	6.53	7.32	7.48	12.93	4.89	5.32	6.81	7.86	9.16	16.08	6.23	6.20
Colobus verus	20/21	5.11	6.99	6.46	11.75	3.65	4.46	5.25	7.19	7.43	14.90	4.62	5.58
Erythrocebus patas	8/27	6.75	9.77	10.39	17.98	3.75	5.40	6.88	10.58	12.17	26.50	4.49	6.72
Gorilla gorilla gorilla	18/18	13.54	19.13	16.44	27.06	10.70	14.83	15.09	23.67	17.40	30.26	11.30	16.80
Homo sapiens	10/10	7.92	8.30	10.01	11.04	6.93	7.43	8.24	9.29	9.97	10.85	7.82	8.40
Hylobates concolor	14/13	6.88	7.12	10.50	12.74	5.30	5.71	7.69	8.14	16.49	19.11	5.14	5.31
Hylobates hoolock	17/34	7.30	7.66	10.32	11.86	5.11	5.59	7.69	8.42	17.09	17.75	5.61	5.92
Hylobates klossi	7/13	6.32	6.46	9.77	9.64	4.70	4.75	7.64	7.71	16.63	15.72	5.43	5.41
Hylobates lar carpenteri	18/20	6.52	7.03	9.65	11.13	4.75	5.36	6.98	7.92	15.79	18.32	4.77	5.35
Hylobates syndactylus syndactylus	24/27	7.94	8.59	11.66	13.39	5.89	6.67	8.45	9.72	17.61	20.94	6.06	6.68
Macaca fascicularis fascicularis	16/11	6.30	8.79	10.10	16.25	3.59	5.26	6.21	9.73	10.67	24.09	4.53	6.43
Macaca fuscata fuscata	25/11	6.10	8.88	8.77	14.26	4.09	5.56	6.58	10.20	9.59	19.57	5.25	6.46
Macaca hecki	13/11	6.66	9.68	10.21	15.68	4.37	5.85	7.03	11.00	11.85	24.36	5.23	7.82
Macaca mulatta mulatta	25/20	5.70	8.34	7.77	13.21	3.64	5.15	6.10	8.67	8.13	16.97	4.65	6.42
Macaca nemestrina nemestrina	13/18	7.34	11.10	10.92	21.29	4.81	6.83	7.42	12.29	12.24	28.89	5.77	7.80

Species													
Macaca nigra	14/18	6.33	9.73	10.01	17.41	4.23	6.01	6.79	11.55	11.38	29.73	5.24	7.95
Macaca sinica	15/25	6.00	7.77	9.26	13.84	3.49	4.86	5.79	9.05	9.96	19.69	4.90	6.38
Macaca tonkeana	10/11	6.91	9.52	10.70	15.32	4.67	6.23	7.77	11.07	12.72	24.66	5.76	7.63
Mandrillus leucophaeus leucophaeus	17/22	7.58	13.17	10.06	28.73	5.07	8.96	8.60	17.89	11.70	47.97	6.24	11.02
Miopithecus talapoin talapoin	12/25	3.45	4.23	5.05	7.81	2.22	2.85	3.71	4.83	6.70	11.35	3.01	3.56
Nasalis larvatus	8/7	6.23	7.53	8.39	15.87	3.93	5.75	6.90	9.23	10.42	22.96	5.19	6.75
Papio anubis	3/6	8.91	13.86	14.66	26.60	5.41	8.44	8.83	16.10	15.95	35.43	6.96	11.57
Papio cynocephalus kindae	12/9	6.10	8.75	7.00	17.14	3.85	5.60	6.46	9.72	9.13	25.57	4.87	7.16
Papio hamadryas	5/20	8.20	13.25	11.12	24.05	5.00	8.78	7.85	14.14	11.17	30.64	6.35	10.66
Papio ursinus	3/4	8.58	13.93	11.83	33.56	5.21	8.77	8.71	17.00	12.12	46.53	6.12	10.72
Pan paniscus	12/12	8.85	10.37	10.92	14.38	6.48	7.70	9.37	11.44	11.24	15.57	6.89	8.88
Pan troglodytes troglodytes	25/14	11.63	14.06	15.11	19.17	9.22	11.52	11.94	15.30	15.26	21.72	9.48	11.81
Pongo pygmaeus pygmaeus	19/20	12.18	16.12	17.19	26.78	8.95	13.20	13.33	17.50	15.95	27.00	10.41	14.26
Presbytis comata	7/6	5.68	6.19	7.30	10.98	4.05	4.32	5.94	6.75	9.23	13.93	5.00	5.00
Presbytis cristata pyrrhus	20/20	6.29	6.95	8.80	13.55	4.38	4.92	6.77	8.12	10.94	18.57	5.80	6.39
Presbytis entellus thersites	6/8	6.38	8.05	9.20	15.85	4.81	5.75	7.29	9.58	10.82	19.08	5.96	6.87
Presbytis melalopohos chrysomelas	15/16	5.42	5.64	6.77	9.93	3.80	3.99	5.82	6.40	8.15	13.95	4.93	4.51
Presbytis obscura obscura	25/14	5.69	6.45	7.43	11.59	3.67	4.66	6.14	7.62	8.51	15.10	5.36	5.75
Presbytis pileata shortridgei	6/7	6.89	7.78	10.12	14.57	4.62	5.41	6.91	9.09	10.52	19.85	5.85	6.66
Presbytis potenziana	9/13	6.08	6.03	7.53	10.71	4.02	4.44	6.71	7.25	9.56	13.56	5.63	5.33
Presbytis rubicunda chrysomelas	18/15	5.37	5.46	6.84	9.41	3.75	4.07	5.69	5.89	8.48	12.73	4.46	4.20
Presbytis vetulus	22/22	5.50	6.63	7.90	12.80	3.88	6.99	6.11	7.82	9.34	16.96	5.04	6.04
Pygathrix nemaeus nigripes	7/13	5.79	6.63	7.29	12.57	4.02	5.32	5.82	8.04	8.54	16.59	4.98	6.27
Simias concolor	9/7	5.47	6.50	6.96	11.41	3.85	5.07	5.54	7.63	8.07	13.91	4.77	5.57
Theropithecus gelada	6/22	7.25	11.75	11.33	24.56	4.69	7.61	7.88	14.89	12.27	39.62	5.79	9.32
Platyrrhines													
Alouatta belzebul	9/10	4.99	6.76	6.13	10.91	3.90	5.71	5.29	8.03	6.78	12.28	4.42	6.06
Alouatta caraya	9/10	5.90	7.74	8.03	12.00	4.20	5.72	6.22	8.58	9.56	14.24	4.32	5.60
Alouatta fuscata	10/7	5.09	6.95	7.40	12.09	4.09	6.07	5.78	8.13	8.73	13.18	4.26	5.83
Alouatta palliata aequatorialis	20/13	5.36	6.88	8.04	11.74	4.12	5.45	6.46	8.16	9.14	13.79	4.60	5.81
Alouatta pigra	9/6	5.01	6.27	7.68	11.42	4.16	5.17	6.73	8.48	10.20	13.94	4.61	5.92
Alouatta seniculus seniculus	18/20	5.61	6.74	7.88	10.65	4.69	6.31	6.64	8.64	8.96	13.42	4.80	6.65
Aotus trivirgatus lemurinus	12/13	2.82	2.87	4.07	4.49	2.08	2.22	3.13	3.19	4.80	5.62	2.71	2.69
Ateles geoffroyi vellerosus	20/17	5.30	6.04	7.33	10.23	4.40	4.77	5.71	6.38	7.50	11.43	5.14	5.17
Ateles paniscus chamek	14/10	5.66	6.06	7.91	10.80	4.32	4.99	5.94	6.60	7.73	12.12	4.95	5.29
Cacajao calvus	7/11	7.14	8.22	11.70	14.05	5.52	6.74	7.08	8.15	10.82	13.58	5.85	7.26
Callicebus moloch discolor	6/12	2.88	2.96	4.18	4.51	2.08	2.04	2.89	2.84	4.04	4.36	2.75	2.75
Callicebus torquatus lugens	10/10	3.30	3.26	4.51	4.61	2.32	2.36	3.46	3.50	4.10	4.28	3.09	3.18
Cebuella pygmaea pygmaea	22/13	1.86	1.89	3.19	3.16	0.91	0.90	1.63	1.64	2.99	3.06	1.35	1.34
Cebus apella libidinosis	21/11	6.33	7.78	9.72	13.82	4.82	6.23	6.32	7.49	9.70	13.67	5.48	6.45

Species	n(F/M)	Mandibular						Maxillary					
		MD		Hgt		BL		MD		Hgt		BL	
		F	M	F	M	F	M	F	M	F	M	F	M
Cebus olivaceus	26/27	6.23	7.11	9.67	14.22	5.00	6.00	6.10	7.49	9.12	14.35	5.57	6.20
Chiropotes satanas chiropotes	18/13	6.03	6.31	10.24	11.83	5.01	5.34	6.24	6.59	9.35	11.18	5.16	5.70
Lagothrix lagothricha peopigaii	8/13	6.10	7.01	7.38	11.50	4.71	5.86	5.78	7.43	8.05	14.23	5.39	6.54
Pithecia pithecia	9/14	4.00	4.69	6.60	8.30	3.18	3.70	4.26	5.00	6.77	8.93	3.84	4.60
Saguinus fuscicollis nigrifrons	15/19	2.85	2.77	5.31	5.21	2.42	2.36	2.83	2.77	5.47	5.66	2.30	2.31
Saguinus midas niger	25/31	2.68	2.69	5.04	4.96	2.28	2.24	2.84	2.80	5.42	5.38	2.24	2.31
Saguinus oedipus geoffroyi	20/14	2.79	2.82	5.06	4.96	2.17	2.23	2.90	2.90	5.65	5.65	2.33	2.42
Saimiri oerstedii oerstedii	9/21	3.05	3.46	4.07	6.78	2.02	2.85	2.86	3.81	4.50	7.58	2.96	3.30
Saimiri sciureus macrodon	3/6	3.19	3.52	4.40	6.84	2.42	2.95	3.25	3.94	5.02	7.10	2.83	3.26

Appendix 9.2

The following table shows mating systems (Matesys.), competition levels (Comp.) and competition classifications for primates. Mating system classifications are multi-male/multi-female (MM), single-male/multi-female (SM), monogamous (M), solitary (S), or polyandrous (P). Competition intensity (Int) and frequency (Freq) are dichotemised into 'high' and 'low' classes. For coalitionary competition (Coal), each sex is classified as coalitions present (yes) or not (no). Strepsirhines were not classified into competition levels. Complete references are available in Plavcan and van Schaik (1992, 1997b) and Plavcan et al. (1995). Mating system classifications for strepsirhines are taken from Kappeler (1993).

Species	Matesys.	Comp.	Male			Female		
			Coal.	Int.	Freq.	Coal.	Int.	Freq.
Catarrhines								
Cercocebus agilis	MM	4	no	high	high	yes	high	high
Cercocebus albigena	MM	4	no	high	high	yes	high	high
Cercocebus torquatus	MM	4	no	high	high	yes	high	high
Cercopithecus aethiops	MM	4	no	high	high	yes	high	high
Cercopithecus ascanius	SM	3	no	high	low	no	high	high
Cercopithecus cephus	SM	3	no	high	low	—	—	—
Cercopithecus diana	SM	3	no	high	low	—	—	—
Cercopithecus lhoesti	SM	3	no	high	low	—	—	—
Cercopithecus mitis	SM	3	no	high	low	no	high	high
Cercopithecus mona	SM	3	no	high	low	no	high	high
Cercopithecus neglectus	SM	3	no	high	low	—	—	—
Cercopithecus nictitans	SM	3	no	high	low	—	—	—
Cercopithecus pogonias	SM	3	no	high	low	—	—	—
Cercopithecus preussi	SM	3	no	high	low	—	—	—
Cercopithecus wolfi	SM	3	no	high	low	—	—	—
Colobus angolensis	MM	4	no	high	high	no	low	high
Colobus badius	MM	4	yes	high	high	no	low	high
Colobus guereza	SM	3	no	high	low	no	low	high
Colobus polykomos	MM	4	no	high	high	no	low	high
Colobus satanas	MM	4	no	high	high	—	—	—
Colobus verus	MM	4	no	high	high	—	—	—
Erythrocebus patas	SM	3	no	high	low	no	high	high
Gorilla gorilla	SM	3	no	high	low	no	low	high
Hylobates concolor	M	1	no	high	low	no	high	high
Hylobates hoolock	M	1	no	high	low	no	high	low
Hylobates klossi	M	1	no	high	low	no	high	low

Species								
Hylobates lar	M	1	no	high	low	no	high	low
Hylaboates syndactylus	M	1	no	high	low	no	high	low
Macaca fascicularis	MM	4	no	high	high	yes	high	high
Macaca fuscata	MM	4	no	high	high	yes	high	high
Macaca hecki	MM	4	no	high	high	—	—	—
Macaca mulatta	MM	4	no	high	high	yes	high	high
Macaca nemestrina	MM	4	no	high	high	yes	high	high
Macaca nigra	MM	4	no	high	high	—	—	—
Macaca sinica	MM	4	no	high	high	yes	high	high
Macaca tonkeanna	MM	4	no	high	high	—	—	—
Mandrillus leucophaeus	MM	4	no	high	high	—	—	—
Miopithecus talapoin	MM	3	no	high	high	—	—	—
Nasalis larvatus	MM	4	no	high	high	no	low	high
Pan paniscus	MM	2	yes	high	high	no	low	high
Pan troglodytes	MM	2	yes	high	high	yes	high	high
Papio cynocephalus	MM	4	no	high	high	no	high	high
Papio hamadryas	MM	4	no	high	high	—	low	low
Pongo pygmaeus	S	3	no	high	low	no	—	—
Presbytis comata	SM	3	no	high	low	no	low	high
Presbytis cristata	SM	3	no	high	low	no	low	high
Presbytis entellus	MM	4	no	high	high	no	low	high
Presbytis melalophos	SM	3	no	high	low	no	low	high
Presbytis obscura	SM	3	no	high	low	—	low	low
Presbytis pileatus	SM	3	no	high	low	—	—	—
Presbytis potenziani	M	1	no	high	low	—	—	—
Presbytis rubicunda	SM	3	no	high	low	no	—	—
Presbytis vetulus	SM	3	no	high	low	no	low	high
Pygathrix nemaeus	SM	3	no	high	low	no	low	high
Simias concolor	SM	3	no	high	low	no	low	high
Theropithecus gelada	MM	4	no	high	high	yes	high	high

Species	Matesys.	Comp.	Male			Female		
			Coal.	Int.	Freq.	Coal.	Int.	Freq.
Platyrrhines								
Alouatta belzebul	MM	4	no	high	high	—	—	—
Alouatta caraya	MM	4	no	high	high	—	—	—
Alouatta fusca	MM	4	no	high	high	—	—	—
Alouatta palliata	MM	4	no	high	high	no	high	high
Alouatta pigra	MM	4	no	high	high	—	—	—
Alouatta seniculus	MM	4	no	high	high	no	low	high
Aotus trivirgatus	M	1	no	high	low	no	high	low
Ateles geoffroyi	MM	2	yes	high	high	no	low	high
Ateles paniscus	MM	2	yes	high	high	no	low	high
Brachyteles arachnoides	MM	2	yes	high	high	no	low	low
Cacajao calvus	SM	3	no	high	low	—	—	—
Callicebus moloch	M	1	no	low	low	no	low	low
Callicebus torquatus	M	1	no	low	low	no	low	low
Cebuella pygmaea	M	1	no	low	low	no	low	low
Cebus apella	MM	4	no	high	high	yes	high	high
Chiropotes satanas	M	1	no	low	low	—	—	—
Lagothrix lagothricha	MM	2	no	low	high	—	—	—
Pithecia pithecia	MM	2	—	—	—	—	—	—
Saguinus fuscicollis	P	1	no	low	high	no	high	low
Saguinus midas	P	1	no	low	high	no	high	low
Saguinus oedipus	P	1	no	low	high	no	high	high
Saimiri oerstedii	MM	3	no	high	low	no	low	high
Saimiri sciureus	MM	3	no	high	low	yes	high	high

Strepsirhines

Arctocebus calabarensis	S	—	no	low	low	no	low	no	low	low
Avahi laniger	M	—	no	low	low	no	low	no	low	low
Chierogaleus medius	S	—	no	low	low	no	low	no	low	low
Eulemur coronatus	MM	—	no	high	high	no	high	no	high	high
Eulemur fulvus	MM	—	no	high	high	no	high	no	high	high
Eulemur macrodon	MM	—	no	high	high	no	high	no	high	high
Eulemur mongoz	MM	—	no	low	low	no	low	no	low	low
Eulemur rubriventor	MM	—	no	low	low	no	low	no	low	low
Galago alleni	S	—	no	low	low	no	low	no	low	low
Galago demidovii	S	—	no	low	low	no	low	no	low	low
Galago elegantulus	S	—	no	low	low	no	low	no	low	low
Galago moholi	S	—	no	low	low	no	low	no	low	low
Galago senegalensis	S	—	no	low	low	no	low	no	low	low
Galago zanzibaricus	S	—	no	low	low	no	low	no	low	low
Hapalemur aureus	M	—	no	low	low	no	low	no	low	low
Indri indri	M	—	no	low	low	no	low	no	low	low
Lemur catta	MM	—	no	high	high	no	high	no	high	high
Lepilemur leucopus	S	—	no	low	low	no	low	no	low	low
Lepilemur mustelinus	S	—	no	low	low	no	low	no	low	low
Lepilemur ruficaudatus	S	—	no	low	low	no	low	no	low	low
Loris tardigradus	S	—	no	low	low	no	low	no	low	low
Microcebus murinus	S	—	no	low	low	no	low	no	low	low
Microcebus rufus	S	—	no	low	low	no	low	no	low	low
Nyctocebus coucang	S	—	—	—	—	—	—	—	—	—
Nyctocebus pygmaeus	S	—	—	—	—	—	—	—	—	—
Otolemur crassicaudatus	S	—	no	low	low	no	low	no	low	low
Otolemur garnetti	S	—	no	low	low	no	low	no	low	low
Perodicticus potto	S	—	no	low	low	no	low	no	low	low
Propithecus diana	MM	—	no	high	high	no	high	no	high	high
Propithecus verreauxi	MM	—	no	high	high	no	high	no	high	high
Varecia variegata	MM	—	no	high	high	no	low	no	low	low

Part 3
Comparative socioecology and social evolution

Editor's introduction

Why have the diverse social systems of primates evolved? Are the causal variables the same for different taxa? These questions are explored in the final six chapters. Here, problems specific to several different taxa are identified and causality is proposed. Where the underlying mechanisms producing social system variation have yet to be determined, the routes to further understanding are highlighted.

The contributors have tackled some of the more problematic issues in primate socioecology. Two radiations, those of the Malagasy primates and of the neotropical monkeys, pose intriguing and difficult questions about adaptive arrays. These groups are of particular interest in that they both represent a number of adaptive types and exhibit a range of social systems. We are only now approaching a sufficient knowledge base to address questions about their social structure in a broad evolutionary and ecological context. Kappler takes on the lemuroids in general in relation to social dynamics (Chapter 10), while Strier challenges the generality of the 'cercopithecine' model for social system evolution using examples from the platyrrhines (Chapter 11). It is of particular importance to be able to examine the existing models for intergroup and intragroup competition for food and mates in these groups of species in the light of an existing paradigm drawn primarily from terrestrial Old World monkeys and apes. As the authors in this section point out, this theoretical paradigm based on female resource competition (*sensu* Wrangham, 1980) needs reappraisal in the light of our growing knowledge of non-cercopithecoid foraging and reproductive strategies.

An energetic model of time budgeting among the papionines is further developed by Williamson and Dunbar (Chapter 12), in order to tease apart the complex interactions between local ecological variation and group size and structure. Interestingly, one of the best studied groups of primates, the baboons, continues to suggest new models or new approaches to understanding the role of ecology in primate social system evolution. Bean's approach to the socioecology of the great apes (Chapter 13) is also to address questions of energetics and time budgeting, but here specifically focusing on evolved behavioural sex differences in relation to reproductive costs and constraints. Foley (Chapter 14) reconstructs the ecology and elements of social structure for the hominid radiation. Such reconstructions remain speculative, in that the taxa of interest are not subject to direct observation, and furthermore they may have existed under conditions that are no longer represented by modern ecologies and environments. However, an analysis of extinct primate taxa should highlight one major problem in comparative biology: that the endpoints represented by extant species are the remnants of lineages with the potential for different rates of evolution, different selective pressures, and variable outcomes in the form of living species. Thus, living species may represent less about ancestor–descendant reconstructions than about that lineage's past radiation. This should be a cautionary tale for those engaging in phylogenetic subtraction techniques where socioecological patterns within a now primarily extinct lineage are unknown.

Finally, the evolutionary ecology of modern humans is addressed by Mace and Holden in Chapter 15, in which a number of important hypotheses about human marriage systems are firmly rooted in a comparative, evolutionary perspective. Such work remains controversial, with anthropologists reluctant to apply Darwinian models to human social complexity, but the careful generation of hypotheses and their robust statistical examination should provide one route forward for those with an evolutionary interest.

10 Lemur social structure and convergence in primate socioecology

PETER M. KAPPELER

Introduction

The study of social relationships lies at the core of socioecology, which attempts to explain social behaviour as an adaptation to ecological factors. Studies of primates have featured prominently in this field (Emlen and Oring, 1977; Rubenstein and Wrangham, 1986; Standen and Foley, 1989), partly because this comparatively small mammalian order exhibits stunning variation in both social organisation (i.e. size and composition) and structure of individual societies (Smuts *et al.*, 1987). This chapter focuses on social structure, which is defined by the complex network of behavioural interactions among members of a society (Hinde, 1976). Differences in the patterning and nature of these interactions give rise to particular social relationships between pairs of individuals. Consistent features of these dyadic relationships, in turn, can be used to characterise the social structure of a society or even a taxon.

The specific aim is to portray the social structure of lemur societies and examine the evolutionary forces that have shaped them. This focus on lemurs (Lemuriformes) is particularly interesting from a comparative perspective because the living primates of Madagascar represent the end-points of an adaptive radiation following a single colonisation event more than 50 million years ago (Yoder *et al.*, 1996), and, thus, offer an opportunity to examine patterns of convergent evolution (Kappeler and Ganzhorn, 1994). In addition, phylogenetic reconstructions revealed that group living, i.e. the permanent association of more than two adults, evolved at least twice independently among lemurs, compared to only once among anthropoids (Kappeler, 1998). Furthermore, lemurs deviate in several basic features of their behavioural ecology and life history from most anthropoids (see below, and van Schaik and Kappeler, 1996). A framework is provided for this comparison between lemurs and anthropoids by first outlining the range of variation observed in anthropoid social structure, and by briefly summarising current theories that attempt to explain it.

Patterns in anthropoid social structure

Four types of interactions can be used to broadly characterise social relationships among members of a society. Populations and species differ with respect to the establishment and maintenance of spatial proximity (affinitive behaviour), the exchange of affiliative behaviour and agonistic support, and in the nature of dominance relations among members of particular dyads (Bernstein and Williams, 1986; Cheney, Seyfarth and Smuts, 1986; de Waal and Luttrell, 1989). These aspects of social relationships are not independent of a species' social organisation, and do not vary independently from each other, although they may be organised differently in different taxa (de Waal, 1989; Foley and Lee, 1989; Cheney, 1992).

Relationships between individuals reflect behavioural strategies that have been selected because they maximise inclusive fitness (Crook and Gartlan, 1966; van Schaik, 1989). Sex has emerged as a major organising principle in the analysis of social structure due to the fundamental sex differences in mechanisms of maximising inclusive fitness (Trivers, 1972; Emlen and Oring, 1977). Because female reproductive success, in particular, is limited by ecological factors, the adaptive basis of female social relationships has traditionally been the focus of primate socioecology (Wrangham, 1980, 1987; van Schaik, 1983; van Schaik and van Hooff, 1983). Social relationships among males and between the sexes, on the other hand, are primarily shaped by sexual selection (Smuts, 1987b; Cowlishaw and Dunbar, 1991; Plavcan and van Schaik, 1992; Smuts and Smuts, 1993; van Hooff and van Schaik, 1994; Brereton, 1995; Clutton-Brock and Parker, 1995; van Schaik, 1996). Thus, both ecological and social factors are ultimately responsible for the observed variation in social relationships. However, the relative importance attributed to these factors, and especially that of their components, is still a controversial topic (Wrangham, 1987; Dunbar, 1988; van Schaik, 1996; Sterck, Watts and van Schaik, 1997).

Female–female relationships

Because success in feeding competition is most closely tied to variation in female fitness, socioecological models of female social relationships have concentrated on causes and consequences of feeding competition (Wrangham, 1980; van Schaik, 1989; Sterck *et al.*, 1997). The nature of feeding competition is shaped by the distribution of resources and it can occur within and between groups. When food patches, relative to group size, are clumped, monopolisable and of intermediate size, contest competition among females is expected, whereas scramble competition predominates over other types of patches (van Schaik, 1989). Existing socioecological models disagree over the identity of primary competitors. According to one hypothesis, related females form coalitions to defend access to preferred food sources against other such coalitions (Wrangham, 1980). Alternatively, group formation is seen as a response to predation risk and within-group competition as an inevitable consequence (van Schaik, 1983).

Whatever the ultimate causes of sociality, each female in a group-living species will experience a mix of contest and scramble competition within and between groups (van Schaik, 1989). The consequences of a given competitive regime for social relationships with other females can be summarised by four inter-related variables: philopatry, nepotism, tolerance and despotism, which probably vary continuously (Sterck *et al.*, 1997). However, it may nevertheless be useful to categorise this variation in order to facilitate broad comparisons. Accordingly, four main categories of female relationships have been identified (van Schaik, 1989; Sterck *et al.*, 1997).

In resident–nepotistic groups, females are philopatric and establish stable, linear and nepotistic hierarchies with despotic dominance relations as a result of strong within-group contest competition. In contrast, in dispersal–egalitarian groups, females regularly transfer between groups, forming neither stable linear hierarchies nor coalitions as a result of weak within-group contest competition. When there are additional strong contests between groups, however, resident–egalitarian groups are formed, characterised by female philopatry, a lack of decided agonistic relationships, and coalitions with relatives. Finally, when both within-group and between-group contests are marked, resident–nepotistic–tolerant groups will develop, in which philopatry is combined with decided relationships within a stable hierarchy, regular coalitions, but also pronounced tolerance by dominants.

Male–male relationships

Male relationships are determined by competition for access to receptive females. As a result, associations of several males and alliances are less common than among females (van Schaik, 1996). Thus, male relationships are typically characterised by competition and intolerance, resulting in clear dominance relations that are frequently age dependent (Bercovitch, 1991; Cowlishaw and Dunbar, 1991; van Hooff and van Schaik, 1994). Intolerance, threats and physical combat are therefore common among males, whereas affiliative behaviour is rarely observed (Cords, 1987; Plavcan and van Schaik, 1992, 1997; van Hooff and van Schaik, 1994). Competition among males is more intense than that among females because the contested resources are more valuable and cannot be shared. As such, and because most males change group membership several times during their lives, coalitions among males are also relatively rare (Noe and Sluijter, 1990; van Hooff and van Schaik, 1992; Alberts and Altmann, 1995). In a few species with pronounced male bonding, on the other hand, males are philopatric and develop highly differentiated relationships, including grooming bonds and coalitionary behaviour (van Hooff and van Schaik, 1992, 1994). These males are also characterised by promiscuous mating and presumably intense sperm competition (Harcourt, 1996). Thus, female distribution and the resulting nature of intrasexual selection are proposed to determine male social relationships to a large extent.

Male–female relationships

Male–female relationships are ultimately shaped by sexual selection and coercion (Smuts and Smuts, 1993; van Schaik, 1996). These forces may be responsible for the evolution of permanent association between the sexes found in the majority of primates (van Schaik and Kappeler, 1997). The relationships resulting from permanent association reflect a compromise between the behavioural strategies of males and females that aim at maximising the number and quality of offspring, respectively (Trivers, 1972). Because primate females cannot be forced into copulation, they can influence the evolution of male behaviour through their sexual behaviour. Males may therefore improve their chances of obtaining matings by providing females with certain services, such as agonistic support, which in turn affects their relationships (van Schaik and van Noordwijk 1989; Noe, van Schaik and van Hooff, 1991; Noe and Hammerstein, 1994). Conflicts between the sexes are frequently resolved on the basis of differences in agonistic power (Clutton-Brock and Parker, 1995). Because of the consequences of intrasexual selection, males typically have greater agonistic power and are therefore able to dominate and coerce females in their group (Smuts, 1987a; Smuts and Smuts, 1993; Clutton-Brock and Parker, 1995).

Male–female relations among anthropoids are highly variable, both within and among species. They are influenced, among other things, by the duration of male residence in a group, the respective rank in the same-sex dominance hierarchy, the degree of paternal certainty, the risk of infanticide and the degree of sexual dimorphism (Hamilton, 1984; Smuts and Smuts, 1993; Wright, 1993; Gubernick, 1994; van Schaik, 1996).

Previous comparative studies of lemur social systems focused either on social organisation (Kappeler, 1997) or on male–female relationships and their consequences (Richard, 1987; Kappeler, 1993a, 1993b; Jolly, 1998). These studies revealed that certain types of social organisation (e.g. single-male, multi–female groups) do not appear to have evolved among lemurs as modal types and that the average size and composition of their groups deviate from those of most anthropoids. These differences suggest a lack of convergence between the social systems of anthropoids and lemurs. Inter-sexual relationships among lemurs also differ from those among an-thropoids in that female dominance is widespread. A new hypothesis attributes this lack of convergence to recent ecological changes in Madagascar permitting formerly nocturnal pair-living taxa to form partly or predominantly diurnal groups (van Schaik and Kappeler, 1996). However, the consequences for social relationships of this possible tran-sition have not yet been analysed. Because specific predictions for such a disequilibrium situation are difficult to formulate, the available infor-mation on lemur social relationships is summarised first before their pos-sible determinants are discussed and contrasted with those identified for anthropoids.

Lemur social structure

Group living has evolved in five extant lemur genera (Kappeler, 1998). This section summarises the data on social relationships within and between the sexes in these taxa. Because long-term field studies of lemur social behav-iour are sparse, results are also included of some captive studies. Informa-tion on size, composition and cohesion of these groups is summarised in Kappeler (1997) and briefly presented in Table 10.1.

Lemur groups contain on average only between one and five females (Richard and Dewar, 1991; Kappeler, 1997). As a result, the basis for the social structure of most lemur groups is radically different from that of the better known anthropoids, which often contain more females. For example, it is almost impossible for such a small average number of lemur females to be organised into several matrilines. Furthermore, because some female transfer between groups has been reported for all group-living lemur taxa (Kappeler, 1997), the effects of genetic relatedness among females on their relationships are presently difficult to evaluate. Moreover, group-living lemurs are characterised by, on average, even adult sex ratios (Kappeler, 1997), which may affect both male–male and male–female rela-tionships. This is a sharp contrast to anthropoids, where sex ratios are often female biased. Finally, social communication among group-living lemurs relies to a much larger extent on olfaction than in most anthropoids

Table 10.1. *Summary of lemur life history and social traits*

Taxon	FM	Activity	Group size
Lemuridae			
Eulemur coronatus	1687	C	5.5
Eulemur fulvus fulvus	2550	(C)	9.1
Eulemur fulvus rufus	2251	C	9.5
Eulemur fulvus sanfordi	2147	C	7.7
Eulemur macaco macaco	2552	C	7.0
Eulemur mongoz	1658	C	3.1
Eulemur rubriventer	1960	C	3.4
Hapalemur aureus	1500	C	3
Hapalemur griseus	892	C	2.6
Hapalemur simus	1300	C	7
Lemur catta	2678	D	16.9
Varecia variegata rubra	3473	D	5.0
Varecia variegata variegata	3548	D(C)	11.0
Indridae			
Propithecus diadema candidus	—	D	4.8
Propithecus diadema diadema	6360	D	—
Propithecus diadema edwardsi	5895	D	6.0
Propithecus diadema perrieri	—	D	4.5
Propithecus tattersalli	3167	D	4.1
Propithecus verreauxi coquereli	3757	D	5.5
Propithecus verreauxi verreauxi	3525	D	6.3

Mean female body mass (FM, in grams), activity (N, nocturnal; C, cathemeral; D, diurnal) and mean foraging party size observed in the largest census are summarised for each taxon. After Kappeler (1997).

(Schilling, 1979). Behavioural mechanisms mediating and defining social structure may therefore differ fundamentally from the more visually and acoustically oriented anthropoids. These caveats and possible constraints on social structure should be kept in mind for the subsequent species-by-species summaries.

Female–female relationships among lemurs

Gentle lemurs (*Hapalemur griseus, H. simus* and possibly *H. aureus*) are regularly found in groups with two or more adult females (Kappeler, 1997). One captive study reported complete intolerance between two female *H. griseus* (Petter, Albignac and Rumpler, 1977), but detailed studies of the social structure of these cathemeral bamboo-specialists are lacking. They are therefore not considered further below.

Ruffed lemurs (*Varecia variegata*) live in dispersed communities consisting of four to five adult males and females, but stable pairs have also been observed. However, community members are rarely all in one location simultaneously (Rigamonti, 1993); in one study, individuals spent more than 50% of their time more than 20 m from the nearest conspecific (Morland, 1991; see also Pereira, Seeligson and Macedonia, 1989). Nevertheless, coalitions of females, without help from males, defended their communal home range against females of other communities (Morland, 1991; Rigamonti, 1993). Females are probably also the philopatric sex (White *et al.*, 1993). Interestingly, most disputes between communities occurred during the lactation season, when female energetic requirements are highest (Morland, 1991).

Female ruffed lemurs have affiliative relationships that are differentiated between individuals (Morland, 1991), but genetic relatedness is unknown. During the cool season, when communities break up into stable subgroups, rates of female interactions decrease (Morland, 1991). Huddling, which is common in other lemurs, is only rarely observed and never between females (Pereira *et al.*, 1989). Rates of aggression between females peaked in the two months prior to mating (Morland, 1993). In captivity, one adult female evicted both her mother and her adolescent sister from the group prior to the mating season after she and her mother nursed each other's offspring and jointly cared for them in the previous birth season (Pereira, Klepper and Simons, 1987). The two had also regular affiliative interactions and the daughter outranked her mother before the eviction (White *et al.*, 1992). The same mother was later outranked by another daughter (White, 1991).

The genus *Eulemur* contains five species. Three species (*E. fulvus, E. macaco* and *E. coronatus*) are mostly found in multi-male, multi-female groups. Two species (*E. mongoz* and *E. rubriventer*), whose social structure remains poorly studied (but see Overdorff, 1996; Curtis, 1997), typically form family groups with single females. Therefore, only the relationships of the group-living species are considered here.

It is difficult to make generalisations about social relationships among brown lemur (*E. fulvus*) females. In one captive study, associations were at the level expected by chance (Kappeler, 1993c). Rates of agonistic interactions can be high, possibly because they have only one, infrequently used, submissive signal (Pereira and Kappeler, 1997). As a result, the majority of agonistic interactions remained undecided, i.e. neither party exhibited only submissive behaviour (Hausfater, 1975), and dominance relations, if they exist at all, are difficult to discern (Harrington, 1975; Pereira and Kappeler, 1997). About half of all agonistic interventions were by adult females and supported male and female kin, but interventions were generally rare and occurred in only about 7% of conflicts (Pereira and Kappeler, 1997). While all females in a group had both agonistic and affiliative interactions (Kappeler, 1993c), most dyads could be characterised as either friends or adversaries (Pereira and Kappeler, 1997), depending on the nature of the majority of interactions. Female adversaries may try to evict each other from a group, using episodic targeted aggression (Vick and Pereira, 1989; Fornasieri and Roeder, 1992; Gresse *et al.*, 1994). After relatively rare failed evictions, however, former adversaries have been observed to huddle and groom again (Pereira and Kappeler, 1997). Reconciliation following conflicts is rare but more frequent than expected by chance (Kappeler, 1993d). Females typically remain in their natal groups, but singletons and pairs of females have regularly been observed (Kappeler, 1997). Unfortunately, the life histories of such individuals remain obscure.

Social relationships among black lemur females (*E. macaco*) have been rarely studied. In one captive group, no clear dominance relations could be discerned among four females (Fornasieri and Roeder, 1992). One introduced unrelated female was initially peripheralised and targeted for aggression but later integrated into the group. Females participate in intergroup encounters and may leave their natal group, as suggested by one documented transfer (Colquhoun, 1993).

Social relationships among female crowned lemurs (*E. coronatus*) are poorly known because only one small captive group has been studied in detail (Kappeler, 1993c). Crowned lemurs lack clear submissive signals. As a result, only 63% ($n = 27$) of agonistic interactions in a mother–daughter pair were decided. In all but one, the adult daughter exhibited submissive behaviour towards her mother. Both females more often had a male as nearest neighbour than each other, and they groomed each other at frequencies expected by chance. One year, when both had female infants, they targeted each other's infants for intense aggression when the offspring were about seven months old, but continued to groom each other during that period (Kappeler, unpublished data).

Ring-tailed lemurs (*Lemur catta*) form the largest groups among lemurs, with four to five adult females on average. Female ringtails typically remain in their natal group (Sussman, 1991) but may be evicted (Vick and Pereira, 1989). Females maintain unambiguous dyadic dominance relations based on submissive signalling throughout the year, partly because the vast majority of their agonistic interactions are clearly decided (Pereira and Kappeler, 1997). However, the resulting hierarchy is neither transitive nor stable (Pereira and Kappeler, 1997). Rank reversals occurred primarily during premating and birth seasons, when instances of episodic targeted aggression are also most likely (Pereira, 1993a).

During episodes of targeted aggression, characterised by persistent spontaneous and unprovoked aggression towards particular individuals, females typically evict related females from their group (Pereira, 1993a). With few exceptions, victims were not supported by their mother or sisters. Agonistic intervention was also extremely rare during other conflicts; less than 1% were characterised by third-party intervention by females on behalf of female kin (Kappeler, 1993c; Pereira and Kappeler, 1997). Agonistic interactions were not followed by affiliative interactions more than expected by chance; i.e. there is no apparent reconciliation or other post-conflict behavioural mechanism to reduce social tension (Kappeler, 1993d). While female–female dyads had the highest absolute rate of conflict, these were still at rates expected simply on the basis of proximity (Kappeler, 1993c). However, females groomed each other less than expected, given their relatively high spatial affinity.

At the dyadic level, female relationships among ringtails were characterised by the combination of above-average rates of agonistic interactions and below-average rates of affiliative interactions, or vice versa (Kappeler, 1993c). Dyads of the first type were more likely to exhibit targeted aggression (Pereira and Kappeler, 1997). Groups initially composed of two matrilines exhibited regular targeted aggression until all members of the lower-ranking matriline were evicted (Taylor and Sussman, 1985; Vick and Pereira, 1989; Pereira, 1993a; cf. Sauther and Sussman, 1993) or the group fissioned (Hood and Jolly, 1995). Evicted females may form or join other groups (Pereira and Izard, 1989), where they may even lactate spontaneously for unrelated infants.

Verreaux's sifakas (*Propithecus verreauxi*) live in groups of two to six adults with a highly variable sex ratio (Richard, 1985; Brockman, 1994). Whereas the majority of females are philopatric, they can leave their natal group to found new groups (Richard, Rakotomanga and Schwartz, 1993). Targeted aggression with subsequent eviction has also been reported (Jolly, 1998). Whenever two or more adult females were present in a study group, they generally exhibited lower rates of aggression and higher rates of affiliative behaviour towards each other than towards males (Richard, 1974a; Brockman, 1994). In one well-studied stable group, grooming frequencies among females were seven times higher than between males and females, but rates of agonistic interactions were identical for both types of dyads (Brockman, 1994). Dominance relations among females were clear and transitive in some groups, but not in others, in which females engaged in protracted agonistic interactions with unpredictable outcome (Richard, 1974a; Brockman, 1994). Some triadic agonistic interactions, but no coalitions among females, have been reported (Richard, 1974a).

In golden-crowned sifakas (*P. tattersalli*), female social relationships are highly variable among groups (Meyers, 1993). In three studied groups with two adult females, the female–female dyad had the highest, intermediate or lowest nearest neighbour index. Grooming relations between females were found to correspond to the strength of their spatial association and ranged from 0.09 to 0.71 affiliative interactions/dyad per day. In all three groups, however, both females had clear dominance relations over an entire year. In one group, the lower-ranking female helped carry the infant of the dominant female on a regular basis. In another group, the subordinate female was expelled by the dominant female (and a male) while both had young infants. The expelled female subsequently lost her infant and later disappeared herself.

In *P. diadema*, most groups contain two adult females (Wright, 1995). Females preferentially associate with either a male or the other female (Hemingway, 1994). Patterns of affiliative interactions among these large lemurs have not yet been reported. Over nine years, two females emigrated from their respective natal groups, whereas only one remained and reproduced at least twice (Wright, 1995). One of the emigrating females may have been evicted by another resident female (Vick and Pereira, 1989).

Male–male relationships among lemurs

Male–male relationships in a wild *Varecia* population are characterised by low levels of interaction. Aggression was limited primarily to the two months spanning the short breeding season (Morland, 1993). It was less frequent than that between females or the sexes and did not result in wounding. Affiliative interactions between males were also rare throughout the year; only one out of 13 dyads exhibited above-average affiliation (Morland, 1991). Grooming among males can be relatively frequent, but perhaps only when they are closely related, as in one captive study (Blanckenhorn, 1990). More typically, males may be unrelated as a result of natal dispersal and subsequent transfers (White, Balko and Fox, 1993).

Social relationships among *Eulemur fulvus* males are characterised by a lack of dominance relations, low rates of conflict and regular affiliative interactions (Kappeler, 1993c; Pereira and Kappeler, 1997). Males have relatively high proportions of decided conflicts, but consistent asymmetries in propensities to submit to other males are rare (Pereira and Kappeler, 1997). However, in small groups, marked dominance relations among resident males have been observed (Colquhoun, 1987; Roeder and Fornasieri, 1995). During the mating season, males can target particular rivals for aggression and even evict them from a group (Colquhoun, 1987; Vick and Pereira, 1989).

Social relationships between *E. macaco* males have not been studied. The males appear to experience intense mating competition, however, because rates of aggression, injuries and roaming behaviour were observed to increase during the mating season (Colquhoun, 1993).

Information on social relationships between *E. coronatus* males is only available from one father–son pair (Kappeler, 1993c). The pair had very low rates of aggression, and all their conflicts were undecided. They associated closely and groomed each other more often than expected by their close association. Mating competition was reduced by the fact that both females in their group came into oestrus on the same day in two successive years.

Male *Lemur catta* maintain decided dominance relations year-round (Budnitz and Dainis, 1975; Pereira and Kappeler, 1997), but rank reversals are common, especially around the brief mating season when the male hierarchy may temporarily break down during days of female oestrus (Jolly, 1967; Sauther, 1991). Reversals are also common after immigration of new males. Males usually transfer every three to four years, often in pairs or trios (Jones, 1983; Sussman, 1992). Most of them join neighbouring groups during the birth season. Rates of conflict and the proportion of undecided agonistic interactions are highest among males, but targeted aggression has not been observed (Vick and Pereira, 1989). Adult males also provide little and receive no agonistic support (Pereira and Kappeler, 1997). Some males residing in the same group may not exchange affiliative behaviour over years (Kappeler, 1993c), but some peripheral males associate and groom each other relatively frequently (Gould, 1997).

In a substantial proportion of *P. verreauxi* groups, there is only one resident adult male. In multi-male groups, males have low rates of agonistic interactions outside the mating season. It appears that one male received no aggression from the others, but it is not known whether males have decided relationships (Richard, 1992). Males groom each other regularly, sometimes more often than females (Richard, 1974a). It may be difficult for males to develop long-term relationships, however, because a large proportion of them transfer between groups every year (Richard *et al.*, 1993). During the mating season, males in some groups engage in extended chases and fierce fights, often involving visiting rivals from other groups (Richard, 1974b, 1985, 1992).

Male *P. tattersalli* have clearly decided dominance relationships year-round (Meyers, 1993). Independent of association levels, two pairs of adult males had relatively high rates of affiliative interactions (0.5 and 1.52 interactions/dyad per day, respectively). Males transfer between groups during the mating season; in one case, three males transferred together. Males participate in conflicts between groups, but nothing is known about mating competition among them.

In *P. diadema*, young males leave their natal groups and some experience severe resistance from resident males in their new groups (Wright, 1995). However, one resident male frequently groomed and huddled with a recent immigrant, event though the immigrant killed two likely offspring of the resident just after his immigration.

Male–female relationships among lemurs

Male–female pairs of ruffed lemurs exhibit the strongest affinitive and affiliative relationships of all dyads (Morland, 1991). Male and female *Varecia* also co-operate in caring for young and appear to co-ordinate related activities. In particular, males guard and defend parked infants during maternal absence (Pereira *et al.*, 1987). However, males do not support females in intergroup conflicts (Morland, 1991). Outside the breeding season, adult female ruffed lemurs invariably dominate males (Kaufman, 1991; Morland, 1993; Raps and White, 1995). The vast majority of conflicts are decided, and virtually all are won by females, who receive spontaneous submission from males in about a third of all conflicts (Raps and White, 1995). During their brief oestrus, females are also aggressive towards males (Foerg, 1982; Morland, 1993), but they have been observed to emit submissive vocalisations and to be chased by males during behavioural oestrus (Morland, 1993). Matings are mainly observed between pairs with stable affiliative relationships before and after the breeding season, but one female also solicited a copulation from a non-group male (Morland, 1993), and males in another population are reported to roam about for several days during the mating season (White *et al.*, 1993).

Male–female relationships appear to form the basic social unit in many *Eulemur* groups. In brown lemurs (*E. fulvus*), particular male–female pairs often maintain close proximity and groom each other more than all other same-sexed conspecifics together (Kappeler, 1993c). In addition, most observed agonistic support among brown lemurs is exchanged between such friendly males and females (Pereira and Kappeler, 1997). On some occasions, male brown lemurs support attempts by females to evict others from groups (Vick and Pereira, 1989). These pair bonds are neither exclusive nor always clear-cut, however, and nothing is yet known about their reproductive consequences. Female brown lemurs may mate with one or several males, but mating competition among males is often inconspicuous (Harrington, 1975). Agonistic relations between male and female brown lemurs are characterised by a lack of decided dominance relations (Pereira *et al.*, 1990; Roeder and Fornasieri, 1995). Consistent dyadic agonistic asymmetries are rare and sex has no apparent effect on their direction (Pereira and Kappeler, 1997). Males and females jointly participate in between-group encounters (Harrington, 1975).

Little is known about intersexual relationships in *E. macaco*. Female dominance has been reported (Colquhoun, 1993; Roeder and Fornasieri, 1995), but without supporting data. In contrast, no evidence for female dominance was found in another study of the same captive group (Fornasieri and Roeder, 1992). A firm conclusion about intersexual dominance relations must await the results of detailed field studies. Association and grooming patterns between the sexes have not been detailed, but only one male per group has been observed mating (Colquhoun, 1993).

In *E. coronatus*, male–female pairs associate significantly more often than same-sexed pairs, but tend to groom each other less than expected from this high affinity (Kappeler, 1993c). Grooming is observed in all male–female dyads, and a significantly larger proportion of grooming interactions is initiated by males rather than females (Kappeler, 1989). Agonistic relations between the sexes are characterised by high proportions of undecided conflicts, during which males also direct aggressive behaviour towards females. However, whereas males exhibit submissive behaviour in decided conflicts, this is only in response to female aggression (Kappeler, 1993a).

Unconditional female dominance is the salient feature of intersexual relationships among *L. catta* (Jolly, 1966; Kappeler, 1990; Pereira and Kappeler, 1997), and is affected mainly by spontaneous male submission (Pereira *et al.*, 1990; Kappeler, 1993a). Rates of intersexual conflict vary seasonally, peaking during mating and birth seasons (Jolly, 1967; Budnitz and Dainis, 1975; Pereira and Weiss, 1991), especially during immigration attempts by non-natal males. In many groups, the highest-ranking male occupies a central position with respect to interactions with females. He is more often in close proximity to all females and interacts affiliatively with them more frequently than do all other males (Budnitz and Dainis, 1975; Sauther, 1991; Kappeler, 1993c). Such central males are also often the first males to mate (Pereira and Weiss, 1991; Sauther, 1991; Sussman, 1992), but females also mate with other group males and regularly with non-group males (Sauther, 1991). Non-central males interact less frequently with females and spend much time at the periphery of the group. Males do not support females in between-group conflicts (Jolly *et al.*, 1993) and do not intervene in other male–female or female–female agonistic interactions (Pereira and Kappeler, 1997). Males initiate the vast majority of grooming interactions with females (Kappeler, 1993c), but females may be responsible for maintaining proximity (Gould, 1996).

In *P. verreauxi*, females direct more aggressive behaviour towards males than towards females and typically receive submissive behaviour in return (Richard, 1974a; Brockman, 1994). Females may even target individual males for prolonged intense aggression and eventually evict them from groups (Richard, 1992). However, on a few occasions, males have been observed to direct aggressive behaviour towards females, and females were observed to give submissive signals to males (Brockman, 1994). In captive pairs of *P. verreauxi*, females showed more aggressive behaviour, but no submissive behaviour has been observed (Kubzdela, Richard and Pereira, 1992). Males are also more responsible than females for maintaining proximity (Kubzdela *et al.*, 1992; Brockman, 1994). Some males show pronounced preferences for proximity to particular females, and these preferences were also reflected in the relative frequency of grooming and other affiliative interactions (Richard, 1974a; Brockman, 1994). Males and females also spend much of their resting time in physical contact (Richard, 1974a). Most females mate with more than one male, regularly including roaming non-residents (Richard, 1992; Brockman, 1994).

Intersexual relationships in *P. tattersalli* are characterised by clear-cut female dominance (Meyers, 1993). Only a single female was once seen exhibiting submissive behaviour towards a male during the mating period. Spontaneous male submission towards females is common. Affinitive and affiliative relationships between the sexes vary between groups, however. In one group, the single male rarely associated and interacted with the two females (0.07 affiliative interactions/dyad per day). In other groups, in which one of the two males interacted most frequently with both females or where one male–female pair was responsible for most interactions, affiliative interactions occurred at similar rates (0.25 interactions/dyad per day). There is no evidence for promiscuous mating, but only a few copulations have been observed.

In *P. diadema*, male–female pairs are reported to form the strongest social bonds (Hemingway, 1994), but there is considerable variation among groups. It has been suggested that females dominate males (Wright, 1988), but this does not prevent male infanticide (Wright, 1995). Only one male, usually the oldest one, has been observed to mate with oestrous females (Wright, 1995).

Discussion

The most important conclusion from this chapter is that female–female, male–female and to some extent male–male social relationships among group-living lemurs appear to differ in many aspects from those of the better-known anthropoids. This may either be due to the fact that few anthropoids live in groups that are as small as those of lemurs, or to real differences. Because the two extant lemur families (Lemuridae and Indridae) that independently gave rise to group-living species, as revealed by phylogenetic reconstructions (Kappeler, 1998), exhibit few parallels in basic aspects of social structure, the second possibility appears more likely.

It must be stressed, however, that many conclusions and classifications of behavioural attributes are preliminary, because only one or a few groups have been studied, sometimes for relatively short periods and/or under captive conditions that constrain important social variables. In addition, an entire genus (*Hapalemur*) and several other species remain essentially unstudied, and comparisons across some species are further hampered by varying methodology. Therefore only the most salient and robust similarities and differences are discussed here, followed by an assessment of determinants, with the hope that this will stimulate additional focused research.

Social relationships among lemur females cannot easily be classified within existing categories (Table 10.2). They show parallels with most anthropoids in that female philopatry appears to be the rule and females jointly defend their ranges. This, together with the observation that group fissioning, rather than individual voluntary dispersal, occurs may indicate that between-group competition is a significant force in lemur social evolution. On the other hand, the degree of exclusive home range use can vary dramatically within species (Jolly *et al.*, 1993; Sauther and Sussman, 1993), suggesting that other factors, such as certain ecological conditions, also promote female philopatry. In addition, the egalitarian dominance relations expected for female-resident species with high between-group competition (van Schaik, 1989; Isbell, 1991) are uncommon among lemurs.

In some lemur species, mothers and daughters maintain relationships characterised by frequent proximity, grooming and occasional agonistic support, but such relationships rarely extend to sisters or other close relatives (see also Jolly, 1998). In fact, cohesion between females in most taxa is low and coalitionary aggression generally rare or absent, attributed to a lack of sufficient visual acuity necessary for distinguishing fighting conspecifics (Pereira, 1995). As a result of rare coalitions in species in which two or more matrilines do coexist, there is no maternal rank inheritance (Pereira, 1995). Furthermore, in anthropoid groups with female philopatry, mothers usually outrank daughters (but see, for example, Borries, 1993), whereas this does not appear to be the case among lemurs.

The regular occurrence of targeted aggression and subsequent eviction of group members in most lemur taxa provides a major qualitative difference to anthropoids (Vick and Pereira, 1989; Jolly, 1998). The patterning of targeted aggression suggests that most groups contain only females from a single matriline (Taylor, 1986), a hypothesis that has yet to be tested with

Table 10.2. *Female–female social relationships among group-living lemurs*

	Hs	Vv	Ef	Em	Ec	Lc	Pv	Pt	Pd
Philopatry									
obligatory	?	Y?	Y?	Y?	Y?	Y	Y	Y?	?
conditional?	?	Y?	Y?	?	N	Y?	?	Y	
TA and eviction	?	Y	Y	Y	Y?	Y	Y	Y	Y?
Fissioning	?	?	?	?	?	Y	?	?	?
Cohesion	?	LO	LO	LO	LO	HI	LO	LO	LO
Grooming bonds	?	N	N	?	N	Y	Y	N?	N?
Reconciliation	?	?	Y	?	?	N	?	?	?
Aggression frequency	?	VAR	HI	?	?	HI	LO	VAR	?
Submissive signalling	?	Y	N	?	N	Y	Y	Y?	?
Dominance stability	?	N	NA	?	?	N	?	Y?	?
Dominance transitivity	?	?	NA	?	?	N	Y?	?	?
Dominance steepness	?	D	NA	?	?	D	E	D?	?
Nepotism	?	N	N	?	?	N	N	N	?
Coalitions	?	N?	N	?	?	N	N?	?	?
Intergroup coalitions	?	Y	Y	Y	Y	Y	Y	Y	Y

For *Hapalemur simus* (Hs), *Varecia variegata* (Vv), *Eulemur fulvus* (Ef), *E. macaco* (Em), *E. coronatus (Ec)*, *Lemur catta* (Lc), *Propithecus verreauxi* (Pv), *P.* tattersalli (Pt) and *P. diadema* (Pd) fundamental aspects of social behavior are summarised. TA refers to targeted aggression; cohesion and aggression frequency are classified relative to other types of dyads. Submissive signalling refers to the existence of formal signals of subordination; dominance stability to the long-term stability of dominance relations; dominance transitivity to the linearity of dominance relations; and dominance steepness describes the continuum of the quality of dominance relations between despotic and egalitarian. Y, yes; N, no; ?, not known; VAR, variable among groups; HI, high; LO, low; D, despotic; E, egalitarian; NA, not applicable.

genetic data. The ultimate causes of this unusual phenomenon remain elusive, but it has been suggested that targeted aggression is related to unusually competitive regimes as a result of seasonally harsh ecological conditions (Vick and Pereira, 1989; Pereira, 1993a).

Low spatial cohesion observed among lemurs, often reflected by the formation of temporary subgroups, indicates that within-group contest competition among females is high and/or that predation risk is low. The latter possibility is not supported by data on the predation of lemurs or by their anti-predator behaviour (Goodman, O'Connor and Langrand, 1993; Macedonia, 1993; Goodman, 1994; Rasoloarison et al., 1995), so that within-group feeding competition may be intense. However, the stable, transitive despotic dominance hierarchies expected under these conditions are apparently not realised among lemurs. Therefore, the conclusion that most lemurs do not fit easily into existing theories calls for both alternative theoretical explanations and additional research on lemurs. This would then allow us to arrive at a comprehensive understanding of the evolution

Table 10.3. *Male–male social relationships among group-living lemurs*

	Hs	Vv	Ef	Em	Ec	Lc	Pv	Pt	Pd
Philopatry									
obligatory	?	N	N	N	N	N	N	N	N
conditional	?	?	?	?	?	N	N	?	N
TA and eviction	?	?	Y	Y	Y?	N	N?	?	?
Association	?	LO	LO	LO	LO	LO	LO	LO	LO
Grooming frequency	?	LO	HI	?	HI?	LO	HI	HI?	?
Aggression frequency	?	LO	LO	?	LO	HI	LO	HI	?
Dominance stability	?	?	NA	?	?	N	?	Y?	?
Dominance transitivity	?	?	NA	?	?	N	Y?	?	?
Dominance steepness	?	D	NA	?	?	D	E	D?	?
Coalitions	?	N?	N	?	?	N	N?	?	?
Intergroup coalitions	?	N	Y	Y	Y	N	Y	Y	Y

See Table 10.2 for abbreviations.

of female social relationships among haplorhine *and* strepsirhine primates.

Lemur males show a few similarities with their anthropoid cousins (Table 10.3). As with most anthropoids, there is no evidence for male philopatry because males typically leave their natal group as young adults and compete for access to receptive females. Permanent all-male bands, which are also rare in anthropoids (e.g. Cords, 1988; Sommer, 1994), have not been observed. Lemur males are also as inconsistent as anthropoids in the degree of participation in between-group encounters (Cheney, 1987).

In contrast to many anthropoids, rates of aggression within groups can be low for most of the year, indicating that the primary function of aggression is mediating access to females. It is striking in this context that many lemur males also seek copulations outside their current group. Some typical morphological correlates of contest competition for fertilisations, such as superior size and weaponry compared to females, are generally lacking, whereas adaptations to scramble (i.e. sperm) competition generally conform to theoretical expectations (Kappeler, 1993b). These behavioural and morphological traits may ultimately be shaped by the pronounced seasonality of reproduction, compared to most anthropoids, and the resulting synchrony among female oestrous periods (Dunbar, 1988; Pereira, 1991).

Lemur males are also capable of establishing and maintaining relationships among each other characterised by frequent proximity, huddling and grooming that are rarely observed among unrelated adult anthropoid males. However, their possible function is still unclear (see Gould, 1997).

Relationships between lemur males and females converge with those of

Table 10.4. *Male–female social relationships among lemurs*

	Hs	Vv	Ef	Em	Ec	Lc	Pv	Pt	Pd
TA and eviction	?	N	Y	Y?	?	N	Y	Y?	?
Association	?	HI	HI	?	HI	LO	LO	VAR	?
Grooming frequency	?	HI	HI	?	HI?	LO	HI	VAR	?
Aggression frequency	?	LO	LO	?	?	HI	VAR	?	?
Female dominance	?	Y	N	Y?	Y	Y	Y?	Y	Y?
Coalitions	?	N?	Y	?	?	N	N?	?	?
Intersexual pair bonds	?	Y	Y	?	Y	N	N	N?	?
Extragroup matings	?	Y	?	Y?	?	Y	Y	?	Y?

See Table 10.2 for abbreviations.

anthropoids in that intersexual association evolved in response to the risk of infanticide (van Schaik, 1996; van Schaik and Kappeler, 1997). Furthermore, mating frequency and partner choice appear to be largely under female control.

Intersexual relationships among group-living lemurs are also unusual by lacking male dominance over females in dyadic agonistic interactions (Table 10.4). The evolution of female dominance has been functionally related to selection for female feeding priority during gestation and lactation (Jolly, 1984; Young, Richard and Aiello, 1990), but the central predictions of this energetic stress hypothesis were not supported in recent comparative studies of the energetic costs of reproduction across prosimians (Tilden and Oftedal, 1995, 1997; Kappeler, 1996). Studies of behavioural development and reproductive strategies (Richard, 1992; Pereira, 1995) suggested that female dominance may be part of male reproductive strategies, with males deferring to females as part of courtship, but its ultimate function still remains elusive.

Female dominance may be a relict of a pair-living phase, in that co-dominance and female feeding priority are not uncommon among other pair-living primates (Wright, 1993; van Schaik and Kappeler, 1996; see also Jolly, 1998). In those taxa with regular nocturnal activity, many adult lemurs appear to associate and interact primarily with one particular member of the opposite sex, but field data on the dynamics and postulated reproductive consequences of such pair bonds are still lacking. Primarily diurnal taxa, on the other hand, appear to have social structures in which a single male interacts disproportionately more often with most or all females (see, for example, Janson, 1984), but the dynamics of these systems are also poorly understood.

These differences and similarities indicate that the social structures of lemur societies have no counterparts among anthropoids, even though lemurs exhibit great heterogeneity and no two genera appear to have evolved identical social systems. Two explanations for this lack of convergence are possible. First, it has been proposed that lemur social systems are in an evolutionary transition towards anthropoid states, but that a lack of time, in combination with physiological constraints, in particular of the visual system, have impeded an 'anthropoid' evolutionary trajectory (Pereira, 1995; van Schaik and Kappeler, 1996). These possible transitions from nocturnal pair living to diurnal group living in the past (2000?) years are thought to be ultimately due to ecological factors, such as altered patterns of interspecific competition and predation following the quarternary extinctions in Madagascar. Based on this scenario, one should find: (1) social traits among group-living lemurs that are commonly found among pair-living species, and (2) corresponding differences between cathemeral and diurnal species. The occurrence of targeting aggression and eviction, female dominance and female reproductive competition is indeed compatible with the first of these predictions. Radical differences in social structure between closely related sympatric species (Pereira and Kappeler, 1997) with different activity patterns are compatible with the second prediction. Additional specific tests of this hypothesis are therefore indicated.

Second, it is possible that the observed systems represent adaptations to current ecological conditions in Madagascar, but that yet unidentified selective forces besides feeding competition, predation and sexual selection have a profound impact on social relationships among lemur males and females. These forces should be sought among aspects and determinants of lemur life histories, which are unusual, e.g. with respect to metabolic and growth strategies (Pereira, 1993b, 1995; van Schaik and Kappeler, 1993; Kappeler, 1995; Schmid and Ganzhorn, 1995; Schmid, 1996). One such explanation, focusing on the significance of the energetic costs of female reproduction (Jolly, 1984), has generated much research on lemur life histories but is unsupported by empirical tests. However, the increasing number of field studies combining behavioural, ecological, physiological and genetic sampling may generate alternative hypotheses by illuminating more detailed patterns in the natural histories and life histories of lemurs.

Acknowledgements

I thank P. Lee for the invitation to contribute to this volume, M. Pereira and C. van Schaik for inspiring discussions, and J. Ganzhorn, E. Heymann,

P. Lee, A. Koenig, J. Kuester, C. van Schaik and D. Zinner for their comments on the manuscript.

References

Alberts, S. and Altmann, J. (1995). Balancing costs and opportunities: dispersal in male baboons. *American Naturalist* **145**, 279–306.

Bercovitch, F.B. (1991). Social stratification, social strategies, and reproductive success in primates. *Ethology and Sociobiology* **12**, 315–33.

Bernstein, I.S. and Williams, L.E. (1986). The study of social organization. In *Comparative Primate Biology*, Vol. 2A, ed. G. Mitchell and J. Erwin, pp. 195–216. New York: A.R. Liss.

Blanckenhorn, W.U. (1990). A comparative study of tolerance and social organization in captive lemurs. *Folia Primatologica* **55**, 133–41.

Borries, C. (1993). Ecology of female social relationships: hanuman langurs (*Presbytis entellus*) and the van Schaik model. *Folia Primatologica* **61**, 21–30.

Brereton, A. (1995). Coercion–defence hypothesis: the evolution of primate sociality. *Folia Primatologica* **64**, 207–14.

Brockman, D. (1994). Reproduction and mating system of Verreaux's sifaka, *Propithecus verreauxi* at Beza Mahafaly, Madagascar. PhD thesis, Yale University.

Budnitz, N. and Dainis, K. (1975). *Lemur catta*: ecology and behavior. In *Lemur Biology*, ed. I. Tattersall and R.W. Sussman, pp. 219–35. New York: Plenum Press.

Cheney, D.L. (1987). Interactions and relationships between groups. In *Primate Societies*, ed. B.B. Smuts, D.L. Cheney, R.M. Seyfarth, R.W. Wrangham and T.T. Struhsaker, pp. 267–81. Chicago: University of Chicago Press.

Cheney, D.L. (1992). Intragroup cohesion and intergroup hostility – the relation between grooming distributions and intergroup competition among female primates. *Behavioral Ecology*, **3**, 334–45.

Cheney, D.L., Seyfarth, R.M. and Smuts, B.B. (1986). Social relationships and social cognition in non-human primates. *Science* **234**, 1361–6.

Clutton-Brock, T.H. and Parker, G.A. (1995). Sexual coercion in animal societies. *Animal Behaviour* **49**, 1345–65.

Colquhoun, I.C. (1987). Dominance and 'fall fever': the reproductive behavior of male brown lemurs (*Lemur fulvus*). *Canadian Review of Physical Anthropology* **6**, 10–19.

Colquhoun, I.C. (1993). The socioecology of *Eulemur macaco macaco*: a preliminary report. In *Lemur Social Systems and their Ecological Basis*, ed. P.M. Kappeler and J.U. Ganzhorn, pp. 11–23. New York: Plenum Press.

Cords, M. (1987). Forest guenons and Patas monkeys: male–male competition in one-male groups. In *Primate Societies*, ed. B.B. Smuts, D.L. Cheney, R.M. Seyfarth, R.W. Wrangham and T.T. Struhsaker, pp. 98–111. Chicago: University of Chicago Press.

Cords, M. (1988). Mating systems of forest guenons: a preliminary review. In *A Primate Radiation: Evolutionary Biology of the African Guenons*, ed. A.

Gauthier-Hion, F. Bourliere, J.-P. Gauthier and J. Kingdon, pp. 323–39. Cambridge: Cambridge University Press.

Cowlishaw, G. and Dunbar, R.I.M. (1991). Dominance rank and mating success in male primates. *Animal Behaviour*, **41**, 1045–56.

Crook, J.H. and Gartlan, J.C. (1966). Evolution of primate societies. *Nature* **210**, 1200–3.

Curtis, D.J. (1997). The mongoose lemur (*Eulemur mongoz*): a study in behavior and ecology. PhD thesis, University of Zürich.

de Waal, F.B.M. (1989). Dominance 'style' and primate social organization. In *Comparative Socioecology*, ed. V. Standen and R.A. Foley, pp. 243–63. Oxford: Blackwell.

de Waal, F.B.M. and Luttrell, L.M. (1989). Toward a comparative socioecology of the genus *Macaca*: different dominance styles in rhesus and stumptail monkeys. *American Journal of Primatology*, **19**, 83–110.

Dunbar, R.I.M. (1988). *Primate Social Systems*. Ithaca: Cornell University Press.

Emlen, S.T. and Oring, L.W. (1977). Ecology, sexual selection, and the evolution of mating systems. *Science* **197**, 215–23.

Foerg, R. (1982). Reproductive behavior in *Varecia variegata*. *Folia Primatologica* **38**, 108–21.

Foley, R.A. and Lee, P.C. (1989). Finite social space, evolutionary pathways, and reconstructing hominid behavior. *Science* **243**, 901–6.

Fornasieri, I. and Roeder, J.J. (1992). Marking behaviour in two lemur species (*L. fulvus* and *L. macaco*): relation to social status, reproduction, aggression and social change. *Folia Primatologica* **59**, 137–48.

Goodman, S. M. (1994). The enigma of anti-predator behavior in lemurs: evidence of a large extinct eagle on Madagascar. *International Journal of Primatology* **15**, 129–34.

Goodman, S.M., O'Connor, S. and Langrand, O. (1993). A review of predation on lemurs: implications for the evolution of social behavior in small, nocturnal primates. In *Lemur Social Systems and Their Ecological Basis*, ed. P.M. Kappeler and J.U. Ganzhorn, pp. 51–66. New York: Plenum Press.

Gould, L. (1996). Male–female affiliative relationships in naturally occurring ringtailed lemurs (*Lemur catta*) at the Beza-Mahafaly Reserve, Madagascar. *American Journal of Primatology* **39**, 63–78.

Gould, L. (1997). Intermale affiliative behavior in ringtailed lemurs (*Lemur catta*) at the Beza-Mahafaly Reserve, Madagascar. *Primates* **38**, 15–30.

Gresse, M., Lacour, M., Fornasieri, I. and Roeder, J.J. (1994). Targeting aggression in *Lemur fulvus albifrons*. In *Current Primate Biology II*, ed. J.J. Roeder, B. Thierry, J. Anderson and N. Herrenschmidt, pp. 233–9. Strasbourg: University Louis Pasteur.

Gubernick, D. (1994). Biparental care and male–female relations in mammals. In *Infanticide and Parental Care*, ed. S. Parmigiani and F. vom Saal, pp. 427–63. Chur, Switzerland: Harwood.

Hamilton, W.J. (1984). Significance of paternal investment by primates to the evolution of adult male–female associations. In *Primate Paternalism*, ed. D. Taub, pp. 309–35. New York: Van Nostrand Reinhold.

Harcourt, A.H. (1996). Sexual selection and sperm competition in primates: what are male genitalia good for? *Evolutionary Anthropology*, **5**, 121–9.

Harrington, J.E. (1975). Field observations of social behavior of *Lemur fulvus fulvus*. In *Lemur Biology*, ed. I. Tattersall and R.W. Sussman, pp. 259–79. New York: Plenum Press.

Hausfater, G. (1975). Dominance and reproduction in baboons: a quantitative analysis. *Contributions to Primatology* **7**, 1–150.

Hemingway, C. (1994). Spatial relations in three groups of the diademed sifaka, *Propithecus diadema edwardsi* in southeastern Madagascar. *American Journal of Physical Anthropology* **18**, 105.

Hinde, R.A. (1976). Interactions, relationships and social structure. *Man* **11**, 1–17.

Hood, L. and Jolly, A. (1995). Troop fission in female *Lemur catta* at Berenty Reserve, Madagascar. *International Journal of Primatology* **16**, 997–1015.

Isbell, L.A. (1991). Contest and scramble competition: patterns of female aggression and ranging behavior among primates. *Behavioral Ecology* **2**, 143–55.

Janson, C.H. (1984). Female choice and mating system of the brown capuchin monkey *Cebus apella* (Primates: Cebidae). *Zeitschrift für Tierpsychologie* **65**, 177–200.

Jolly, A. (1966). *Lemur Behavior*. Chicago: University of Chicago Press.

Jolly, A. (1967). Breeding synchrony in wild *Lemur catta*. In *Social Communication among Primates*, ed. S.A. Altman, pp. 3–14. Chicago: University of Chicago Press.

Jolly, A. (1984). The puzzle of female feeding priority. In *Female Primates: Studies by Woman Primatologists*, ed. M.F. Small, pp. 197–215. New York: A.R. Liss.

Jolly, A. (1998). Pair-bonding, female aggression, and the evolution of lemur societies. *Folia Primatologica* **69**, 1–13.

Jolly, A., Rasamimanana, H.R., Kinnaird, M.F. *et al.* (1993). Territoriality in *Lemur catta* groups during the birth season at Berenty, Madagascar. In *Lemur Social Systems and their Ecological Basis*, ed. P.M. Kappeler and J.U. Ganzhorn, pp. 85–110. New York: Plenum Press.

Jones, K.C. (1983). Inter-troop transfer of *Lemur catta* males at Berenty, Madagascar. *Folia Primatologica* **40**, 145–60.

Kappeler, P.M. (1989). Agonistic and grooming behavior of captive crowned lemurs (*Lemur coronatus*) during the breeding season. *Human Evolution* **4**, 207–15.

Kappeler, P.M. (1990). Female dominance in *Lemur catta*: more than just female feeding priority? *Folia Primatologica* **55**, 92–5.

Kappeler, P.M. (1993a). Female dominance in primates and other mammals. In *Perspectives in Ethology*, Vol. 10: *Behaviour and Evolution*, ed. P.P.G. Bateson, P.H. Klopfer and N.S. Thompson, pp. 143–58. New York: Plenum Press.

Kappeler, P.M. (1993b). Sexual selection and lemur social systems. In *Lemur Social Systems and their Ecological Basis*, ed. P.M. Kappeler and J.U. Ganzhorn, pp. 223–40. New York: Plenum Press.

Kappeler, P.M. (1993c). Variation in social structure: the effects of sex and kinship on social interactions in three lemur species. *Ethology* **93**, 125–45.

Kappeler, P.M. (1993d). Reconciliation and post-conflict behaviour in ringtailed lemurs, *Lemur catta* and redfronted lemurs, *Eulemur fulvus rufus*. *Animal Behaviour* **45**, 901–15.

Kappeler, P.M. (1995). Life history variation among nocturnal prosimians. In *Creatures of the Dark: the Nocturnal Prosimians*, ed. L. Alterman, M.K. Izard and G.A. Doyle, pp. 75–92. New York: Plenum Press.

Kappeler, P.M. (1996). Causes and consequences of life history variation among strepsirhine primates. *American Naturalist* **148**, 868–91.

Kappeler, P.M. (1997). Determinants of primate social organization: comparative evidence and new insights from Malagasy lemurs. *Biological Reviews* **72**, 111–51.

Kappeler, P.M. (1998). Convergence and nonconvergence in primate social systems. In *Primate Communities*, ed. J. Fleagle, C. Janson and K. Reed. Cambridge: Cambridge University Press.

Kappeler, P.M. and Ganzhorn, J.U. (1994). The evolution of primate communities and societies in Madagascar. *Evolutionary Anthropology* **2**, 159–71.

Kaufman, R. (1991). Female dominance in semifree-ranging black-and-white ruffed lemurs, *Varecia variegata*. *Folia Primatologica* **57**, 39–41.

Kubzdela, K., Richard, A.F. and Pereira, M.E. (1992). Social relations in semi-free-ranging sifakas (*Propithecus verreauxi coquereli*) and the question of female dominance. *American Journal of Primatology* **28**, 139–45.

Macedonia, J.M. (1993). Adaptation and phylogenetic constraints in the antipredator behavior of ringtailed and ruffed lemurs. In *Lemur Social Systems and their Ecological Basis*, ed. P.M. Kappeler and J.U. Ganzhorn, pp. 67–84. New York: Plenum Press.

Meyers, D. (1993). The effects of resource seasonality on behavior and reproduction in the golden-crowned sifaka (*Propithecus tattersalli* Simons, 1988) in three Malagasy forests. PhD thesis, Duke University.

Morland, H.S. (1991). Preliminary report on the social organization of ruffed lemurs (*Varecia variegata variegata*) in a northeast Madagascar rainforest. *Folia Primatologica* **56**, 157–61.

Morland, H.S. (1993). Reproductive activity of ruffed lemurs (*Varecia variegata variegata*) in a Madagascar rainforest. *American Journal of Physical Anthropology* **91**, 71–82.

Noe, R. and Hammerstein, P. (1994). Biological markets: supply and demand determine the effect of partner choice in cooperation, mutualism and mating. *Behavioral Ecology and Sociobiology* **35**, 1–12.

Noe, R. and Sluijter, A.A. (1990). Reproductive tactics of male savanna baboons. *Behaviour* **113**, 117–70.

Noe, R., van Schaik, C.P. and van Hooff, J. (1991). The market effect: an explanation for pay-off asymmetries among collaborating animals. *Ethology* **87**, 97–118.

Overdorff, D.J. (1996). Ecological correlates to social structure in two lemur species in Madagascar. *American Journal of Physical Anthropology*, **100**, 487–506.

Pereira, M.E. (1991). Asynchrony within estrous synchrony among ringtailed lemurs (Primates: Lemuridae). *Physiology and Behaviour*, **49**, 47–52.

Pereira, M.E. (1993a). Agonistic interaction, dominance relation, and ontogenetic trajectories in ringtailed lemurs. In *Juvenile Primates: Life History, Development, and Behavior*, ed. M.E. Pereira and L.A. Fairbanks, pp. 285–305. New York: Oxford University Press.

Pereira, M.E. (1993b). Seasonal adjustment of growth rate and adult body weight in ringtailed lemurs. In *Lemur Social Systems and their Ecological Basis*, ed. P.M. Kappeler and J.U. Ganzhorn, pp. 205–22. New York: Plenum Press.

Pereira, M.E. (1995). Development and social dominance among group-living primates. *American Journal of Primatology* **37**, 143–75.

Pereira, M.E. and Izard, M.K. (1989). Lactation and care for unrelated infants in forest-living ringtailed lemurs. *American Journal of Primatology* **18**, 101–8.

Pereira, M.E. and Kappeler, P.M. (1997). Divergent systems of agonistic relationship in lemurid primates. *Behaviour* **134**, 225–74.

Pereira, M.E., Kaufman, R., Kappeler, P.M. and Overdorff, D.J. (1990). Female dominance does not characterize all of the Lemuridae. *Folia Primatologica* **55**, 96–103.

Pereira, M.E., Klepper, A. and Simons, E.L. (1987). Tactics of care for young infants by forest-living ruffed lemurs (*Varecia variegata variegata*): ground nests, parking, and biparental guarding. *American Journal of Primatology* **13**, 129–44.

Pereira, M.E., Seeligson, M.L. and Macedonia, J.M. (1989). The behavioral repertoire of the black and white ruffed lemur, *Varecia variegata variegata* (Primates, Lemuridae). *Folia Primatologica* **51**, 1–32.

Pereira, M.E. and Weiss, M.L. (1991). Female mate choice, male migration, and the threat of infanticide in ringtailed lemurs. *Behavioral Ecology and Sociobiology* **28**, 141–52.

Petter, J.J., Albignac, R. and Rumpler, Y. (1977). *Faune de Madagascar 44: Mammifères Lemuriens (Primates Prosimien)*. Paris: ORSTOM and CNRS.

Plavcan, M.J. and van Schaik, C.P. (1992). Intrasexual competition and canine dimorphism in primates. *American Journal of Physical Anthropology* **87**, 461–77.

Plavcan, M.J. and van Schaik, C.P. (1997). Interpreting hominid behavior on the basis of sexual dimorphism. *Journal of Human Evolution* **32**, 345–74.

Raps, S. and White, F.J. (1995). Female social dominance in semi-free ranging ruffed lemurs, *Varecia variegata*. *Folia Primatologica* **65**, 163–8.

Rasoloarison, R., Rasolonandrasana, B.P.N., Ganzhorn, J.U. and Goodman, S.M. (1995). Predation on vertebrates in the Kirindy forest, Western Madagascar. *Ecotropica* **1**, 59–65.

Richard, A. (1974a). Intra-specific variation in the social organization and ecology of *Propithecus verreauxi*. *Folia Primatologica* **22**, 178–207.

Richard, A. (1974b). Patterns of mating in *Propithecus verreauxi verreauxi*. In *Prosimian Biology*, ed. R.D. Martin, G.A. Doyle and A.C. Walker, pp. 49–74. London: Duckworth.

Richard, A.F. (1985). Social boundaries in a Malagasy prosimian, the sifaka (*Propithecus verreauxi*). *International Journal of Primatology* **6**, 553–68.

Richard, A.F. (1987). Malagasy prosimians: female dominance. In *Primate Societies*, ed. B.B. Smuts, D.L. Cheney, R.M. Seyfarth, R.W. Wrangham and T.T. Struhsaker, pp. 25–33. Chicago: University of Chicago Press.

Richard, A.F. (1992). Aggressive competition between males, female-controlled polygyny and sexual monomorphism in a Malagasy primate, *Propithecus verreauxi*. *Journal of Human Evolution* **22**, 395–406.

Richard, A.F. and Dewar, R.E. (1991). Lemur ecology. *Annual Review of Ecology and Systematics* **22**, 145–75.

Richard, A.F., Rakotomanga, P. and Schwartz, M. (1993). Dispersal by *Propithecus verreauxi* at Beza Mahafaly, Madagascar: 1984–1991. *American Journal of*

Primatology **30**, 1–20.

Rigamonti, M.M. (1993). Home range and diet in red ruffed lemurs (*Varecia variegata rubra*) on the Masoala peninsula, Madagascar. In *Lemur Social Systems and their Ecological Basis*, ed. P.M. Kappeler and J.U. Ganzhorn, pp. 25–40. New York: Plenum Press.

Roeder, J.-J. and Fornasieri, I. (1995). Does agonistic dominance imply feeding priority in lemurs? A study in *Eulemur fulvus mayottensis. International Journal of Primatology* **16**, 629–42.

Rubenstein, D.I. and Wrangham, R.W. (1986). *Ecological Aspects of Social Evolution*. Princeton, NJ: Princeton University Press.

Sauther, M.L. (1991). Reproductive behavior of free-ranging *Lemur catta* at Beza Mahafaly Special Reserve, Madagascar. *American Journal of Physical Anthropology* **84**, 463–77.

Sauther, M.L. and Sussman, R.W. (1993). A new interpretation of the social organization and mating system of the ringtailed lemur (*Lemur catta*). In *Lemur Social Systems and their Ecological Basis*, ed. P.M. Kappeler and J.U. Ganzhorn, pp. 111–22. New York: Plenum Press.

Schilling, A. (1979). Olfactory communication in prosimians. In *The Study of Prosimian Behavior*, ed. G.A. Doyle and R.D. Martin, pp. 461–542. New York: Academic Press.

Schmid, J. (1996). Oxygen consumption and torpor in mouse lemurs (*Microcebus murinus* and *M. myoxinus*): preliminary results of a study in western Madagascar. In *Adaptations to the Cold: Tenth International Hibernation Symposium*, ed. F. Geiser, A. Hulbert and S. Nicol, pp. 47–54. Armidale: University of New England Press.

Schmid, J. and Ganzhorn, J.U. (1995). Resting metabolic rates of *Lepilemur ruficaudatus. American Journal of Primatology* **38**, 169–74.

Smuts, B.B. (1987a). Gender, aggression, and influence. In *Primate Societies*, ed. B.B. Smuts, D.L. Cheney, R.M. Seyfarth, R.W. Wrangham and T.T. Struhsaker, pp. 400–12. Chicago: University of Chicago Press.

Smuts, B.B. (1987b). Sexual competition and mate choice. In *Primate Societies*, ed. B.B. Smuts, D.L. Cheney, R.M. Seyfarth, R.W. Wrangham and T.T. Struhsaker, pp. 385–99. Chicago: University of Chicago Press.

Smuts, B.B., Cheney, D.L., Seyfarth, R.M., Wrangham, R.W. and Struhsaker, T.T. (1987). *Primate Societies*. Chicago: Chicago University Press.

Smuts, B.B. and Smuts, R.W. (1993). Male aggression and sexual coercion of females in nonhuman primates and other mammals: evidence and theoretical implications. *Advances in the Study of Behavior* **22**, 1–63.

Sommer, V. (1994). Infanticide among the langurs of Jodhpur: testing the sexual selection hypothesis with a long-term record. In *Infanticide and Parental Care*, ed. S. Parmigiani and F. vom Saal, pp. 155–98. Chur, Switzerland: Harwood.

Standen, V. and Foley, R. (1989). *Comparative Socioecology: the Behavioural Ecology of Humans and other Animals*. Oxford: Blackwell Scientific Publications.

Sterck, E., Watts, D. and van Schaik, C.P. (1997). The evolution of female social relationships in nonhuman primates. *Behavioral Ecology and Sociobiology* **41**, 291–309.

Sussman, R.W. (1991). Demography and social organization of free-ranging *Lemur catta* in the Beza Mahafaly Reserve, Madagascar. *American Journal of Physical Anthropology* **84**, 43–58.

Sussman, R.W. (1992). Male life history and intergroup mobility among ringtailed lemurs (*Lemur catta*). *International Journal of Primatology* **13**, 395–414.

Taylor, L.L. (1986). Kinship, dominance and social organization in a semi-free ranging group of ringtailed lemurs (*Lemur catta*). PhD thesis, Washington University.

Taylor, L.L. and Sussman, R.W. (1985). A preliminary study of kinship and social organization in a semi-free ranging group of *Lemur catta*. *International Journal of Primatology* **6**, 601–14.

Tilden, C. and Oftedal, O. (1995). The bioenergetics of reproduction in prosimian primates: is it related to female dominance? In *Creatures of the Dark*, ed. L. Alterman, G. Doyle and M. Izard, pp. 119–31. New York: Plenum Press.

Tilden, C. and Oftedal, O. (1997). Milk composition reflects pattern of maternal care in prosimian primates. *American Journal of Primatology* **41**, 195–211.

Trivers, R.L. (1972). Parental investment and sexual selection. In *Sexual Selection and the Descent of Man*, ed. B. Campbell, pp. 136–79. Chicago: Aldine.

van Hooff, J. and van Schaik, C.P. (1992). Cooperation in competition: the ecology of primate bonds. In *Coalitions and Alliances in Humans and other Animals*, ed. A.H. Harcourt and F.B.M. de Waal, pp. 357–89. Oxford: Oxford University Press.

van Hooff, J. and van Schaik, C.P. (1994). Male bonds: affiliative relationships among nonhuman primate males. *Behaviour* **130**, 309–37.

van Schaik, C.P. (1983). Why are diurnal primates living in groups? *Behaviour* **87**, 120–44.

van Schaik, C.P. (1989). The ecology of social relationships amongst female primates. In *Comparative Socioecology*, ed. V. Standen and R.A. Foley, pp. 195–218. Oxford: Blackwell.

van Schaik, C.P. (1996). Social evolution in primates: the role of ecological factors and male behaviour. *Proceedings of the British Academy* **88**, 9–31.

van Schaik, C.P. and Kappeler, P.M. (1993). Life history, activity period and lemur social systems. In *Lemur Social Systems and their Ecological Basis*, ed. P.M. Kappeler and J.U. Ganzhorn, pp. 241–60. New York: Plenum Press.

van Schaik, C.P. and Kappeler, P.M. (1996). The social systems of gregarious lemurs: lack of convergence with anthropoids due to evolutionary disequilibrium? *Ethology* **102**, 915–41.

van Schaik, C.P. and Kappeler, P.M. (1997). Infanticide risk and the evolution of male–female bonding in primates. *Proceedings of the Royal Society: Biological Sciences B* **264**, 1687–94.

van Schaik, C.P. and van Hooff, J. (1983). On the ultimate causes of primate social systems. *Behaviour* **85**, 91–117.

van Schaik, C.P. and van Noordwijk, M.A. (1989). The special role of male *Cebus* monkeys in predation avoidance and its effect on group composition. *Behavioral Ecology and Sociobiology* **24**, 265–76.

Vick, L.G. and Pereira, M.E. (1989). Episodic targeting aggression and the histories of *Lemur* social groups. *Behavioral Ecology and Sociobiology* **25**, 3–12.

White, F.J. (1991). Social organization, feeding ecology, and reproductive strategy of ruffed lemurs, *Varecia variegata*. In *Proceedings of the XIII Congress of the*

International Primatological Society, ed. A. Ehara, T. Kimura, O. Takenaka and M. Iwamoto, pp. 81–4. Amsterdam: Elsevier.

White, F.J., Balko, E.A. and Fox, E.A. (1993). Male transfer in captive ruffed lemurs, *Varecia variegata variegata*. In *Lemur Social Systems and their Ecological Basis*, ed. P.M. Kappeler and J.U. Ganzhorn, pp. 41–50. New York: Plenum Press.

White, F.J., Burton, A.S., Buchholz, S. and Glander, K.E. (1992). Social organization of free-ranging ruffed lemurs, *Varecia variegata variegata*: mother–adult daughter relationship. *American Journal of Primatology* **28**, 281–7.

Wrangham, R.W. (1980). An ecological model of female-bonded primate groups. *Behaviour* **75**, 262–300.

Wrangham, R.W. (1987). Evolution of social structure. In *Primate Societies*, ed. B.B. Smuts, D.L. Cheney, R.M. Seyfarth, R.W. Wrangham and T.T. Struhsaker, pp. 282–97. Chicago: University of Chicago Press.

Wright, P.C. (1988). Social behavior of *Propithecus diadema edwardsi* in Madagascar. *American Journal of Physical Anthropology* **75**, 289.

Wright, P.C. (1993). Variations in male–female dominance and offspring care in non-human primates. In *Sex and Gender Hierarchies*, ed. B. Miller, pp. 127–45. Cambridge: Cambridge University Press.

Wright, P.C. (1995). Demography and life history of free-ranging *Propithecus diadema edwardsi* in Ranomafana National Park, Madagascar. *International Journal of Primatology* **10**, 835–54.

Yoder, A., Cartmill, M., Ruvolo, M., Smith, K. and Vilgalys, R. (1996). Ancient single origin for Malagasy primates. *Proceedings of the National Academy of Sciences*, USA **93**, 5122–6.

Young, A.L., Richard, A.F. and Aiello, L.C. (1990). Female dominance and maternal investment in strepsirhine primates. *American Naturalist* **135**, 473–88.

11 *Why is female kin bonding so rare? Comparative sociality of neotropical primates*

KAREN B. STRIER

Introduction

Some female primates remain in their natal groups for life. Female kin in these cohesive, matrilineal groups form affiliative bonds that are expressed through proximity, grooming, and agonistic support against members of other matrilines and unrelated males. Kin-bonded females generally play key roles in determining group movements, defending resources during intergroup encounters, and repelling or assisting extra-group males in their attempts to immigrate or oust resident males from positions in the group (Wrangham, 1980). Male-biased dispersal often, but not always, covaries with the occurrence of female kin bonding, contributing to the avoidance of inbreeding and corresponding to the lower levels of nepotism among males in multi-male groups (Moore, 1984; Pusey and Packer, 1987).

The prevalence of female kin bonding and male-biased dispersal among many of the best studied Old World monkeys has fuelled an enduring myth in which primates with different dispersal and social systems were treated as deviant exceptions to this otherwise 'typical' primate pattern (Moore, 1984; Strier, 1994a). Long-term studies on a greater diversity of primate taxa have since shifted this perspective, however, and now the distribution of female kin bonding and male-biased dispersal across primates is more appropriately recognised as a reflection of phylogenetic or ecological affinities instead of a universal standard against which all interspecific and intraspecific variability is compared (Moore, 1992). Thus, female kin bonding with male-biased dispersal among Strepsirhines can be seen as consistent with a primitive mammalian condition, whereas the absence of female kin bonding among extant hominoids can be considered a derived condition, reflecting both their monophyletic antiquity (Lee, 1994) and the relaxation of predation pressures on these large-bodied primates. High costs of feeding competition due to a dietary reliance on ripe fruit further preclude chimpanzees and orangutans from forming the cohesive groups necessary for the maintenance of female kin bonds, while indifference towards co-operative resource defence leads female gorillas to bond with influential males rather than one another (Wrangham, 1980; van Schaik, 1989).

The rarity of female kin bonding among the New World primates has been more difficult to explain (Table 11.1). Even among those New World primates in which femle kin groups may occur (e.g. Peruvian *Saimiri*: Mitchell, 1994; Mitchell, Boinski and van Schaik, 1991; *Cebus*: Robinson and Janson, 1987; Fedigan, 1993; and *Callithrix*: Garber, 1994; Ferrari and Digby, 1996), behavioural differences distinguish them from those of most cercopithecines. Phylogenetic and ecological models account for some of these distinctions, while others make sense only when life histories, reproductive patterns and demography are considered. This chapter examines New World primate sociality in an effort to understand why female kin bonding is so rare.

Phylogenetic perspectives

Molecular data place the platyrrhine–catarrhine branch point at 35 million years ago (Schneider *et al.*, 1993). Whether female kin bonding with male-biased dispersal characterised the societies of ancestral haplorhines, or whether it was secondarily derived among the Old World cercopithecine monkeys during their much more recent radiation is difficult to determine (Strier, 1990, 1994a; Di Fiore and Rendall, 1994; Lee, 1994; Rendall and Di Fiore, 1996). Nonetheless, phylogenetic analyses do not support female kin bonding as an ancestral trait among platyrrhines. Instead, female-biased dispersal with male kin bonding or concomitant dispersal by males and weak kinship bonds among females emerge as the probable ancestral conditions in at least two phyletic groups.

Callitrichinae ancestors

The subfamily Callitrichinae includes the marmosets (*Callithrix*), pygmy marmosets (*Cebuella*), tamarins (*Saguinus*), golden lion tamarins (*Leontopithecus*), and Goeldi's monkeys (*Callimico*). Extant callitrichines are divided into two tribes: Callimini, represented by the monotypic genus, *Callimico*; and Callitrichini, represented by the other four genera. Compared to other Platyrrhines, the Callitrichinae are small-bodied primates

Table 11.1. *Dispersal and evidence of bonding among neotropical primates*

Genus (references)	Dispersal	Evidence of female bonding	Variation in female bonds	Evidence of male bonds
Alouatta[1,2,3]	Both sexes	None	—	Tolerance of young male kin
Aotus[4]	Both sexes	None	—	—
Ateles[5,6]	Female biased	None	—	Co-operation in monitoring females
Brachyteles[5,6]	Female biased	None	—	Co-operation in monitoring females, strong affiliations + tolerance
Cacajao[4,7]	?	?	Groups fragment	?
Callicebus[4]	Both sexes	None	—	—
Callimico[8]	Both sexes	None	—	—
Callithrix[8,9]	Both sexes	Daughters frequently tolerated in natal group	Reproduction may be inhibited	Males migrate in pairs; co-operative infant care
Cebuella[8]	Both sexes	None	—	—
Cebus[7,10–15]	Male biased	Yes, see text and Table 11.3	Variation in group cohesiveness, female resource defence, strength of female–alpha male bond; females dominate some males	Variations in male influxes; superstructure societies; low male aggression
Chiropotes[4,7]	?	?	Groups fragment	?
Lagothrix[5,6,16]	Female biased	None	—	Weak
Leontopithecus[8,17,18]	Both sexes	Daughters occasionally tolerated in natal group	Reproduction usually inhibited	Males migrate in pairs; co-operative infant care
Pithecia[4,7]	?	?	Groups fragment	?
Saguinus[8,19,20]	Both sexes	Daughters sometimes tolerated in natal group	Reproduction often inhibited	Males migrate in pairs; co-operative infant care

| *Saimiri*[21-24] | Variable sex biased | Variable, see text and Table 11.2 | Females dominate males | Strong affiliations with female dispersal; cohort migration alliances |

References: (1) Crockett and Eisenberg, 1987; (2) Glander, 1992; (3) Crockett and Pope, 1993; (4) Robinson *et al.*, 1987; (5) Nishimura, 1994; (6) Strier, 1994b; (7) Kinzey and Cunningham, 1994; (8) Garber, 1994; (9) Ferrari and Digby, 1996; (10) Fedigan, 1993; (11) Janson, 1986; (12) Janson, 1985; (13) O'Brien, 1991; (14) Izawa, 1994a; (15) O'Brien and Robinson, 1993; (16) Stevenson *et al.*, 1994; (17)Dietz and Baker, 1993; (15) Baker *et al.*, 1993; (19) Goldizen *et al.*, 1996; (20) Savage *et al.*, 1996; (21) Mitchell *et al.*, 1991; (22) Boinski and Mitchell, 1994; (23) Boinski, 1994; (24) Mitchell, 1994.

with average female body weights ranging from 126 g in *Cebuella pygmaea* to 587 g in *Leontopithecus rosalia* (Ford, 1994). They have high reproductive rates characteristic of species that evolved in secondary and edge forests (Ross, 1991; Rylands, 1996), rapid postnatal infant growth, and extensive co-operative infant care in which group members share the energetic burden of carrying infants. Reproductive rates are further increased among callitrichins, with all four genera characteristically producing twins, and occasionally triplets. The two marmoset genera, which annually produce two litters instead of one, have the highest re-productive rates of all (Garber, 1994).

Callitrichine social and mating systems are highly variable, both between genera and within populations of the same species (Rylands, 1996). Monogamy, polygyny, polygamy and polyandry have been described in wild groups of *Saguinus* (Goldizen *et al.*, 1996; Heymann, 1996; Savage *et al.*, 1996). *Cebuella* groups tend to contain a single adult male and female, while *Leontopithecus* groups often contain multiple males and females (Garber, 1994). Multi-male, multi-female *Callithrix* groups appear to be extended families spanning up to three generations (Ferrari and Digby, 1996). Male and female dominance hierarchies are evident in multi-male, multi-female groups, and reproductive opportunities may be limited for subordinates. The dominant male accounts for most observed matings during female periovulatory periods in *Leontopithecus* (Baker *et al.*, 1993) and is thought to be the primary breeder in *Callithrix* (Ferrari and Lopes, 1989; Coutinho and Corrêa, 1995), but multiple males have been observed to mate with the same female on the same day in *C. jacchus* (Rylands, 1996) and *Saguinus mystax* (Garber *et al.*, 1993). Regardless of paternity, both dominant and subordinate males contribute equally to infant care. The tendency of males to migrate in pairs from the same natal cohort may facilitate their co-operative investment in offspring (Garber, 1994).

Subordinate female callitrichins experience varying degrees of reproductive inhibition. In captivity, *Callithrix* and *Saguinus* females cease to ovulate in the presence of a dominant, reproductively active female (Abbott, 1989), whereas social suppression of reproduction appears to inhibit reproduction in ovulating subordinate *Leontopithecus* females (Garber, 1994), while the situation for *Cebuella* is still unclear (French, 1997; Carlson, Ziegler and Snowdon, 1997). Field studies on a variety of callitrichins have reported the presence of multiple breeding females (*Callithrix jacchus*: Scanlon, Chalmers and Monteiro da Cruz, 1988; Digby and Ferari, 1994; Ferrari and Digby, 1996; *C. arita*: Coutinho and Corrêa, 1995; *Leontopithecus rosalia*: Dietz and Baker, 1993; *Saginus fuscicollis*: Goldizen *et al.*, 1996; *S. oedipus*: Savage *et al.*, 1996), suggesting that the inhibition of reproduction is strongly affected by ecological and demographic conditions.

Callitrichin females attain a secure breeding position through dominance, which is related to aggression in *Leontopithecus*, age in *Saguinus*, and inheritance in *Callithrix* (Garber, 1994). *Saguinus* females may also stay in their natal groups to inherit the dominant breeding position, which in *S. fuscicollis* is maintained for an average of three years (Goldizen *et al.*, 1996). Subordinate females remaining in their natal groups may be released from reproductive inhibition when exposed to novel (Widowski *et al.*, 1990) or immigrant males in *Saguinus* (Savage *et al.*, 1996) and *Leontopithecus* (Baker *et al.*, 1993), or when opportunities for dispersal are severely limited in *Callithrix* (Ferrari and Digby, 1996). In contrast to cohort migration among males, all callitrichin females that disperse from their natal groups do so alone (Garber, 1994).

Using these and other distinguishing features of the five extant genera, Garber (1994) has modelled the hypothetical callitrichine and callitrichin ancestors. He regards the single breeding pairs of *Cebuella* as derived from primitive, flexible multi-male, multi-female breeding and social groups, and the three-week postpartum delay in co-operative infant care found exclusively in *Callimico* as primitive. The callitrichin ancestor retained the flexible multi-male, multi-female breeding and social group and the production of a single litter annually, but also expressed twinning, reproductive suppression mediated through hormonal or behavioural means, and co-operative infant care at birth. Paired male migration, with the potential for long-term male kin bonds, also characterised the ancestral callitrichin, whereas solitary dispersal by both sexes occurred in the ancestral callitrichine.

The production of two annual litters of twins in the marmosets could create intense intragroup competition for resources, including food and infant carriers. *Cebuella* maintains small groups (two to nine individuals) by expelling offspring of both sexes through directed aggression (Soini, 1988). *Callithrix* group sizes grow larger (9–15 individuals), due, at least in part, to tolerance towards females waiting to inherit a breeding position or share breeding opportunities (Ferrari and Digby, 1996). These female kin bonds may differ from those of Old World monkeys, however, because the hormonal inhibition of reproduction is a powerful selective force (Rylands, 1996). Reproductive inhibition in *Callithrix* makes female dispersal particularly sensitive to local demographic and ecological conditions affecting natal group size and population density (Digby and Ferrari, 1994). Sexually mature daughters may remain in their natal groups because they lack alternative dispersal options, but not all daughters can inherit their mother's position as dominant breeder, and opportunities for females to become established breeders in non-natal groups are rare or non-existent. The fact that female *Callithrix*, like other female callitrichins, disperse as solitaries also implies weaker kin bonds among them than among males migrating as pairs.

Atelinae ancestors

The subfamily Atelinae includes four extant genera, howling monkeys (*Alouatta*), woolly monkeys (*Lagothrix*), spider monkeys (*Ateles*) and muriquis (*Brachyteles*). *Alouatta* is the sole representative of the Tribe Alouattini; the other three genera comprise the Tribe Atelini. Average adult female weights range from the 4605 g in *Alouatta caraya* to 9450 g in *Brachyteles arachnoides* (Ford, 1994), making them the largest New World primates. Both *Brachyteles* and *Ateles* are distinguished by slow reproductive rates due to delayed maturation and the birth of single infants at roughly three-year intervals (Chapman and Chapman, 1990; Strier, 1996a). *Alouatta* and *Lagothrix* life histories are more like those of Old World monkeys than apes, distinguishing them from the other atelines and other non-Callitrichinae genera (Ross, 1991).

One of the most striking similarities among the four ateline genera is their flexible grouping patterns, which vary both interspecifically and intraspecifically in response to ecological and demographic conditions. *Alouatta* groups tend to be the most cohesive in the subfamily (Strier, 1992), although they fragment into smaller feeding aggregates at localised food patches (Leighton and Leighton, 1982) and may split into subgroups for up to four-week stretches (Chapman, 1988). *Ateles* groups tend to be the most consistently fluid, with females adjusting their party sizes in response to the size of fruit patches, and males, for the most part, monitoring the movements of females. *Brachyteles* groups at different field sites have been described as cohesive (Lemos de Sá, 1991), facultatively cohesive (Strier, 1989), fluid with females and males encountering one another only during the mating season (Milton, 1984), and fluid with fluctuating mixed-sex subgroups (Strier *et al.*, 1993). *Lagothrix* groups are reported as being cohesive in Colombia (Nishimura, 1990, 1994; Stevenson *et al.*, 1994) and more fluid in the Brazilian Amazon (Peres, 1994).

A second, and possibly related, similarity among the atelines is the routine dispersal of females from their natal groups (Strier, 1994b). Female dispersal should preclude opportunities for female kin bonding unless females migrate as cohorts or immigrate into the same groups that older members of their matriline previously joined. But, even in small, isolated populations of *Brachyteles*, in which female dispersal options are limited to one or two non-natal groups such that sisters end up together, bonds between mothers and daughters are severed when daughters disperse (Strier, 1997).

Flexible associations among dispersing females result in weak social relationships between female atelines. Dominance hierarchies among *Alouatta* females appear to be age graded, with young females achieving alpha rank in the process of immigration (Glander, 1980; Jones, 1980). The basis for dominance relationships among female *Ateles* and *Lagothrix* is poorly understood, but agonistic interactions over food resources have been described for both genera (Chapman and Lefebvre, 1990; Stevenson *et al.*, 1994). *Brachyteles* females at low densities appear to maintain egalitarian relationships or avoid one another altogether (Strier, 1990).

Males are philopatric in the three atelin genera, but male kin bonds range from egalitarian to hierarchical depending on the relative degree of cohesiveness among females and whether or not males can dominate females (Strier, 1994b). In *Alouatta*, males as well as females disperse, resulting in both single-male groups and multi-male groups with weak kin bonds except in age-graded groups (Crockett and Eisenberg, 1987; Glander, 1992; Crockett and Pope, 1993). Genetic studies of multi-male *Alouatta seniculus* groups indicate that they are reproductively single-male groups, with paternity for the most part attributed to the dominant male (Pope, 1990).

Whether male philopatry or male dispersal occurred in the ancestral ateline remains uncertain, but comparisons among extant genera support fluid multi-male, multi-female groups, with female dispersal (Rosenberger and Strier, 1989). Neither female dispersal nor flexible grouping patterns are consistent with female kin bonds in this subfamily. Furthermore, as Lee (1994) notes, inbreeding avoidance would prevent the transition to female kin groups once male philopatry becomes established, as appears to have been the case among ancestral atelins.

Ecological perspectives

Ecological models predict female kin bonding when predation pressures make the maintenance of cohesive groups beneficial and when defensible resources favour co-operation among females during intragroup or intergroup contest competition (van Schaik, 1989). Mixed species associations may be an alternative solution to the problem of predators when group size is limited by feeding competition (Terborgh and Janson, 1986), whereas scramble feeding competition and the energetic constraints imposed by folivory can lead to egalitarian relationships or context-specific aggression among females in cohesive groups irrespective of whether or not they are kin (van Schaik, 1989).

Callitrichine and ateline ecology

The absence of female kin bonding emphasised in phylogenetic analyses of callitrichines and atelines is only partially consistent with ecological predictions. Vulnerability to predation may account for the cohesive groups of the small-bodied callitrichines (Caine, 1993), while the relaxation of predation pressures may account for the tendencies of the large-bodied atelines to adjust their grouping patterns in response to the size of food patches. The fruit, invertebrate prey, and plant exudates eaten by calli-thrines include defensible resources that could lead to female kin bonding unless, as van Schaik (1989, p. 212) has noted, 'female dispersal is enforced by some other factor'. The social and hormonal mechanisms that can inhibit reproduction in subordinate callitrichines may exemplify one such 'other factor'. Both the individualistic (non-kin) hierarchies among some female *Saguinus* and the nepotistic (kin) hierarchies among female *Callithrix* (Garber, 1994) are consistent with the potential compromises between the ecological benefits of female kin bonding and the reproductive benefits of female dispersal to avoid female–female competition.

Population differences in ateline grouping patterns co-vary as expected with differences in diet, but dispersal and social systems among members of cohesive groups also appear to reflect phylogenetic constraints on behaviour. For example, the fluid, *Ateles*-like grouping patterns of Brazilian *Lagothrix* are consistent with their *Ateles*-like dietary reliance on plant matter (99%) and, in particular, fruit (83%; Peres, 1994), whereas the cohesive groups of Colombian *Lagothrix* have been attributed to a higher proportion of arthropods (23%) in their diet (Stevenson *et al.*, 1994). Yet, despite low levels of intergroup feeding competition (Nishimura, 1990, 1994), Colombian *Lagothrix* exhibit aggressive, intragroup contest competition in which non-lactating females and immatures are excluded from fruit patches (Stevenson *et al.*, 1994). Such *intra*group contest competition over fruit should favour nepotistic and hierarchical relationships according to the criteria proposed to account for female kin bonding in other primates (van Schaik, 1989). Explaining the absence of female kin bonding in Colombian *Lagothrix* may require reconsidering the importance of *inter*group competition as a selective force on female bonding (Wrangham, 1980), as well as distinguishing between components of female kin bonding that respond to ecological conditions, such as grouping patterns (Strier, 1992; Kinzey and Cunningham, 1994), and those that are phylogenetically conservative, such as sex-biased philopatry (Lee, 1994).

Ecological contrasts among Saimiri

Ecological variables related to the defensibility of food sources are consistent with differences between the dispersal and social systems of Peruvian and Costa Rican squirrel monkeys, formerly classified as *Saimiri sciureus* and *S. oerstedi,* respectively (Mitchell *et al.,* 1991), but now considered to be discrete populations of the same species (Costello *et al.,* 1993). Predation is a serious threat to both populations, and both live in large, cohesive groups that may split up at feeding trees before reuniting to travel or forage for insects (Boinski and Mitchell, 1994; Kinzey and Cunningham, 1994). Both populations devote comparable proportions of their feeding time to animal and plant material, but Peruvian *Saimiri* include a higher diversity of fruits in their diet and feed at larger fruit patches than Costa Rican *Saimiri.* Neither population experiences strong intergroup contest competition. High rates of aggression over fruit observed in Peruvian *Saimiri* imply high levels of intragroup contest competition, but aggression over fruit resources has not been observed within groups of Costa Rican *Saimiri* (Mitchell *et al.,* 1991).

Female-biased dispersal and the absence of female hierarchies or coalitions observed among Costa Rican *Saimiri* correspond to the lack of contest competition in their small, non-defensible fruit patches, whereas female philopatry and the stable linear hierarchies and coalitions observed among Peruvian *Saimiri* correspond to the predicted benefits associated with female kin bonding when resources utilised by group members can be monopolised (Mitchell *et al.,* 1991). The only close associations observed among female Costa Rican *Saimiri* were limited to mothers carrying small infants, presumably to deter predators (Boinski, 1987). Relationships among female Peruvian *Saimiri* resemble those among other female-bonded primates, but their heterosexual relationships, with female dominance over males, and high rates of male migration are more similar to female-bonded lemurs (Richard and Dewar, 1991) than to female-bonded Old World monkeys (Table 11.2).

Saimiri males from Peru, like male ring-tailed lemurs (*Lemur catta;* Sussman, 1992) and male callitrichines (see above), routinely disperse and maintain long-term alliances with same-aged peers from their natal groups (Mitchell, 1994). The fact that one or two males appear to account for most consortships during the breeding system may result in close genetic

Table 11.2. *Saimiri population variability*[a]

	Costa Rica	Peru
Competitive regime	Scramble	Contest
Dispersing sex	Female biased	Male biased
Female–female relationships	Weak affiliations; egalitarian	Strong affiliations; hierarchical
Male–male relationships	Strong affiliations; hierarchical during breeding season	Agonistic support among migration alliances; hierarchical year-round
Female–male relationships	Weak affiliations; egalitarian	Female dominance

[a]References cited in text.

relationships among sons who form migration alliances (Mitchell, 1994), but even if male cohorts were sired by different fathers, they would still share some genetic ties due to the kinship among their mothers. Philopatric males in Costa Rican *Saimiri* maintain strong affiliative, egalitarian bonds, except during the annual breeding season, when their hierarchical relationships resemble those that Peruvian males maintain year round (Boinski, 1994).

The divergent social and dispersal systems exhibited by *Saimiri* could have arisen from an ancestral system in which both sexes dispersed (*contra* Boinski and Mitchell, 1994). Ancestral female *Saimiri* may have dispersed individually, as exhibited by extant *Saimiri* inCosta Rica. Ancestral male *Saimiri* may have formed kin-based migration alliances for competition with unrelated males that are similar to those of extant *Saimiri* in Peru (Mitchell, 1994) and that extended into the affiliative bonds exhibited today among philopatric male *Saimiri* in Costa Rica (Boinski, 1994). Once female kin bonding becomes ecologically advantageous due to the defensibility of prized food resources, as in extant *Saimiri* in Peru, male migration alliances may become potential threats to females over contested food and feeding sites (van Hooff and van Schaik, 1992), particularly when reproduction is seasonally constrained (Strier, 1996b). The fact that male migration alliances in Peruvian *Saimiri* are not effective at overriding female dominance (Mitchell, 1994) supports the possibility that female dominance over males arose in response to male migration alliances, and is consistent with the proposal that female kin bonding (and male philopatry) are derived conditions in this genus.

The Cebus *anomaly*

Capuchin monkeys (*Cebus*) are consistently characterised by male-biased dispersal with female philopatry, making them anomalies among New World primates. For this reason, as well as because of similarities in body size, dietary categories and ranging patterns, *Cebus* has often been compared to the Old World monkey genus *Cercopithecus* (Robinson and Janson, 1987; Fedigan, 1993). *Cebus* populations are highly variable (Table 11.3), but some resemble other female kin-bonded primates in their affiliative female bonds (*C. capucinus*: Fedigan, 1993), agonistic support and inheritance of maternal ranks that affect the outcome of within-group contest competition over food (*C. olivaceus*: O'Brien, 1991; O'Brien and Robinson, 1993), tendency to fission along matrilineal lines (Colombian *C. apella*: Izawa, 1994a), and active participation of females in intergroup encounters over food (Peruvian *C. apella* and *C. albifrons*: Janson, 1986).

Some of the population variation exhibited by female-bonded *Cebus* can be explained by ecological differences. For example, the fact that *C. capucinus* females seldom participate in co-operative resource defence may reflect low levels of intergroup feeding competition among 'opportunistic, extractive foragers' (Fedigan, 1993) that also shift their diets rather than their grouping patterns in response to seasonal fluctuations in their food supply (Chapman, 1990; Kinzey and Cunningham, 1994). Similarly, females may increase access to food for themselves and their offspring through affiliative bonds and preferential mating with alpha males whenever within-group contest competiton is strong (Peruvian *C. apella*: Janson, 1985; *C. olivaceus*: O'Brien, 1991). Low levels of feeding competition in large, indefensible fruit patches may account for the absence of affiliative or mating preferences that females exhibit towards alpha males in *C. albifrons* (Janson, 1986), as well as in *C. capucinus* females (Fedigan, 1993) and *C. apella* females in Colombia (Izawa, 1994b).

Population variation in female kin bonding among *Cebus* can also be attributed to their unique life history traits and reproductive patterns. Unlike other female-bonded primates of similar body size, *Cebus* have low reproductive rates due to long interbirth intervals (Ross, 1991) and large relative brain size, which may increase interbirth intervals (Fedigan and Rose, 1995). High postnatal investment in offspring, particularly during the first year of life, is consistent with reports of allomothering (Robinson and Janson, 1987; O'Brien and Robinson, 1993), and high infant mortality (O'Brien and Robinson, 1993; Izawa, 1994b).

Compared to *Saimiri*, which also have low reproductive rates (Ross, 1991), *Cebus* infants may be more vulnerable to lethal aggression from

Table 11.3. *Cebus population variation*

	C. albifrons[1,2]	C. apella (Peru)[1-4]	C. apella (Colombia)[5,6]	C. capucinus[7,8]	C. olivaceus[2,9,10]
Competitive regime (within group)	Contest + scramble	Contest + scramble	Contest + scramble	Contest + scramble	Contest + scramble
Between group resource defence	Females + males	Females + alpha male	Females + males	None (?)	Females + alpha male (?)
Dispersing sex	Male biased	Male biased	Male biased	Male biased; influxes aseasonal	Male biased
Female–female realtionships	Hierarchical	Hierarchical	Hierarchical	Hierarchical + affiliative	Hierarchical + affiliative
Male–male relationships	Tolerant, hierarchical multi-male mating	Aggressive, hierarchical; alpha male mating	Tolerant, hierarchical; multi-male mating	Tolerant, hierarchical; multi-male mating	Aggressive, hierarchical; alpha male mating
Female–male relationships	Hierarchical; alpha female dominates some males	Hierarchical; alpha female dominates some males; affiliative with alpha male	Hierarchical; alpha female dominates some males	Hierarchical; alpha female dominates some males	Hierarchical; alpha female dominates some males; affiliative with alpha male
Reproductive seasonality	4 months	4 months	4 months	Year-round, most during 6 months	4 months, most during 2 months

References: (1) Janson, 1986; (2) Robinson and Janson, 1987; (3) Janson, 1984; (4) Janson, 1985; (5) Izawa, 1994a; (6) Izawa, 1994b; (7) Fedigan, 1993; (8) Fedigan and Rose, 1995; (9) O'Brien, 1991; (10) O'Brien and Robinson, 1993.

unrelated males because reproduction in *Cebus* is less tightly synchronised (Strier, 1996b). Reports of aggression towards unrelated immature *C. apella* (Janson, 1984) and the coincidence of infant deaths or disappearances with turn-overs in male rank and group membership in *C. apella* (Izawa, 1994b), *C. olivaceus* (O'Brien, 1991), and *C. capucinus* (Fedigan, 1993) are consistent with predictions that male–male competition, including interference with infant survival, will be more intense whenever female reproduction is not seasonally constrained (Strier, 1996b).

Sexual dimorphism in body size (Ford, 1994) and canine size (Plavcan and van Schaik, 1992) contributes to the ability of male *Cebus* to harm infants, as well as females (Smuts, 1987). Kin provide females with reliable allies against the threat of aggression that unrelated males, as well as unrelated females, may pose to themselves and their infants (Smuts and Smuts, 1993). Kin bonds may enable alpha *Cebus* females to establish dominance over all but the alpha and, more rarely, beta males (*C. apella*: Janson, 1985; *C. olivaceus*: O'brien, 1991; *C. capucinus*: Fedigan, 1993), and facilitate coalitions among females that can influence the outcome of male–male competition for status and group membership (*C. olivaceus*: O'Brien, 1991; *C. apella*: Izawa, 1994b).

Conclusions

In the nearly three decades since Wrangham (1980) proposed an ecological model of female bonding in primates, increasingly fine-grained analyses have been employed to understand the determinants, corollaries and mechanisms of female kin bonding among primates. The importance of predation (van Schaik and van Hooff, 1983), female transfer (Moore, 1984), constraints on group size (Terborgh and Janson, 1986) and cohesiveness (Kinzey and Cunningham, 1994), and varying within-group competitive regimes (van Schaik, 1989; van Hooff and van Schaik, 1992) on female bonding are now widely recognised. The interacting effects of phylogeny (Strier, 1990; Di Fiore and Rendall, 1994; Lee, 1994; Rendall and Di Fiore, 1996), demography (Moore, 1992), sexual and social selection pressures (Smuts and Smuts, 1993), and life histories (van Schaik, 1989) on the ecological distribution of female primate kin bonds require further exploration.

Ecological variables by themselves are insufficient to explain the variation in New World primate social and dispersal systems. Crucial links between foraging strategies, dispersal patterns, and male and female social relationships emerge from considerations of the interacting effects of phylogenetic constraints, such as those that impede shifts from male to female kin bonding in atelines, reproductive constraints, such as those that inhibit reproduction in subordinate callitrichins, and social constraints such as those arising from reproductive seasonality in *Saimiri* and male aggression in *Cebus*.

Female kin bonding is far from ubiquitous among primates. Approaching questions about the evolution of female kin bonding from the perspective of it being a highly specialised phenomenon subject to phylogenetic, demographic and reproductive constraints will accelerate efforts to understand what defines female kin bonding, and why it is so rare.

Acknowledgements

Research for this manuscript was supported by NSF grant BNS 8959298 and a sabbatical award from the University of Wisconsin–Madison. The author is grateful to Anne Carlson for her valuable comments on an earlier version of this manuscript.

References

Abbott, D.H. (1989). Social suppression of reproduction in primates. In *Comparative Socioecology: the Behavioural Ecology of Humans and Other Mammals*, ed. V. Standen and R.A. Foley, pp. 285–304. Oxford: Blackwell Scientific.

Baker, A.J., Dietz, J.M. and Kleinman, D.G. (1993). Behavioural evidence for monopolization of paternity in multi-male groups of golden lion tamarins. *Animal Behaviour* **46**, 1091–103.

Boinski, S. (1987). Birth synchrony in squirrel monkeys. (*Saimiri oerstedi*). *Behavioral Ecology and Sociobiology* **21**, 393–400.

Boinski, S. (1994). Affiliation patterns among male Costa Rican squirrel monkeys. *Behaviour* **130**, 191–209.

Boinski, S. and Mitchell, C.L. (1994). Male residence and association patterns in Costa Rican squirrel monkeys. (*Saimiri oersedti*). *American Journal of Primatology* **34**, 157–69.

Caine, N.G. (1993). Flexibility and co-operation as unifying themes in *Saguinus* social organization and behavior: the role of predation pressures. In *Marmosets and Tamarins*, ed. A.B. Rylands, pp. 200–19. New York: Oxford University Press.

Carlson, A.A., Ziegler, T.E. and Snowdon, C.T. (1997). Ovarian function of pygmy marmoset daughters (*Cebuella pygmaea*) in intact and motherless families. *American Journal of Primatology* **43**, 347–55.

Chapman, C.A. (1988). Patch use and patch depletion by the spider and howling monkeys of Santa Rosa National Park, Costa Rica. *Behaviour* **105**, 99–116.

Chapman, C.A. (1990). Ecological constraints on group size in three species of neotropical primates. *Folia Primatologica* **55**, 1–9.

Chapman, C.A. and Chapman, L.J. (1990). Reproductive biology of captive and free-ranging spider monkeys. *Zoo Biology* **9**, 1–9.

Chapman, C.A. and Lefebvre, L. (1990). Manipulating foraging group sizes: spider monkey food calls at fruiting trees. *Animal Behaviour* **39**, 891–6.

Costello, R.K., Dickinson, C., Rosenberger, A.L., Boinski, S. and Salay, F.S. (1993). Squirrel monkey (genus *Saimiri*) taxonomy: a multidisciplinary study of the biology of species. In *Species, Species Concepts, and Primate Evolution*, ed. B. Kimbel and L. Martin, pp. 177–237. New York: Plenum.

Coutinho, P.E.G. and Corrêa, H.K.M. (1995). Polygyny in a free-ranging group of buffy-tufted-ear marmosets, *Callithrix aurita. Folio Primatologica* **65**, 25–9.

Crockett, C.M. and Eisenberg, J.F. (1987). Howlers: variations in group size and demography. In *Primate Societies*, ed. B.B. Smuts, D.L. Cheney, R.M. Seyfarth, R.W. Wrangham and T.T. Struhsaker, pp. 54–68. Chicago: University of Chicago Press.

Crockett, C.M. and Pope, T.R. (1993). Consequences of sex differences in dispersal for juvenile red howler monkeys. In *Juvenile Primates: Life History, Development, and Behavior*, ed. M.E. Pereira and L.A. Fairbanks, pp. 104–18. New York: Oxford University Press.

Di Fiore, A. and Rendall, D. (1994). Evolution of social organization: a reappraisal for primates by using phylogenetic methods. *Proceedings of the National Academy of Science, USA* **91**, 9941–5.

Dietz, J.M. and Baker, A.J. (1993). Polygyny and female reproductive success in golden lion tamarins, *Leontopithecus rosalia. Animal Behaviour* **46**, 1067–78.

Digby, L.J. and Ferrari, S.F. (1994). Multiple breeding females in free-ranging groups of *Callithrix jaccus. International Journal of Primatology* **15**, 389–97.

Fedigan, L. (1993). Sex differences and intersexual relations in adult white-faced capuchins (*Cebus capucinus*). *International Journal of Primatology* **14**, 853–78.

Fedigan, L.M. and Rose, L.M. (1995). Interbirth interval variation in three sympatric species of neotropical monkey. *American Journal of Primatology* **37**, 9–24.

Ferrari, S.F. and Digby, L.J. (1996). wild *Callithrix* groups: stable extended families? *American Journal of Primatology* **38**, 19–27.

Ferrari, S.F. and Lopes, M.A. (1989). A re-evaluation of the social organisation of the Callitrichidae, with reference to the ecological differences between genera. *Folia Primatologica* **52**, 132–47.

Ford, S.M. (1994). Evolution of sexual dimorphism in body weight in platyrrhines. *American Journal of Primatology* **34**, 221–44.

French, J.A. (1997). Proximate regulation of singular breeding in callitrichid primates. In *Cooperative Breeding Mammals*, ed. N.G. Solomon and J.A. French, pp. 34–75. New York: Cambridge University Press.

Garber, P.A. (1994). Phylogenetic approach to the study of tamarin and marmoset social systems. *American Journal of Primatology* **34**, 199–219.

Garber, P.A., Encaración, F., Moya, L. and Puetz, J.P. (1993). Demography and reproductive patterns in moustached tamarin monkeys (*Saguinus mystax*): implications for reconstructing platyrrhine mating systems. *American Journal of Primatology* **29**, 235–54.

Glander, K.E. (1980). Reproduction and population growth in free-ranging mantled howling monkeys. *American Journal of Physical Anthropology* **53**, 25–36.

Glander, K.E. (1992). Dispersal patterns in Costa Rican mantled howling monkeys. *International Journal of Primatology* **13**, 415–36.

Goldizen, A.W., Mendelson, J., van Vlaardingen, M. and Terborgh, J. (1996). Saddle-back tamarin (*Saguinus fuscicollis*) reproductive strategies: evidence from a thirteen-year study of a marked population. *American Journal of Primatology* **38**, 57–83.

Heymann, E.W. (1996). Social behavior of wild moustached tamarins, *Saguinus mystax*, at the Estación Biológica Quebrada Blanco, Peruvian Amazonia. *American Journal of Primatology* **38**, 101–13.

Izawa, K. (1994a). Group division of wild black-capped capuchins. *Field Studies of New World Monkeys, La Macarena Colombia* **9**, 5–14.

Izawa, K. (1994b). Social changes within a group of wild black-capped cpuchins, IV. *Field Studies of New World Monkeys, La Macarena Colombia* **9**, 5–14.

Janson, C.H. (1984). Female choice and mating system of the brown capuchin monkey *Cebus apella* (Primates: Cebidae). *Zeitschrift fur Tierpsychologie* **65**, 177–200.

Janson, C. (1985). Aggressive competition and individual food consumption in wild brown capuchin monkeys (*Cebus apella*). *Behavioral Ecology and Sociobiology* **18**, 125–38.

Janson, C.H. (1986). The mating systems as a determinant of social evolution in capuchin monkeys (*Cebus*). In *Primate Ecology and Conservation*, Vol. 2, ed. J.G. Else and P.C. Lee, pp. 169–79. Cambridge: Cambridge University Press.

Jones, C.B. (1980). The functions of status in the mantled howler monkey. *Alouatta palliata* Grav: intraspecific competiton for group membership in a folivorous neotropical primate. *Primates* **21**, 389–405.

Kinzey, W.G. and Cunningham, E.P. (1994). Variability in platyrrhine social organization. *American Journal of Primatology* **34**, 185–98.

Lee, P.C. (1994). Social structure and evolution. In *Behaviour and Evolution*, ed. P.J.B. Slater and T.R. Halliday, pp. 266–303. Cambridge: Cambridge University Press.

Leighton, M. and Leighton, D.R. (1982). The relationship of size of feeding aggregate to size of food patch: howler monkeys (*Alouatta palliata*) feeding in *Trichilia cipo* fruit trees on Barro Colorado Island. *Biotropica* **14**, 81–90.

Lemos de Sá, R.M. (1991). Poplação de *Brachyteles arachnoides* (Primates, Cebidae) da Fazenda Esmerald, Rio Casca, Minas Gerais. In *A Primatologia no Brasil-3*, ed. A.B. Rylands and A.T. Bernardes, pp. 235–8. Belo Horizonte, Brasil: Fundação Biodiversitas.

Milton, K. (1984). Habitat, diet and activity patterns of free-ranging woolly spider monkeys (*Brachyteles arachnoides* E. Geoffroy 1806). *International Journal of Primatology* **5**, 491–514.

Mitchell, C.L. (1994). Migration alliances and coalitions among adult male south American squirrel monkeys (*Saimiri sciureus*). *Behaviour* **130**, 169–90.

Mitchell, C.L., Boinski, S. and van Schaik, C.P. (1991). Competitive regimes and female bonding in two species of squirrel monkeys (*Saimiri oerstedi* and *S. sciureus*). *Behavioral Ecology and Sociobiology* **28**, 55–60.

Moore, J. (1984). Female transfer in primates. *International Journal of Primatology* **5**, 537–89.

Moore, J. (1992). Dispersal, nepotism, and primate social behavior. *International Journal of Primatology* **13**, 361–78.

Nishimura, A. (1990). A sociological and behavioral study of woolly monkeys, *Lagothrix lagotricha*, in the upper Amazon. *The Science and Engineering Review of Doshisha University* **31**, 87–121.

Nishimura, A. (1994). Social interaction patterns of woolly monkeys (*Lagothrix lagotricha*): a comparison among the atelins. *The Science and Engineering Review of Doshisha University* **35**, 235–54.

O'Brien, T.G. (1991). Female–male social interactions in wedge-capped capuchin monkeys: benefits and costs of group living. *Animal Behaviour* **41**, 555–68.

O'Brien, T.G. and Robinson, J.G. (1993). Stability of social relationships in female wedge-capped capuchin monkeys. In *Juvenile Primates: Life History, Development, and Behavior*, ed. M.E. Pereira and L.A. Fairbanks, pp. 197–210. New York: Oxford University Press.

Peres, C.A. (1994). Diet and feeding ecology of gray woolly monkeys (*Lagothrix lagotricha cana*) in Central Amazonia: comparisons with other atelines. *American Journal of Primatology* **15**, 333–72.

Plavcan, J.M. and van Schaik, C.P. (1992). Intrasexual competition and canine dimorphism in anthropoid primates. *American Journal of Physical Anthropology* **87**, 461–77.

Pope, T. (1990). The reproduction consequences of male cooperation in the red howler monkey: paternity exclusion in multi-male and single-male troops using genetic markers. *Behavioral Ecology and Sociobiology* **27**, 439–46.

Pusey, A.E. and Packer, C. (1987). Dispersal and philopatry. In *Primate Societies*, ed. B.B. Smuts, D.L. Cheney, R.M. Seyfarth, R.W. Wrangham and T.T. Struhsaker, pp. 250–66. Chicago: University of Chicago Press.

Rendall, D. and Di Fiore, A. (1996). The road less traveled: phylogenetic perspectives in primatology. *Evolutionary Anthropology* **4**, 43–52.

Richard, A.F. and Dewar, R.E. (1991). Lemur ecology. *Annual Review of Ecology and Systematics* **22**, 145–75.

Robinson, J.G. and Janson, C.H. (1987). Capuchins, squirrel monkeys, and atelines: socioecological convergence. In *Primate Societies*, ed. B.B. Smuts, D.L. Cheney, R.M. Seyfarth, R.W. Wrangham and T.T. Struhsaker, pp. 69–82. Chicago: University of Chicago Press.

Robinson, J.G., Wright, P.C. and Kinzey, W.G. (1987). Monogamous cebids and their relatives: intergroup calls and spacing. In *Primate Societies*, ed. B.B. Smuts, D.L. Cheney, R.M. Seyfarth, R.W. Wrangham and T.T. Struhsaker, pp. 44–53. Chicago: University of Chicago Press.

Rosenberger, A.L. and Strier, K.B. (1989). Adaptive radiation of the ateline primates. *Journal of Human Evolution* **18**, 717–50.

Ross, C. (1991). Life history patterns of New World monkeys. *International Journal of Primatology* **12**, 481–502.

Rylands, A.B. (1996). Habitat and the evolution of social and reproductive behavior in callitrichidae. *American Journal of Primatology* **38**, 5–18.

Savage, A., Giraldo, L.H., Soto, L.H. and Snowdon, C.T. (1996). Demography, group composition, and dispersal in wild cotton-top tamarin (*Saguinus oedipus*) groups. *American Journal of Primatology* **38**, 85–100.

Scanlon, C.E., Chalmers, N.R. and Monteiro de Cruz, M.A.O. (1988). Changes in the size, composition and reproductive condition of wild marmoset groups (*Callithrix jacchus jacchus*) in north east Brazil. *Primates* **29**, 295–305.

Schneider, H., Schneider, M.P.C., Sampaio, I. *et al.* (1993). Molecular phylogeny of the New World monkeys (*Platyrrhini, Primates*). *Molecular Phylogenetics and Evolution* **2**, 225–42.

Smuts, B.B. (1987). Gender, aggression, and influence. In *Primate Societies*, ed. B.B. Smuts, D.L. Cheney, R.M. Seyfarth, R.W. Wrangham and T.T. Struhsaker, pp. 400–12. Chicago: University of Chicago Press.

Smuts, B.B. and Smuts, R.W. (1993). Male aggression and sexual coercion of females in nonhuman primates and other mammals: evidence and theoretical implications. *Advances in the Study of Behavior* **22**, 1–63.

Soini, P. (1988). The pygmy marmoset, genus *Cebuella*. In *Ecology and Behavior of Neotropical Primates*, Vol. 2, ed. R.A. Mittermeier, A.B. Rylands, A.F. Coimbra-Filho, and G.A.B. da Fonseca, pp. 79–129. Washington, DC: World Wildlife Fund.

Stevenson, P.R., Quiñones, M.J. and Ahumada, J.A. (1994). Ecological strategies of woolly monkeys (*Lagothrix lagotricha*) at Tinigua National Park, Colombia. *American Journal of Primatology* **32**, 123–40.

Strier, K.B. (1989). Effects of patch size on feeding associations in muriquis (*Brachyteles arachnoides*). *Folia Primatologica* **52**, 70–7.

Strier, K.B. (1990). New World primates, new frontiers: insights from the woolly spider monkey, or muriqui (*Brachyteles arachnoides*). *International Journal of Primatology* **11**, 7–19.

Strier, K.B. (1992). Atelinae adaptations: behavioral strategies and ecological constraints. *American Journal of Physica Anthropology* **88**, 515–24.

Strier, K.B. (1994a). Myth of the typical primate. *Yearbook of Physical Anthropology* **37**, 233–71.

Strier, K.B. (1994b). Brotherhoods among atelins. *Behaviour* **130**, 151–67.

Strier, K.B. (1996a). Male reproductive strategies in new world primates. *Human Nature* **7**, 105–23.

Strier, K.B. (1996b). Reproductive ecology of female muriquis. In *Adaptive Radiations of Neotropical Primates*, ed. M. Norconk, A. Rosenberger and P. Garber, pp. 511–32. New York: Plenum Press.

Strier, K.B. (1997). Behavioral ecology and conservation biology of primates and other animals. *Advances in the Study of Behavior* **26**, 101–58.

Strier, K.B., Mendes, F.D.D., Rímoli, J. and Rímoli, A.O. (1993). Demography and social structure in one group of muriquis (*Brachyteles arachnoides*). *International Journal of Primatology* **14**, 513–26.

Sussman, R.W. (1992). Male life history and intergroup mobility among ringtailed lemurs (*Lemur catta*). *International Journal of Primatology* **13**, 395–413.

Terborgh, J. and Janson, C.H. (1986). The socioecology of primate groups. *Annual Review of Ecology and Systematics* **17**, 111–35.

van Hooff, J.A.R.A.M. and van Schaik, C.P. (1992). Cooperation in competition: the ecology of primate bonds. In *Coalitions and Alliances in Humans and Other Animals*, ed. A.H. Harcourt and F.B.M. de Waal, pp. 357–90. Oxford: Oxford University Press.

van Schaik, C.P. (1989). The ecology of social relationships amongst primate females. In *Comparative Socioecology: the Behavioural Ecology of Humans and*

Other Mammals, ed. V. Standen and R.A. Foley, pp. 195–218. Oxford: Blackwell Scientific.

van Schaik, C.P. and van Hooff, J.A.R.A.M. (1983). On the ultimate causes of primate social systems. *Behaviour* **85**, 91–117.

Widowski, T.M., Ziegler, T.E., Elowson, A.M. and Snowdon, C.T. (1990). The role of males in the stimulation of reproductive function in female cotton-top tamarins, *Saguinus o. oedipus. animal Behaviour* **40**, 731–41.

Wrangham, R.W. (1980). An ecological model of female-bonded primate groups. *Behaviour* **75**, 262–99.

12 Energetics, time budgets and group size

DAISY K. WILLIAMSON AND ROBIN DUNBAR

Introduction

Individual animals seek to maximise their fitness by pursuing strategies that optimise their current opportunities. One of the factors that determines the opportunities available to an individual is the size of group it lives in, because group size constrains the choice of social partners available (Dunbar, 1996). Group size in turn is in part determined by environmental conditions, three sets of which are likely to be especially important. First, resource availability and patchiness will set an upper limit on group size and density. These will be determined by broad climatic and geophysical variables, but indirect competition from both conspecifics and other ecological competitors will also be important. Second, resource quality and thermoregulatory considerations will add an additional constraint on group size through their impact on time budgets. Both resource quality and thermoregulation will be directly influenced by climate. Finally, predation risk (and perhaps direct competition from conspecifics) will set lower limits on group size below which animals will not be able to resist predation or competitive exclusion. Over evolutionary time, these constraints will give rise to cognitive mechanisms designed to facilitate the cohesion of groups of the size that is typical for the species as a whole, given the ecological niche it occupies. These cognitive mechanism will impose upper limits on group size that are species specific.

This chapter presents a general approach to the determinants of group size in primates that helps us to understand why group sizes vary across habitats both within and between species. The general approach is illustrated in Figure 12.1 which shows a linear programming model for baboon (*Papio* spp.) group sizes in relation to one environmental variable (rainfall). In this case, the range of possible group sizes that a species can maintain in a given habitat is determined by three key variables: (1) the minimum permissible group size required to solve some pressing ecological problem such as predation risk; (2) the maximum ecologically tolerable group size set by a combination of resource availability and the species' time budget

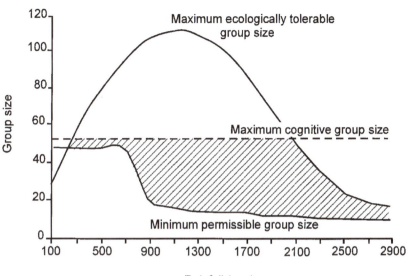

Fig. 12.1 Linear programming model of baboon group sizes, based on the time budgets model given by Dunbar (1996). The realisable range of group sizes is determined by the maximum ecologically tolerable group size (set by habitat-specific time budget constraints), the minimum permissible group size (set by habitat-specific predation risk) and the cognitive constraints characteristic of the species. The values shown are those for habitats with a mean ambient temperature of 25 °C (approximately the middle of the genus's climatic range). Animals cannot survive in habitats in which the minimum permissible group size is larger than the maximum tolerable. Redrawn from Dunbar (1996).

flexibility; and (3) the cognitive constraints on group size imposed by the species-typical ability to handle and maintain relationships. The first two variables are important because they set limits on the viability of populations: if the maximum ecologically tolerable group size lies below the habitat-specific minimum permissible group size (given the local predation risk), the species will be unable to live in that habitat.

Note that animals in a given habitat do not have to live in groups of the maximum tolerable size. Rather, the observed group size will lie within the state space defined by these variables. The optimal group size will depend on whether the animals are trying to maximise birth rates (in which case it will lie close to the minimum line) or maximise survival rates (in which case it will lie close to the line defining the maximum tolerable group size).

Models are presented of maximum group size for two genera of baboons (the gelada, *Theropithecus gelada*, and the common baboons, *Papio* spp.) and the chimpanzee (*Pan* spp.). In each case, these models allow us to place an upper limit on the size of group that the species can maintain in a given habitat. The basic model is presented first and its derivation is explained before differences in parameter values between the genera are discussed. The next step is to use the models to predict maximum group sizes for each taxon and show how the models explain why there is niche separation between the three genera. Finally, some recent criticisms of the model are considered and it is shown that the models produce robust predictions that are upheld by the available data.

The time budgets model

The basic model assumes that there are four activity categories that account for all (or at least most) of an animal's active day. These categories are taken to be feeding, moving, resting and social time. Each of these is assumed to be determined by a suite of variables, some of which are climatic (e.g. rainfall and temperature), some are related to habitat quality, and some are related either to other time budget categories or to behavioural variables such as group size and day journey length. To determine what these sets of determinants are, stepwise regression analyses are carried out for each relevant dependent variable. In each case, the set of possible independent variables considered comprises only those that make biological sense (i.e. those for which a plausible biological causal explanation can be specified). Some of the relationships involved are likely to form a natural cascade. Climate determines habitat quality (high temperatures produce nutritionally poorer quality plants), and habitat quality then determines feeding time (the poorer the forage quality, the more time has to be spent feeding for the animal to ingest the same quantity of nutrients).

For practical reasons, only four measures of climate were used in the original analyses: these were total annual rainfall, rainfall evenness across the year (i.e. seasonality), number of months with less than 50 mm of rainfall, and the mean ambient temperature. Although better measures could have been chosen (e.g. evapotranspiration), the selected variables have the advantage of being widely available for most baboon study sites. Williamson (1997) has since undertaken a principal components analysis of nine different climatic variables for 218 weather stations randomly sampled throughout sub-Saharan Africa, and has shown that all these climatic variables can be reduced to three principal variables (total rainfall, rainfall variability and mean temperature).

Both rainfall (le Houérou and Hoste, 1977; Rutherford, 1980; Deshmuhk, 1984; McNaughton, 1985) and evapotranspiration (Rosenzweig, 1968; Leith and Box, 1972) turn out to be strongly correlated with plant productivity in sub-Saharan Africa. Evapotranspiration is in turn correlated with the number of dry months:

$$AE = 1615.3 - 126.1 \, V$$

($r^2 = 0.62$, $F_{1,47} = 28.8$, $p < 0.01$) where AE is the annual evapotranspiration (mm/year) and V is the number of months in the year with less than 50mm of rainfall (data from Thornthwaite and Mather, 1962).

Table 12.1 gives relevant data for the *Papio* and gelada models. Table 12.2 gives the equations for the main variables for the *Papio* model. The equations given in Table 12.2 are those from Dunbar (1996), and incorporate a number of corrections from the original equations of Dunbar (1992a). Note that two equations are given for social time. One is determined from the stepwise regression analysis from the time budget data for each study site. The other is a direct relationship between social time and group size, based on the assumption that social time (principally allogrooming) is the essential glue needed to bond primate social groups. At least in Old World monkeys and apes, the amount of social time required is directly proportional to group size (Dunbar, 1992a). Individual groups may be forced to compromise on this in order to survive in a particular habitat (this is reflected in the stepwise equation for social time), but doing so places the group at risk of social fragmentation and, ultimately, fission (see below). Full details and the equations for all the models can be found in Dunbar (1992a, 1992b, 1996) and Williamson (1997).

Once these socioecological relationships are determined, it is possible to build a systems model of the kind shown in Figure 12.2. The arrows linking the variables show the putative causal relationships identified by the stepwise regression analyses.

The final step is to use the systems model to determine the upper limit on group size (the *maximum ecologically tolerable group size*, N_{max}) for populations living in different habitats. This is done iteratively by allowing group size to increase until all the spare resting time has been converted into feeding, moving or social time. Note that the maximum tolerable group size is habitat specific, and can vary considerably across a species' geographical range (see Fig. 12.1).

Taxon-specific elements in the model

The systems models derived for the three taxa are broadly similar, but differ in some important respects. This can be illustrated by a comparison

Table 12.1. *Time budgets and climatic data for the baboon and gelada populations*

Site	Percentage of time: Feed	Move	Rest	Social	Day journey (km)	Mean group size	Annual rainfall (mm)	Rainfall diversity[a]	Number of dry months[b]	Mean temperature (°C)
Gelada populations:										
Sankaber, Ethiopia	45.2	20.4	13.8	20.5	2.2	262	1385	—[c]	—[c]	9.8
Gich, Ethiopia	62.3	14.7	5.2	16.0	1.0	112	1515	—[c]	—[c]	7.7
Bole, Ethiopia	35.7	17.4	26.3	18.5	0.6	60	1100	—[c]	—[c]	15.9
Baboon populations										
Mt Assirik, Senegal	23.5	36.9	20.7	18.9	7.9	247	954	0.817	3	30.1
Shai Hills, Ghana	20.3	18.2	61.4	22.7	1.3	23.7	1065	0.855	4	25.9
Bole Valley, Ethiopia	20.5	25.4	35.4	15.9	1.2	19	1105	0.799	6	19.5
Mulu, Ethiopia	40.8	25.0	22.4	14.7	1.1	22	1105	0.799	6	15.8
Awash Falls, Ethiopia	30.9	25.0	30.5	12.2	5.3	71	517	0.858	9	24.7
Budongo, Uganda	59.3	17.6	5.9	16.9	3.8	37.5	1886	0.886	2	22.0
Chololo, Kenya	40.2	33.1	17.4	7.8	5.6	102	597	0.846	8	22.1
Gilgil, Kenya	50.7	30.4	9.6	9.3	4.3[d]	57	730	0.907	4	17.3
Amboseli NP, Kenya	48.0	24.1	20.9	6.7	6.1	46.5	225	0.820	11	23.1
Gombe NP, Tanzania	25.8	19.4	30.2	10.6	2.4[d]	43	1417	0.861	4	23.5
Ruaha NP, Tanzania	47.4	24.2	16.7	4.5	6.8[d]	72	304	0.718	3	21.7
Mikumi NP, Tanzania	36.5	26.1	25.0	5.9	3.4	120	851	0.863	6	24.3
Giants Castle, South Africa	56.6	17.7	16.8	7.7	0.9[d]	11.8	1197	0.866	5	14.6
Cape Point, South Africa	33.5	29.0	26.3	11.3	8.2	85	483	0.886	8	16.5

Sources: gelada populations: Dunbar (1992b); baboon populations: Dunbar (1992a), with corrections from Williamson (1997).

[a]Simpson's evenness index (based on rainfall in each month of the year).

[b]Number of months with less than 50 mm of rainfall.

[c]The gelada model uses percentage of grass cover as a direct measure of primary production instead of rainfall diversity.

[d]Day journey length is estimated from the equation given in Table 12.2; these data were used only in the derivation of equations for time budget elements.

Table 12.2. *Regression equations for the* Papio *time budget and day journey length variables*

Regression equation
$\ln(F) = 6.87 + 4.08\ln(Z) - 0.75\ln(T) - 0.39\ln(V) + 0.16\ln(J)$
$\ln(M) = 2.21 + 0.16\ln(N) + 0.22\ln(V)$
$\ln(R) = 0.97 - 7.92\ln(Z) + 0.60\ln(V)$
*$\ln(S) = -2.28 + 1.32\ln(N) - 0.04(\ln(N))^2$
*$\ln(S) = -1.60 + 0.49\ln(P) - 4.97\ln(Z)$
$\ln(J) = 1.34 + 0.7\ln(N) - 0.47\ln(P)$

F = feeding time (%); M = moving time (%); R = resting time (%); S = social time (%); J = day jouney length (km); Z = Simpson's index of rainfall evenness; T = mean annual temperature (°C); V = number of dry months (< 50 mm rainfall); P = mean annual rainfall (mm).
*Two equations are given for social time. The upper equation is based on the relationship between grooming time and group size for 13 genera of Old World monkeys and apes; the lower one is the conventional stepwise regression for the 18 baboon sites done in the same way as the other equations.

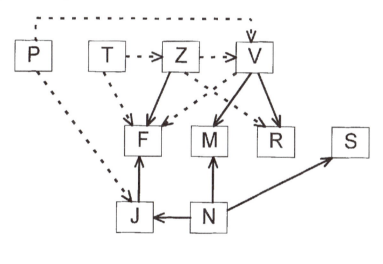

- - > Negative relationship

——> Postive relationship

Fig. 12.2   Flow diagram for the systems model used to predict maximum ecologically tolerable group size. Redrawn from Dunbar (1992c).

of the baboon and gelada models. Ambient temperature is a variable in the equation for feeding time in both models:

$$\ln(F_{gel}) = 5.9 - 0.6\ln(T) - 0.9\ln(Q)$$
$$\ln(F_{bab}) = 6.4 - 0.61\ln(T) + 5.7\ln(Z)$$

where F_{gel} and F_{bab} are the percentage of the day spent feeding by gelada and *Papio* baboons, respectively, T is the mean ambient temperature, Q is the protein content of grass (percentage of dry weight) and Z is Simpson's evenness index of rainfall diversity. The fact that the slope parameter for temperature is identical in the two equations appears to reflect the direct effects of thermoregulation on energy requirements: energy consumption (and hence feeding time requirement) increases as ambient temperature declines.

The variables Q and Z turn out to be surrogates for different vegetation quality parameters (grass and bush layer vegetation, respectively), which are in turn also related to ambient temperature (Dunbar, 1996). When the equations for Q and Z are substituted into the equations for feeding time, we obtain:

$$\ln(F_{gel}) = 14.2 - 8.0\ln(T) + 1.5\ln(T))^2$$
$$\ln(F_{bab}) = -22.2 + 17.5\ln(T) - 3.1\ln(T))^2$$

These equations are mirror images of each other, and this appears to be due to the fact that each taxon's principal food sources (grasses for gelada, bush layer vegetation for baboons) respond in diametrically opposite ways to ambient temperature: within the limits of plant viability, grasses become more abundant and nutritious but bush layer vegetation becomes less abundant as ambient temperature declines (Dunbar, 1994). In contrast to both of these, chimpanzee feeding time equations seem to be more dependent on the factors that determine tree level cover, as might be expected from their ripe-fruit frugivory dietetic style.

Thus, adaptations to specific diets play an important role in determining the form of the equations for feeding time. This can be expected to influence other aspects of the time budget too. Folivores like the gelada may have to travel less far than frugivores like baboons, and therefore need to devote less time to moving. Since the energetic costs of travel are reduced, they may also need to spend less time feeding in order to fuel travel. Similarly, a strictly folivorous diet such as that adopted by many colobines may incur additional resting time costs to facilitate fermentation. Resting time is in fact positively correlated with the percentage of leaf in the diet in primates as a whole (Dunbar, 1988). This appears to be entirely a consequence of the fact that the fermentation processes associated with a high-leaf (and seed-based?) diet require the animal to be resting (van Soest, 1982). This in turn imposes considerable constraints on the time available for social interaction (Dunbar, 1988), which may explain why colobine monkeys have significantly smaller groups than frugivores like baboons and macaques. Indeed, the colobines with the largest naturally occurring groups (*Semnopithecus entellus* and *Procolobus badius*) are also the most frugivorous. The costs of fermentation appear to be sufficiently high that they preclude even social interaction.

Although it is, in principle, possible for animals to effect some savings of time by altering the rate at which they behave (e.g. by feeding faster), only moving time is known to be elastic in this sense. At least in the case of the baboons, moving time remains more or less constant because the animals travel faster when they have longer day journeys to complete: baboons in low rainfall habitats travel at about twice the speed of baboons in wet habitats (Dunbar, 1992a). However, this does not appear to be true of the gelada, for whom moving time is a simple function of day journey length. This probably reflects the contrasting modes of progression characteristic of these two genera: baboons walk between widely spaced food patches, whereas gelada typically shuffle across a more or less continuous grassy sward.

The functional equations that characterise the socioecology of the three model taxa can be used to predict maximum ecologically tolerable group sizes (Fig. 12.3). The data show that gelada only occur in cooler habitats (10–20 °C), whereas baboons tend to favour habitats with higher mean temperatures of 20–30 °C. Consequently, the geographic distribution of gelada and *Papio* baboons has only a limited overlap, with gelada being restricted to the high-altitude grasslands that lie above 1500 m asl on the Ethiopian plateau. The distribution of chimpanzees is a mirror image of that for the baboons. This may reflect dietary differences between the two taxa, with chimpanzees preferring tree-based feeding sites whereas the baboons prefer feeding sites in the shrub/bush layer. The niche separation between the three taxa thus seems to be the result of dietary differences between them, and the way in which their food species respond to climatic variables.

Testing the models

The mark of a good model is that it correctly describes reality, to the point where it can predict what we actually see. Using models to predict the behaviour of the animals is thus an important test of the models' validity. We can test the models presented above in three ways. One is to see

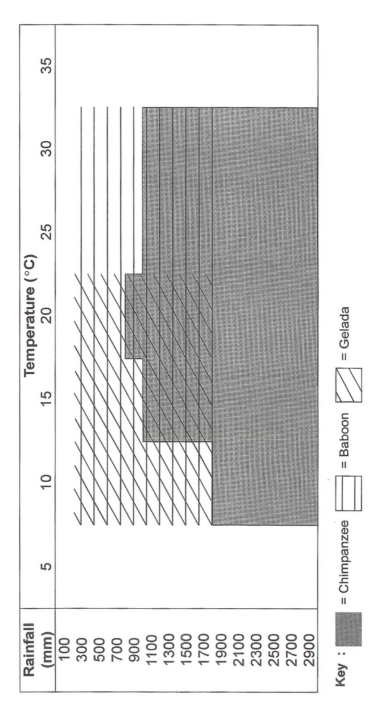

Fig. 12.3   Geographical distributions of the three taxa (gelada, *Papio* baboons and chimpanzees) predicted by each taxon's time budgets model. The distributions represent the combinations of primary climatic variables in which each species can maintain a maximum ecologically tolerable group size that is larger than the minimum permissible.

whether the equations for the time budget variables correctly predict the patterns observed. A second is to use the predicted N_{max} to predict a species' geographical range. The third is to use N_{max} to predict when groups are likely to undergo fission.

Dunbar (1992a) and Williamson (1997) have used the baboon time budgets models (based on a sample of 18 populations) to predict the time budgets of four other populations. In general, both sets of analyses produced significant fits between observed and predicted values, although some equations were better than others. Williamson (1997) tested seven different sets of equations (including those generated by Bronikowski and Altmann, 1996), and found that her own equations, based on a more detailed analysis with an improved dataset and better weather data, provided much the best fit. The predictions generated from the different equations were tested for four independent baboon sites that had not been used to generate the time budget equations: Metahara, Gilgil, Awash and Badi (for methods, see Dunbar, 1992a). Figure 12.4 compares observed and predicted values for the four time budget components in these habitats. Only in one case do they diverge significantly (beyond the 95% confidence limits) and, overall, the predicted values offer a significant fit to the observed values, emphasising the robust nature of the time budget model.

The maximum ecologically tolerable group size can be used to identify the geographical limits to a species' distribution. Figure 12.1 suggests that a species will not be able to survive in habitats in which N_{max} is below some tolerable threshold (e.g. the minimum required to ensure safety from predators). Thus, a simple test of the models is to see how well they predict a species' geographical distribution. It was possible to use background climatic data to predict N_{max} for *Papio* populations in two series of habitats occupying an altitudinal cline at opposite ends of the Ethiopian plateau. Figure 12.5 plots the predicted values of N_{max} for each site against its altitude. The horizontal line marks $N_{max} = 15$, the smallest mean group size observed in any baboon population anywhere in Africa. The results show clearly that baboons do not occur in those habitats where N_{max} falls below 15.

Williamson (1987) selected 70 weather stations at random throughout sub-Saharan Africa (Wernstedt, 1972). The chimpanzee model was then used to determine values of N_{max} for each of these 70 sites in order to predict the presence or absence of chimpanzees at these geographic locations. The model provided a remarkably good fit to the current distribution of chimpanzees ($\chi^2 = 12.451$; df = 1; $p < 0.01$). The only major discrepancy was that the model would expect chimpanzees to inhabit the

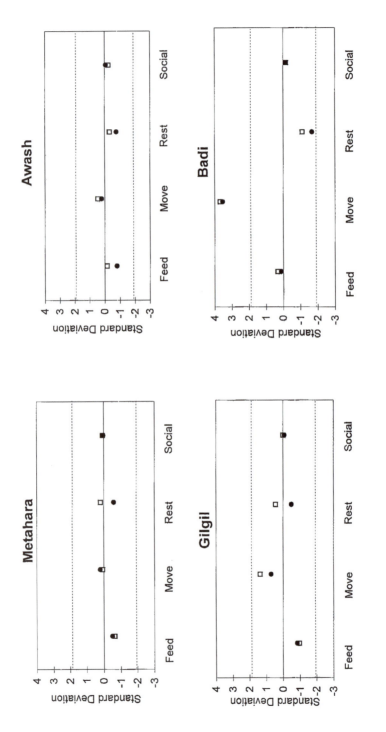

Fig. 12.4 Time budgets predicted from two different sets of equations (Bronikowski and Altmann, 1996 □; and Williamson, 1997 ●), compared with the observed time budgets for each of the four sites.

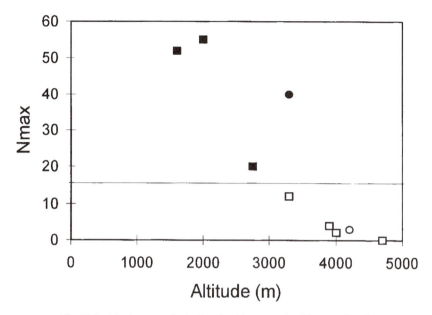

Fig. 12.5   Maximum ecologically tolerable group size, N_{max}, predicted for two series of habitats on the Ethiopian highlands, plotted against altitude. The horizontal line at $N_{max} = 15$ marks the lower limit for population mean group size in *Papio* baboons. Solid symbols indicate habitats where *Papio* baboons occur, open symbols those where they do not. Redrawn from Dunbar (1996).

woodland habitats of the Lake Malawi region, which are currently outside their range.

The model assumes that time budgets ultimately place an upper limit on group size. When the natural demographic processes cause group size to drift above N_{max}, we can expect groups to undergo fission. The final test is whether the models correctly predict this. Dunbar (1992a) showed that, for a sample of 11 *Papio* populations, those cases in which groups were described as habitually fragmenting during foraging (or underwent fission by the end of the study) were significantly more likely to have group sizes that were larger than N_{max} for their habitat than were those groups that did not fragment. More recently, Henzi, Lycett and Piper (1997) showed that the fission patterns of the Drakensberg (South Africa) and Amboseli (Kenya) baboon populations matched the predictions of the *Papio* model extremely well. In the Drakensberg population, the finding that baboon troops invariably undergo fission once they exceed a size of 28 animals comes very close to the maximum ecologically tolerable group size of 27 predicted by the model.

Criticisms of the model

The baboon time budgets model was criticised by Bronikowski and Altmann (1996) and Bronikowski and Webb (1996) on three main grounds: the data (notably the behavioural and climatic data), the statistical methods, and the scale of comparisons (interpopulation versus intrapopulation comparisons). Bronikowski and Altmann (1996) were not solely concerned with re-analysing Dunbar (1992a), but were concerned with the wider issue of whether a few climatic variables can accurately predict changes in resources and the adjustments in time budgets that populations make to such changes. The aim in this section is to clarify the issues relating to data collection and statistical analysis raised by Bronikowski and Altmann (1996).

Statistical methods

To determine the factors that influence baboon time budgets, stepwise multiple regression was used. Stepwise regression finds the set of variables that account for the highest proportion of the observed variance in the time budget data. Bronikowski and Altmann (1996) point out that stepwise regression should be used with caution, particularly with observational data, because it does not always choose the same 'important' explanatory variables between analyses. However, it must also be recognised that the stepwise method is often suggested as the best compromise between finding an optimal equation for predicting future randomly selected datasets from the same population and finding an equation that predicts the maximum variance for a specific data set (Draper and Smith, 1981; Darlington, 1990). Williamson (1997), in an extensive re-analysis, examined correlation matrices before constructing multiple regression equations in order to avoid the confounding effects of intercorrelating variables (collinearity), yet found essentially the same set of equations as Bronikowski and Altmann (1996). Indeed, neither of these sets of equations differed markedly from those originally obtained by Dunbar (1992a).

Such differences as did exist between the equation sets produced by Dunbar (1992a) and Bronikowski and Altmann (1996) turn out, on more detailed analysis, to be due to the fact that different statistical packages use different F-value inclusion criteria or significance criteria (p-value) or both. The more stringent criteria of some stepwise procedures (high F-value to enter) resulted in too few variables entering the predictor set. Importantly, equation sets derived using more stringent criteria (e.g. those of Bronikowski and Altmann, 1996) were much less successful both at predicting time budgets of test populations and in generating realistic values for N_{max} than equations generated using more lenient criteria (e.g. those of Dunbar, 1992a, or Williamson, 1997). Indeed, if a model's ability to predict reality is important, then the original model of Dunbar (1992a) is in fact more robust than Bronikowski and Altmann (1996) assume. Full details of these analyses can be found in Williamson (1997).

Climate data

Bronikowski and Altmann (1996) raise a number of separate issues concerning the climatic data used in the original analyses of Dunbar (1992a). They argue that the model's power was likely to have been weakened by: (a) the use of meteorological data from varied sources of different duration, (b) the fact that year-to-year variation in rainfall may be more important in low-rainfall habitats than in wetter areas, (c) the use of just four crude climatic indices, and (d) the use of Simpson's evenness index to measure within-year rainfall variability.

The original analyses of Dunbar (1992a) used meteorological data from a single year for the baboon site sampled or long-term average meteorological data from a nearby weather station. Preference was given to study site data where possible, on the assumption that spatio-temporal proximity to the study would be of overriding importance. This mixture of long-term and short-term climatic data may have been important because the climatic data were used to index plant productivity and a time lag would be expected between rainfall and its effect on plant productivity (e.g. Western and Lindsay, 1984).

Williamson (1997) collated data from three sources: long-term climate data from a world weather compilation (Anon, 1984); study year data from the nearest official weather station (from government publications); and study year data from the field site itself. A compromise data set, consisting of a mixture of long-term and short-term data, was also compiled, with proximity of the weather data source to the field site as the primary criterion for inclusion. The functional equations between behavioural and environmental variables for the four climatic sources obtained from the core set of 18 baboon populations were broadly similar. These equations were also tested on the subsidiary sample of four baboon populations: the equations based on the mixture of long-term and short-term data, emphasising proximity of the weather station to the field site, were the most accurate in predicting time budgets in the subsidiary sample.

The authors assume that, all else being equal, long-term climatic conditions will best predict the vegetation structure of the habitat: the vegetation broadly available to animals in any one study year will be the cumulative product of the previous years' climate. However, the animals' behavioural responses in any given year are also likely to be determined by current climatic conditions (e.g. current rainfall influencing immediate primary production and current ambient temperature setting energy demands for thermoregulation). Because the data in the long-term climate database included some weather stations not very proximate to the field site and these generally yielded the least successful equation sets, it appears that proximity of the data source to the field site may be more important than the duration of the climatic record.

The success of the 'mixed-climate' dataset in predicting time budgets highlights a second issue raised by Bronikowski and Altmann (1996), who pointed out that, for sites with high annual rainfall (e.g. Budongo forest), year-to-year differences in mean annual rainfall would be slight compared to those for very low-rainfall sites (e.g. Amboseli). While there would be little difference between long-term and short-term climate data for high-rainfall sites, Bronikowski and Altmann expected that study-year climate data would be more accurate than longer term data in low-rainfall sites like Amboseli where year-to-year fluctuations in rainfall would be more exaggerated. However, in fact, the authors' more extensive analyses suggest that the proximity of the weather station to the study site may be more important.

It is worth emphasising that long-term average climate is most likely to set the norm for group size in a given population (mainly because the demographic lags are so long for primates). A population may find itself under significant demographic stress (*sensu* Dunbar, 1992a) in those years in which, for example, rainfall is significantly lower than normal. In such years, the actual group size (the cumulative outcome of past climatic conditions) may exceed the ecologically maximum tolerable group size *for that year*. This may impose considerable strain on the animals, and may ultimately lead to group fission. However, due to demographic lags, it could be as much as a year later before the fission processes actually cause a group to split – by which time climatic conditions may have ameliorated anyway.

Bronikowski and Altmann (1996) also criticised the limited number of climatic variables used as independent variables in the regression analyses. However, Williamson's (1997) detailed analyses of weather data from all over Africa (see above, p. 322) suggest that this particular criticism is misplaced, as these variables are highly intercorrelated and only a small number are robust predictors.

A further issue raised by Bronikowski and Webb (1996) was the choice of index to best characterise rainfall seasonality. Rainfall seasonality has received relatively little attention in the climatic literature. When climatic indices have been used, both their usefulness and accuracy have been criticised (Rutherford and Westfall, 1994; Bronikowski and Webb, 1996). In particular, Bronikowski and Webb (1996) question the value of Simpson's index of evenness (derived from the plant diversity literature: Magurran, 1988) because it does not take the number of dry months, or their pattern, into account. Moreover, variability in rainfall is inversely correlated with the amount that falls. The coefficient of rainfall variability increases from 10–15% in the rainforest to more than 50% in desert habitats (le Houérou and Popov, 1981). The relationship between rainfall variability and the amount that falls appears to be linear for rainfall above 100 mm, but not for rainfall below 100 mm, because the frequency distribution of rainfall is skewed (le Houérou and Popov, 1981). Bronikowski and Webb (1996) argue that this might explain why seasonality indices seemed to characterise the baboon sample as a whole quite well but were not found to characterise Amboseli (a very dry habitat) when considered separately.

There is as yet no one index that is universally useful in all habitat types. Some indices provide insufficient variance between values for statistical analysis (for example when rainfall is very low and/or even). Ultimately, the choice of a rainfall diversity index should be based on how useful the index is in the specific behavioural ecological model tested. Simpson's index was favoured by Dunbar (1992a) to characterise rainfall evenness in a study of the correlation between meteorological and behavioural variables in baboons because it is independent of the quantity of rainfall. In fact, Simpson's index is highly correlated with direct measures of plant productivity in Africa. Climatic data from 187 randomly sampled sub-Saharan stations published in Thornthwaite and Mather (1962) yield:

$$Z = 1471.91 - 185.56 \, (AE)$$

$(r^2 = 0.35, F_{1,185} = 6.84, p = 0.0119)$ where Z = Simpson's index of rainfall diversity and AE = annual evapotranspiration (mm/year), a widely established correlate of primary productivity.

It is worth emphasising in conclusion, however, that the models of Dunbar (1992a) and Williamson (1997) use two measures of rainfall variability (or seasonality): Simpson's index of diversity in monthly rainfall across the year, and the number of months of the year with less than 50 mm of rainfall. Between them, these two measures effectively address both the criteria identified as important by Bronikowski and Altmann (1996).

Conclusion

These analyses suggest that it is possible to model some aspects of the behavioural ecology of baboons and chimpanzees (see also Chapter 14) using a limited set of climatic data. Although the original model of Dunbar (1992a) has been criticised on a number of grounds, more detailed analysis by Williamson (1997) has shown that the results are generally robust as well as predictive in that they yield biological as well as statistically consistent predictions.

Bronikowski and Altmann's (1996) comments do, however, raise an important and interesting issue. The model based on a pan-African sample does not predict particularly well the year-to-year variation in the activity budgets of the Amboseli population of baboons. The models reviewed here are based on analyses of data from populations drawn from throughout the taxon's range. This is likely to give prominence to those climatic and vegetational variables that affect broad patterns of vegetation and behaviour. It may be ecologically naive to expect the same model to represent accurately year-to-year variation within a habitat, because those variables that are the principal determinants of between-habitat variation in vegetation need not necessarily be the same as those that determine within-site year-to-year variation. An interspecific approach will set the broad limits (or overall mean) characteristic of a site, whereas an intra-specific approach will determine the year-specific position within these limits. Within-site analyses will need a second layer of equations to model the fine-detail variation from one year to the next. In fact, the pan-African model predicts extremely well the broad range of group sizes found at Amboseli, particularly when the better estimates of Amboseli climate provided by Bronikowski and Altmann (1996) are used.

These issues aside, the success of the time budget models lies in their ability to predict variance in group size across habitats. This they appear to be able to do with considerable accuracy. Moreover, comparison of the models for different genera throws an interesting light on the ecological differences between the taxa, clearly showing why niche separation exists. Contrary to previous supposition, the gelada occupy a retreat refuge habitat on the Ethiopian highlands not because of ecological competition from the more aggressive *Papio* baboons, but because, as grazers, climatic conditions have forced them up the altitudinal gradient. As global temperatures rise and fall, so the distributions of the two baboon genera move up and down the altitudinal gradient in tandem (Dunbar, 1992c). *Papio* baboons are no more able to survive in the cool, high-altitude habitats favoured by the gelada than the latter are able to survive in the hot, low-altitude habitats preferred by the baboons.

The equations, as they currently stand, include a number of surrogate variables that substitute for crucial biologically causal relationships. An important future step will clearly be to examine the specific causal relationships involved in more detail. Such studies will help substantiate the models and enable us to arrive at a much better understanding of primate behavioural ecology than we have hitherto been able to achieve.

A further direction for new work lies in building models for other taxonomic groups. This will allow us to see how species' dietary and locomotory adaptations alter the form of the equations for feeding and moving. Once we have done this, we may be able to build a general model that incorporates dietary and locomotory style as logistic variables. A start has been made on this by the development of models for three taxa (gelada, baboons and chimpanzees) that differ radically in their ecological adaptations. A similar analysis has also been carried out for gibbons (*Hylobates* spp.), which differ from all three of these taxa in both dietary and locomotor style (Sear, 1994), but a systems model has yet to be built for this genus.

Acknowledgements

DKW is grateful to the Natural Environment Research Council and the Leverhulme Trust (Grant F/25/BK) for financial support.

References

Anon (1984). *Africa, the Atlantic Ocean South of 35 °N and the Indian Ocean (Met. 0.85.d.). Tables of Temperature, Relative Humidity, Precipitation and Sunshine for the World.* London: HMSO.

Bronikowski, A.M. and Altmann, J. (1996). Foraging in a variable environment: weather patterns and the behavioral ecology of baboons. *Behavioral Ecology and Sociobiology* **39**, 11–25.

Bronikowski, A.M. and Webb, C. (1996). Appendix: A critical examination of rainfall variability measures used in behavioral ecology studies. *Behavioral Ecology and Sociobiology* **29**, 27–30.

Darlington, R.B. (1990). *Regression and Linear Models.* New York: Wiley.

Deshmukh, I.K. (1984). A common relationship between precipitation and grassland peak biomass for East and Southern Africa. *African Journal of Ecology* **22**, 181–6.

Draper, N.R. and Smith, H. (1981). *Applied Regression Analysis*, 2nd edition. New York: Wiley.

Dunbar, R.I.M. (1988). *Primate Social Systems*. London: Chapman and Hall.

Dunbar, R.I.M. (1992a). Time: a hidden constraint on the behavioral ecology of baboons. *Behavioral Ecology and Sociobiology* **31**, 35–49.

Dunbar, R.I.M. (1992b). A model of the gelada socioecological system. *Primates* **33**, 69–83.

Dunbar, R.I.M. (1992c). Behavioural ecology of the extinct papionids. *Journal of Human Evolution* **22**, 407–21.

Dunbar, R.I.M. (1994). Ecological constraints on group size in baboons. In *Animal Societies: Individuals, Interactions and Social Organization*, ed. P. Jarman and A. Rossiter, pp. 221–36. Oxford: Blackwell Scientific Publications.

Dunbar, R.I.M. (1996). Determinants of group size: a general model. *Proceedings of the British Academy* **88**, 33–57.

Dunbar, R.I.M. and Nathan, M.F. (1972). Social organisation of the Guinea baboon, *Papio papio*, in Senegal. *Folia Primatologica* **17**, 321–34.

Henzi, S.P., Lycett, J.E., and Piper, S.E. (1997). Fission and troop size in a mountain baboon population. *Animal Behaviour* **53**, 525–35.

le Houérou, H.N. and Hoste, C.H. (1977). Rangeland production and annual rainfall relations in the Mediterranean basin and in the African Sahelo–Sudanian zone. *Journal of Rangeland Management* **30**, 181–9.

le Houérou, H.N. and Popov, G.F. (1981). An eco-climatic classification of inter-tropical Africa (3 maps). *Plant Production and Protection Paper*, Number 31. Rome: FAO.

Leith, H. and Box, E.O. (1972). Evapotranspiration and primary production: C.W. Thornthwaite memorial model. *Publications in Climatology* **25**, 37–46.

Magurran, A.E. (1988). *Ecological Diversity and its Measurement*. Princeton: Princeton University Press.

McNaughton, S.J. (1985). Ecology of a grazing ecosystem: the Serengeti. *Ecological Monographs* **55**, 259–94.

Rosenzweig, M.L. (1968). Net primary productivity of terrestrial communities: prediction from climatological data. *American Naturalist* **102**, 67–74.

Rutherford, M.C. (1980). Annual plant production–precipitation relations in arid and semi-arid regions. *South African Journal of Science* **76**, 53–6.

Rutherford, M.C. and Westfall, R.H. (1994). *Biomes of Southern Africa: an Objective Categorization*. Memoires of the Botanical Survey of Southern Africa, Number 63. Pretoria: National Botanical Institute.

van Soest, P.J. (1982). *The Nutritional Ecology of the Ruminant*. Ithaca, NY: Cornell University Press.

Sear, R. (1994). *A Quantitative Analysis of Gibbon Behavioural Ecology*. MSc thesis, University of London.

Thornthwaite, C.W. and Mather, J.R. (1962). Average climatic water balance of the continents. Part 1: Africa. *Publications in Climatology* **15**(2). Centerton, NJ: Laboratory of Climatology.

Wernstedt, F.L. (1972). *World Climatic Data*. London: Climatic Data Press.

Western, D. and Lindsay, W.K. (1984). Seasonal herd dynamics of a savanna elephant population. *African Journal of Ecology* **22**, 229–44.

Williamson, D.K. (1997). Primate Socioecology: the Development of a Conceptual Model for the Early Hominids. Unpublished PhD thesis, University College, London.

13 *Ecology of sex differences in great ape foraging*

ALLISON BEAN

Sex differences in great ape foraging stragegy vary between populations as well as species. Are these sex differences functions of the energetic constraints of the sexes and thus represent a maximisation of individual fitness goals, or are they a group adaptation that minimises feeding competition?

Male and female energetic constraints

Male and female energetic constraints (the costs of metabolism, activity and reproduction) result from two factors: (1) the degree of sexual dimorphism in body weight that exists between the sexes, and (2) the reproductive state of the individual.

Male primates are often larger than female primates. Large differences in body size have significant energetic consequences. While metabolic rate increases with body size as around body weight$^{0.75}$, commonly the rate of food intake may increase less rapidly with body weight (Clutton-Brock, 1994). Furthermore, heavier bodied animls are further constrained by their greater weight and size and are thus less able to exploit certain areas of food patches such as the outer branches of fruiting trees (Doran, 1993a, 1993b). Conversely, weaker, smaller animals are less able to exploit resources that require greater strength for their processing such as the stripping of bark. Smaller animals are also more likely to be displaced at food sources by larger ones (Wheatley, 1982).

The degree of sexual dimorphism in body mass will therefore affect the amount of sexual segregation within a habitat. A large degree of sexual dimorphism produces males as large-bodied foragers and females as smaller bodied foragers. Males and females are thus under differing selective pressures due to their body size and effectively occupy different niches within the same habitat. This has the advantage of lowering feeding competition between the sexes. Low sexual dimorphism results in greater feeding competition between the sexes because males and females are similar in body size and strength and will therefore overlap in foraging

339

strategy. If feeding competition becomes too high, the sexes may have to segregate by altering foraging behaviour relative to each other in order to lower feeding competiton by exploiting different resources within the same habitat.

Although males are larger than females (so we would expect them to have greater daily energetic costs), female primates suffer the extra energy costs of internal fertilisation, pregnancy, extended lactation and prolonged dependency of young (Sadleir, 1969; Portman, 1970; Pond, 1977). These extra costs often mean that a female must feed for longer than an even larger male (relative to just how much larger that male is). For example, lactating female gelada baboons are found to spend up to 30% more time feeding per day than non-lactating females (Dunbar, 1992). Alternatively, females may select a different high-quality diet that requires extra searching and processing times (Pollock, 1977; Post, Hausfater and McCuskey, 1980; Harrison, 1983; Clutton-Brock, Albon and Guinness, 1984).

Therefore, when the degree of sexual dimorphism that occurs between the sexes is low (such as in bonobos and chimpanzees), female daily energetic costs may overlap with male costs. This is because any sex differences in daily energetic costs cannot be attributed solely to body weight disparity alone, but include the costs of reproductive demands made upon females. The social structure as well as environmental variation between populations may also influence these emergent differences.

Sex differences in foraging, as well as being energetically sensible for the individual, expand the feeding options for dispersed individuals and small groups whose subsistence activities are mostly self-regulated (Galdikas and Teleki, 1981). The exploitation of differing ecological niches between the sexes promotes a reduction in intersexual competition for food and is partly related to an existing energetic separation of the sexes that is the result of a greater degree of sexual dimorphism in body mass. With a difference in the subsistence activities of males and females, the sharing of resources within a group can become adaptive. 'Complementary activities' allow females to benefit from the eating of foods that they normally would not have the time, energy or 'capacity for risk' to attain. Food sharing allows males to 'buy' a female's affection and perhaps secure a mating preference. Among chimpanzees, the selection by a male of the individuals with whom he shares his kill is significant, and 80% of sharing in chimpanzees involves adults of both sexes getting meat from males. Female chimpanzees that are cycling are more successful in getting meat than non-cycling females (Teleki, 1973).

Great ape habitats

The habitat types of great apes vary between populations. A number of African ape populations from different species overlap in habitat, e.g. eastern lowland gorillas (*Gorilla gorilla graueri*) and eastern chimpanzees (*Pan troglodytes schweinfurthii*), and eastern lowland gorillas and central chimpanzees (*Pan troglodytes troglodyes*), and western lowland gorillas (*Gorilla gorilla gorilla*) and central chimpanzees. The habitats of gorillas have a large vertical as well as horizontal range. Gorillas inhabit a range of lowland tropical rainforests up to the high-altitude montane forests of the Virunga Volcanoes (Zaire – now Democratic Republic of Congo, Rwanda and Uganda, e.g. up to 3700 m: Fossey, 1974). Chimpanzees inhabit a broad range of relative aridity, from tropical rainforests (e.g. Budongo, Uganda), through deciduous woodlands (e.g. Gombe, Tanzania) to open savannah (e.g. Mt Assirik, Senegal). Bonobos (*Pan paniscus*) are restricted to the lowland tropical rainforests of Zaire; they have no overlapping ranges with other African apes, as they are geographically isolated from them by the Zaire River. Orangutans (*Pongo pygmaeus*) are limited in distribution to Borneo and Sumatra and so inhabit only a limited range of habitat types compared to the African apes.

Habitat type and habitat quality

Measuring 'habitat type' is difficult because each habitat is different for each individual living within the greater habitat. Indicators of habitat vegetation are 'climatic' variables, such as rainfall, altitude, temperature range, seasonality (number of months with < 50 mm of rainfall per year: see Chapter 12), mean annual temperature and latitude (degrees from the equator: Table 13.1).

An estimate of 'habitat quality' can be made using species' dietary breadth (Table 13.2) because the number of food items in a population's diet is observed to increased as overall food availability decreases (Wrangham, 1977). Where this is the case, food types such as invertebrates (e.g. social insects) are incorporated into the diet in greater proportions. A higher density of terrestrial herbaceous vegetation in a habitat can act as a buffer for a lower density of fruit patches, and thus a lower dependence on fruit, with higher reliance on terrestrial herbaceous vegetation, in a diet is considered to reflect a higher quality habitat.

The habitat type or quality is an important determinant of foraging behaviour within a habitat, particularly for dietary composition and

Table 13.1. *Climatic and habitat type data for great apes*

Subspecies	Study site	Country	Rainfall (mm/year)	Mean altitude (m)	Temperature range (°C)	Latitude	Seasonality
Bornean orangutan	Tanjung Puting	Borneo	3000	30	18–37.5	0°3'S	0
Pongo pygmaeus pygmaeus	Ulu Segama	Borneo	3810	870	22–32	5°3'N	0
	Sepilok Forest	Borneo	3962	152	23–30	5°20'N	0
	Kutai	Borneo	2360	(125)	19–30.5	0°24'N	0
Sumatran orangutan	Ketambe	Sumatra	3229	675	17–34.2	3°40'N	0
Pongo pygmaeus albelii	Ranun River	Sumatra	3000	(1000)	18–42	2°49'N	0
Mountain gorilla	Karisoke (Virunga)	Rwanda	1800	3500	3.8–14.8	1°S	3
Gorilla gorilla beringei	Other Virunga volcano sites	Zaire–Rwanda–Uganda	1700	3150	0–21	0°03'S	3
	Kayonza/Bwindi Forest	Congo–Uganda	(1750)	2058	12–26	0°9'S	3
	Kisoro	Uganda	1676	2957	4.5–17	1°36'S	2
	Kabara	Uganda	1750	1413	4–15	1°20'S	2
Eastern lowland gorillas	Mt Tshiaberimu	Uganda	(1750)	2621	9–19.5	0°9'S	2
Gorilla gorilla graueri	Kahuzi-Biega	Zaire	1700	1954	9.9–24.6	2°25'N	4
	Iterbero	Zaire	1700	950	10.4–17.9	2°25'N	4
	Mt Kahuzi District	Zaire	1548	2384	10.4–17.9	2°25'N	4
	Utu	Zaire	2320	732	20–30	1°12'N	0
Western lowland gorillas	Rio Muni	Equatorial Guinea	2800	550	15.5–33.3	1°65'N	3
Gorilla gorilla gorilla	Belinga	Gabon	2000	850	19–31	1°N	3
	Bai Hokou	C.A.R.	1500	350	19.4–32.5	2°13'N	3
	Lopé	Gabon	1532	(375)	20.1–32.8	0°10'S	3
Bonobos	Lomako	Zaire	1960	390	18.5–32	0°5'S	0
Pan paniscus	Wamba	Zaire	2005	400	12.7–31.4	0°2'N	0
	Yalosidi	Zaire	(1499)	(125)	17.5–21.5	1°7'S	0
	Lake Tumba	Zaire	1398	(125)	22–31	0°5'S	0
	Lilungu/Ikela	Zaire	1653	(125)	22.7–32.6	1°11'S	0

Eastern chimpanzees *Pan troglodytes schweinfurthii*	Gombe	Tanzania	1600	1138	19–28	4°40'S	6
	Mahale	Tanzania	1836	1649	18–27	6°70'S	5
	Ugalla	Tanzania	1012	1300	19–28	5°43'S	8
	Kibale	Uganda	1671	1508	10–31	0°27'N	0
	Budongo	Uganda	1495	1097	28–34	1°43'N	2
	Kahuzi-Biega	Zaire	1700	1954	9.9–24.6	2°25'N	3
Central chimpanzees *Pan t. troglodytes*	Lopé	Gabon	1532	(375)	20.1–32.8	0°10'S	3
	Rio Muni	Equatorial Guinea	2800	550	15.5–33.3	1°65'N	3
Western chimpanzees *Pan t. verus*	Sapo	Liberia	(1850)	(250)	22–28	6°N	1
	Tai Forest	Ivory Coast	1829	203	20–28	5°52'N	1
	Bossou	Equatorial Guinea	2500	600	15.5–33.3	7°39'N	5
	Mt Assirik	Senegal	954	(125)	23–35	12°53'N	9

Values in parentheses are estimated from Pearce and Smith (1994) and *Collin's Atlas of the World*. All other sources are recorded from published literature sources.

All ultimate data sources for this data can be found in Bean (1998).

Table 13.2. *Foraging strategy data for great apes*

Subspecies	Metabolic weight (kg)^0.75	Sexual dimorphism in body mass (M/F)	Study site	Population density (n/km)	Percentage sex ratio	Day range (km)	Home range (km²)	Foraging party size	Number of food items
Pongo pygmaeus pygmaeus	22.23	2.23	Tanjung	2.5	86	0.79	7.75	1.5	317
			Kutai	4	50	0.49	3.25	1.5	80
			Segama	1.5	67	0.5	4.5	1.5	111
Pongo pygmaeus abelii	23.23	2.23	Ketambe	5	133	0.62	1.5	1.5	151
			Ranun	1	96	0.55	3.85	1.5	60
Gorilla gorilla beringei	38.15	1.63	Karisoke	1.1	87	0.75	9.4	10.5	75
			Kabara	2.5	52	0.47	22	16.9	49
			Virunga	1	70	0.69	4–9.7	6.5	78
			Kayonza	0.5	114	0.5	16	8.75	
Gorilla gorilla graueri	37.97	2.19	Kisoro	0.96	53	1.6	10.35		
			Mt Tshiaberimu	0.55	40				
			Mt Kahuzi	0.4	66	0.9	31	17	95
			Kahuzi-Biega	0.35		1.05	14.75	7.3	129
			Utu	0.38	87				20
			Itebero	0.29		2.1			194
Gorilla gorilla gorilla	36.37	2.37	Rio Muni	0.7	51	1.13	6.75		141
			Belinga	0.18		3			107
			Lopé	0.2	50	1.172	8.1	5	221
			Likoula	2.2		0.94	31		
Pan paniscus	15.64	1.36	Bai Hokou	1.14	17	2.3	22.9	4.8	182
			Wamba	2.35	66	2.4	58	13	170
			Lomako	2	69	7.6	22	3.5	113
			Lilungu		50				
			Lake Tumba	1.7	50			3.5	
			Yalosidi		60			4	87

Pan troglodytes schweinfurthii 15.34					1.29	
Gombe	2.6–6	80	3.1	6	4.5	167
Mahale	6	66	4.5	19	4.25	370
Kibale	1	66	4	18.6	3.6	113
Budongo	3.4	77	7.5	32	4.4	118
Filabanga				150		206
Kahuzi	4.5	60	5.3	13.65	4.5	99
Kaskati		35		145		
Ugalla	0.075		10	515		
Pan troglodytes verus 16.16					(1.28)	
Bossou	4.5	67	5.5	228	6	205
Mt Assirik	0.09	83	6	27	3	72
Tai Forest		75	5		6.55	
Pan troglodytes troglodytes 19.84					1.27	
Sapo	0.24			9.7		
Rio Muni	0.31		4	15	11.2	
Lopé	1					174
Bai						114

Value in parentheses is estimated.

Table 13.3. *Dietary composition, dietary breadth and arboreality in great apes*

Subspecies	Study site	Foraging party size	Number of food items	Percentage fruit in diet	Percentage plant parts in diet	Percentage invertebrates in diet
Pongo pygmaeus	Tanjung	1.5	317	61	24	4
	Kutai	1.5	80	53.8	31.22	0.8
	Segama	1.5	111	52	36	0.9
Pongo pygmaeus albelii	Ketambe	1.5	151	80.8	12.4	0.2
	Ranun	1.5	60	62	23.5	3
Gorilla gorilla beringei	Karisoke	10.5	75	1.7	85.8	0.001
	Kabara	16.9	49	6.1	66	
	Virunga	6.5	78	4	70	2
	Visoke			4	67	
	Kayonza	8.75		10	52	
	Kisoro			8	79	
Gorilla gorilla graueri	Mt Tshiaberimu				93	
	Mt Kahuzi	17	95	10	65	
	Kahuzi-Biega	7.3	129	15.5	58	
	Utu		20	20	75	
	Itebero		194	25	41	1
Gorilla gorilla gorilla	Rio Muni		141	38.3		
	Belinga		107	69.2	6.7	1
	Lope	5	221	45.5	33.8	1
	Bai	4.8	182	51	26	1
Pan paniscus	Wamba	13	170	59	27.6	1.5
	Lomako	3.5	113	54	33.6	0.9
	Lake Tumba	3.5		58	54	4
	Yalosidi	4	87	63	27	2
Pan troglodytes schweinfurthii	Gombe	4.5	167	59.4	26.2	11.9
	Mahale	4.25	370	57.7	32	6.4

	Kibale	3.6	113	78	7.2	0.2
	Budongo	4.4	118	90	7	1
	Filabanga		206	69		0.5
	Kahuzi	4.5	99	41	50	7
	Kaskati			35.7		
Pan troglodytes verus	Bossou	6	205	49	41	0.5
	Mt Assirik	3	72	57	35	5
	Tai Forest	6.55				
	Rio Muni	11.2		44.3	39	15.76
	Lope		174	61	27	7
	Bai		114	88	9.7	2.5

For definitions, see Bean (1998).

ranging behaviour (see Bean, 1998, for a review). However, the 'suitability' of the habitat or relative 'resource availability' to the individual forager will further determine foraging behaviour. Habitat type, behavioural variables relative to resource availability and a measure of daily energetic costs incurred by the individual forager may be used to estimate relative ecological costs. For example, a large-bodied ape frugivore is expected to vary foraging party size in response to ecological factors, within a single habitat seasonally (Yamagiwa *et al.*, 1994), or between populations in different habitats. Thus, habitat type, a measure of body mass (e.g. metabolic weight: weight $(kg)^{0.75}$), and foraging party size (Table 13.2) are used to estimate relative ecological costs of the forager as a determinant of foraging behaviour (Bean, in press).

For example, orangutan foraging party size is low; orangutans are almost solitary foragers, they are also heavy (males reach up to 66.3 kg: Leigh and Shea, 1995), arboreally adapted and frugivorous (Table 13.3). Fruits are a limiting resource, however, because they are clumped in patches and are seasonal in production. This 'seasonal, high-quality food patchiness' produces severe competition for food, especially upon smaller animals such as females, which are likely to be displaced by larger males (Rijksen, 1978; Wheatley, 1982; Galdikas, 1988).

Wrangham (1979) has suggested that orangutans are obliged to forage separately by the severe costs incurred from foraging in groups (feeding competiton). These costs can be attributed either to direct competition for access to sites or to indirect costs incurred by having to travel further in groups (Dunbar, 1988). Orangutan habitats are predominantly dipterocarp and are relatively poor in terms of fruit production (Caldecott, 1986a, 1986b). MacKinnon (1974) noted that, unlike chimpanzees that only eat ripe fruit, orangutans will eat fruit at any stage of growth.

Gorillas, whilst being large (males reach up to 128 kg: Leigh and Shea, 1995), are generally terrestrial (Table 13.3). Gorilla foraging party sizes vary with dietary composition (Table 13.3). The more frugivorous the population, the smaller the foraging party size (Tutin and Fernandez, 1984). The same trend is observed for bonobos and chimpanzees; bonobos forage in larger parties than chimpanzees and are less dependent on fruit as a principal food source (Malenky and Wrangham, 1994; Table 13.3). Thus, the higher the expected feeding competition, the lower the expected foraging party size. Foraging party size is therefore expected to be indicative of resource availability.

There is a significant degree of intercorrelation between the variables that describe 'habitat type' (i.e. climatic variables as above; see Table 13.1). Altitude and temperature range largely account for the variation in the other climatic variables, and so habitat type can be described by a composite variable (see Bean, 1998, for a review).

The 'habitat quality' of great ape populations is highly variable. Bonobos and mountain gorillas occupy habitats rich in terrestrial herbaceous vegetation, which is reflected in their diets (Watts, 1984; Wrangham, 1987). The number of food items in the bonobo's diet is less than that in the diet of chimpanzees (Malenky and Wrangham, 1994; Table 13.3). Bonobos use a smaller proportion of food resources available to them. This suggests that the supply of major food is more stable for bonobos and that bonobos do not need to feed on hard-to-process or hard-to-collect food items such as bark and insects. Chimpanzee diet is more varied than the bonobo diet because the habitat of chimpanzees is poorer generally than that of bonobos. Terrestrial herbaceous vegetation is found in far greater quantities in bonobo habitats compared to chimpanzee habitats (Wrangham, 1987; White, 1988). Chimpanzees exploit a wide variety of habitats (Reynolds and Reynolds, 1965; Goodall, 1968; Hladik, 1977; McGrew, Baldwin and Tutin, 1981). They are flexible in feeding and ranging habits and, because of this ability, it might be anticipated that under certain conditions (e.g. a decline in habitat quality), the sexes could adapt to occupy different ecological niches.

Great ape foraging strategy

Foraging strategy can be summarised by the variables day and home range, dietary composition (the percentage of fruit, plant parts and animal matter in the diet), foraging party size, dietary breadth (the number of food items in the diet), arboreality (the percentage of time spent arboreal per day) and activity budget (the percentage of time spent feeding and travelling and time spent neither feeding nor travelling – termed resting time here – see Bean, 1998, for definitions and Dunbar, 1994, for a discussion) (Tables 13.2–13.5). Foraging strategy in great apes is influenced by climatic variables (see Table 13.1) and the degree of sexual dimorphism in body mass (see Table 13.2; see Bean, 1998, for a review).

Evaluating sex differences in great apes

Broad across-species analyses help to identify the primary independent variables that influence sex differences in foraging strategy. Matched-pair t-tests on foraging strategy variables (see above) for males and for females

Table 13.4. Sex differences in great ape foraging behaviour

Subspecies	Study site	Sex	Day range (km)	Home range (km²)	Percentage fruit in diet	Percentage plant parts	Percentage animal (vertebrate or invertebrate)	Percentage of time feeding	Percentage of time travelling	Percentage of time resting	F/T index
Pongo pygmaeus pygmaeus	Tanjung	M	0.86	10 +	44.5	< F	> F	59	18	22	3.3
		F	0.71	5.5		> M	< M	61	17	21	3.6
	Kutait	M	0.68	5 +	58.6	40.6	0.8	40.7	8.7	50	4.7
		F	0.31	1.5	67.1	30.9	1.9	42.7	10.3	41	4.1
	Segama	M	0.75	5 +	49	23.1	2	33	15	52	2.2
		F	0.25	4				32	18	50	1.8
Pongo pygmaeus albelii	Ketambe	M	0.69	2	58	29	2	48	9	43	5.3
		F	0.55	1				40	20	40	2
	Ranun	M	0.6	5.2	62	23.5	3	32.6	19.3	47	1.7
		F	0.5	2.5				35	18.7	44	1.9
Gorilla gorilla beringei	Karisoke	M	0.75	9.4	1.7	85.8	0.01	60.2	5.7	32.5	10.56
		F	0.75	9.4				55.4	6.8	34.7	8.15
	Other Virunga	M	0.35–1.035	4–9.7	4	70	0				
		F	0.35–1.035	4–9.7			> M				
	Kabara	M	0.47	22	6.1	60		41	27.6	31.4	1.49
		F	0.47	22							
Gorilla gorilla graueri	Kahuzi-Biega	M	1.05	14.75	15.5	58		48	9	43	5.3
		F	1.05	14.75				43	9.8	47.2	4.4
	Itebero	M	2.1		25	41	> F				
		F	2.1				< M				
Gorilla gorilla gorilla	Rio Muni	M	1.13	6.75	38.3			45	43	12	1.04
		F	1.13	6.75							
	Lopé	M	1.172	8.1	45.5	32.8	> F	21	39	10	0.5
		F	1.172	8.1			< M				
	Bai	M	2.3	22.9	67	13	2	58	9	30	6.4
		F	2.3	22.9	47	27	0	51	17	26	3
Pan paniscus	Wamba	M	2.4	58	59	27.6		30	13	43	2.3
		F	2.4	< 58				30	13	43	2.3
	Lomako	M	7.6	22	54	33.6	> F	40.4	16.1	31.9	2.5
		F	7.6	< 22			< M	40.4	16.1	31.9	2.5
Pan troglodytes schweinfurthii	Gombe	M	4.1	7.3	59.4	26.2	1.4	56	14	30	4
		F	2.1	4			4.3	60	12.5	27.5	4.8

Mahale	M	6	7.7	57.7	32					
	F	3.24	4							
Kibale	M	5	21.86	78	2.2	< F	62.1	12.1	25.8	5.1
	F	3	18.6			> M	52.4	10	37.6	5.24
Budongo	M	14.4	32	90	7					
	F	9.6	< 32							
Pan troglodytes verus										
Mt Assirik	M	6.68	228	57	35		41	43	14	0.95
	F	6	228				46	28	26	1.6
Tai Forest	M	3	27			< F				
	F	1.9				> M				
Pan troglodytes troglodytes										
Rio Muni	M	4	15	44.3	39					
	F	4	15							

F/T index = foraging index = $\dfrac{\text{time feeding}}{\text{time travelling}}$ (Rodman, 1971).

Table 13.5. Sex differences in great ape activity budgets and daily energetic costs[a]

Subspecies	Study	Sex	Body weight (kg)	BMR (kcal)	Day range (km)	Mean vertical height (km)	Percentage of time feeding per day	Percentage of time travelling per day	Percentage of time resting per day	Percentage of time arboreal per day	Total daily cost of locomotion (kcal)	Total daily cost of feeding (kcal)	Total daily cost of resting (kcal)	Mean daily costs (kcal)[b]	Mean daily costs, female corrected (kcal)[c]
Pongo pygmaeus pygmaeus	Tanjung	M	69	1676	0.85	(0.2265)	59	18	22	(99)	489	220	169	2552	2552
		F	37	1050	0.71	(0.2265)	61	17	21	(95)	236	122	101	1668	2309
	Kutai	M	84	1942	0.68	(0.2265)	40.7	8.7	50	80	272	185	445	2806	2806
		F	38	1071	0.305	(0.2265)	42.7	10.3	41	95	125	88	201	1575	2182
	Segama	M	76.5	1811	0.75	(0.2265)	33	15	52	(85)	437	136	432	2783	2783
		F	38	1071	0.25	(9.2265)	32	18	50	(90)	220	66	245	1765	2445
Pongo pygmaeus albelii	Ketambe	M	70	1694	0.69	(0.2265)	48	9	43	(90)	235	181	334	2442	2442
		F	30	897	0.55	(0.2265)	40	20	40	(95)	194	64.8	164	1320	1828
	Ranun	M	86.2	1980	0.6	(0.2265)	32.6	19.3	47	(85)	604	154	4127	3165	3165
		F	38.3	1078	0.5	(0.2265)	35	18.7	44	(95)	252	72	217	1619	2242
Gorilla g. beringei	Karisoke	M	157.5	3112	0.75	(0.2265)	60.2	5.7	32.5	10	70.38	510	472	4164	4164
		F	87	1994	0.75	(0.2265)	55.4	6.8	34.7	20	48	258	320	2620	3692
Gorilla g. graueri	Kahuzi	M	175.2	3371	1.05	0.002	48	9	43	20	158	454	664	4647	4647
		F	80	1872	1.05	0.003	43	9.8	47.2	30	94	186	41	2193	3090
Gorilla g. gorilla	Bai Hokau	M	169.5	3288	2.3	0.003	58	9	30	24	316	530	411	4545	4545
		F	71.5	1721	2.3	0.002	51	17	26	58	315	197	186	2419	3410
Pan paniscus	Wamba	M	45	1216	2.4	0.0325	30	13	43	80	124	73	240	1653	1653
		F	33.2	968	2.4	0.0325	30	13	43	80	92.3	54	191	1305	1820
	Lomako	M	45	1216	7.6	0.038	40.4	16.1	31.9	94.6	378	98	98	1870	1870
		F	33.2	968	7.6	0.038	40.4	16.1	31.9	94.6	279	72	142	1461	2038
Pan troglodytes schweinfurthii	Gombe	M	39.5	1103	4.1	0.05	56	14	30	32.9	213	119	152	1587	1587
		F	29.8	988	2.1	0.05	60	12.5	27.5	47.8	90	97	125	1300	1847
	Kibale	M	45	1216	5	0.05	62.1	12.1	25.8	37.4	255	151	144	1766	1766
		F	36.5	1039	3	0.075	52.4	10	37.6	68.4	144	103	179	1465	2093
Pan t. verus	Mt Assirik	M	46.4	1244	15	0.05	41	43	14	20	1122	103	80	2519	2519
		F	(36.3)	1035	8	0.05	46	28	26	22	529	90	123	1777	2525

[a] For equations for daily energetic cost estimates see Wheatley (1982) and see Bean (1998) for a review.
[b] Without female reproductive costs included in the calculation.
[c] With female reproductive costs: assuming lactation costs = 1.5 daily energetic costs and gestation costs = 1.1 daily energetic costs (Lee and Bowman, 1995). See Bean (1998) for details of calculations.
Values in parentheses are estimated from the literature. BMR = basal metabolic rate.

show that: females are more arboreal than males ($t = 3.7$, $p = 0.03$); males have longer day ranges ($t = -2.29$, $p = 0.04$) and larger home ranges ($t = -2.9$, $p = 0.013$). It is interesting that when all the species are considered together, intraspecific sex differences are masked by interspecific differences in body mass, female reproductive costs and general activity trends.

The structure of variation in male and female foraging strategies

The structure of variation in male and female foraging strategy was examined using principal components analysis (PCA). In PCA, each variable is expressed as a coefficient per component, termed component loadings. These indicate how much weight per variable is assigned to each component. To determine the 'characteristic' of each component, the foraging strategy variables are labelled climatic, behavioural and energetic (Table 13.6). Summing the coefficients per label, per component allows the percentage weight of each 'component label' per component to be calculated. The resulting weightiest 'component label' is ascribed the characteristic 'component label' of that component.

Statistical Package for the Social Sciences (SPSS®) recommends that three components best represent the data (Table 13.6) – for both males and females. The cumulative results (Table 13.6) show that while over 40% of the variance is accounted for in both male and female foraging strategies (for males 40.7% and for females 46.7%) by component one, the component characteristics vary between the sexes (see Table 13.5). The first component is characterised by climatic variables (38%) for males, whereas for females the component is a complex mix of climatic, energetic and behavioural variables, each contributing about a third of the variance to this component. Component two is characterised by energetic variables in males (41%) but by behavioural variables in females (44%). The third component is equally characterised by climatic and behavioural variables in males (30%), but by climatic variables in females (40%) (Table 13.6). Per variable, the structures of male and female components are quite different (see Table 13.8). For example, percentage time spent arboreal and percentage fruit in diet load negatively into component one in males, but positively in component one for females. Seasonality and altitude load positively into component one for males, but negatively for females. These and further differences between the nature and size of the component loadings per component between the sexes suggest that the structures of the variation in male and female foraging behaviour are not only very complex, but quite different.

Table 13.6. *The structure of variation in male and female great ape foraging behaviour: results of principal components analysis*

Component	Males			Females		
	1	2	3	1	2	3
Total variation[a] (%)	40.7	29.5	12.3	46.7	24.6	12.1
Climatic label[b]	Rainfall Altitude Seasonality	Rainfall Altitude	Altitude	Rainfall Altitude Seasonality	Rainfall Altitude	Altitude Seasonality
CL loading (%)	38	20	30	33	24	40
Energetic label[d]	DEC % Arboreal Day range	DEC Day range		DEC % Arboreal Day range	DEC Day range	
CL loading (%)	31	41	0	35	36	0
Behaviour label[e]	FPS Home range % Fruit	% Fruit Home range	FPS	FPS Home range % Fruit	FPS Home range % Fruit	FPS
CL loading (%)	31	35	30	31	44	28
Component characteristic[f]	Climatic	Energetic	Climate/ behaviour	Climate/ energetic/ behaviour	Behaviour	Climatic

DEC = daily energetic costs; FPS = foraging party size.
[a] Total variation = amount of variation in data set that component accounts for.
[b] Coefficient sum = sum total of variable coefficients/component loadings (see Table 13.8).
[c] Climatic label = sum of rainfall, altitude and seasonality coefficients.
[d] Energetic label = sum of daily energetic costs, percentage time spent arboreal and day range coefficients.
[e] Behaviour label = sum of percentage fruit in diet, foraging party size, and home-range coefficients.
[f] Component characteristic = label that accounts for most of the variation in each component.

Sex differences in foraging strategy between great ape subspecies

An overall comparison of great ape sex differences in foraging strategy (see Table 13.4) results in a lack of apparent significant differences (see above). This is due to the smoothing-out of actual differences between ape populations because of the large range in sexual dimorphism observed across all species. Analysis at the subspecies level (paired sample t-tests) shows sex differences at a closer level of inspection.

On average, orangutan males spend more time feeding, incorporate more insects into their diet ($t = -2.15$; $p = 0.09$), range more widely (day range: $t = 2.67$; $p = 0.05$; home range $t = -4.4$; $p = 0.012$) and have larger daily energetic costs (19%, $t = -4.81$; $p = 0.009$) than females. Female orangutans spend more time in the trees than males ($t = 2.67$; $p = 0.05$). There are no significant differences between the sex differences in foraging behaviour of Sumatran and Bornean orangutans.

Gorillas show no significant sex differences in ranging, but males spend more time feeding ($t = -7.97$; $p = 0.015$) and on the ground and have a greater proportion of insects in their diet than females ($t = -4.14$; $p = 0.054$). Overall, male daily energetic costs are 20% higher than those of females ($t = -4.32$; $p = 0.05$). These sex differences are more marked in low-altitude-living gorillas than in high-altitude-living gorillas.

Bonobos appear to show no significant differences in foraging behaviour and the sexes forage together in relatively large parties. However, although the males have a slightly larger body weight than the females, female daily energetic costs are 8.5% greater than that of males ($t = 17.75$; $p = 0.036$).

Chimpanzee females have smaller day and home ranges and a larger degree of arboreality. Overall, chimpanzee males spend more time resting ($t = -10.53$; $p = 0.009$) and feed less on insects than females. Among the subspecies, females feed longer than males in eastern chimpanzees, but feed for less time in western chimpanzees. As with the gorillas, sex differences in chimpanzee foraging behaviour are more pronounced in the relatively poorer habitats. In particular, male chimpanzees at Mt Assirik rest for twice the time of females each day (Tutin, McGrew and Baldwin, 1983). At Gombe, females spend more time travelling and feeding than males (but less than at Mt Assirik: Wrangham and Smuts, 1980), and at Kibale, males spend more time travelling and feeding than females (Ghiglieri, 1984). Furthermore, hunting is more common at Kibale than at Gombe (Ghiglieri, 1984). Kibale consists of medium-altitude tropical rainforest and Gombe of evergreen forest and woodland, with the countryside being quite broken up. Mt Assirik is particularly harsh, consisting of mainly open savannah with small strips of gallery forest (McGrew *et al.*, 1981).

Predicting sex differences in great ape foraging behaviour

The predictability of sex differences in foraging strategy from independent variables that mark habitat type, relative ecological energetic costs (see above) and degree of sexual dimorphism was examined using multiple linear regression analysis (Bean, 1998).

As the habitat increases in harshness, sex differences in daily energetic costs[1] and foraging index[2] *decrease*, but sex differences in day range *increase* as the sexes begin to overlap in daily energy requirement and expenditure. A harsher habitat means that food is generally less available and/or harder to process/find and so feeding competition is greater. Why sex differences in daily cost decrease is not immediately apparent. If there is a degree of feeding specialisation, then harshness of an environment might need a definition for each sex, thus the habitat is relatively harsher for one sex than for the other.

In a harsh habitat, the individual will have to forage further per day (and cover a larger area seasonally and annually) to find food items that are more spread out. Or, if the food is of low quality, the individual must forage further per day in order to consume a greater quantity of food. Feeding competition will be intense and the sexes might be expected to alter their ranging patterns relative to each other in order to minimise direct competition.

Ecological energetic costs represent the relative exploitation of a habitat type and decrease with an increase in body mass, but also decrease with a decrease in foraging party size and are a result of low foraging efficiency. As ecological energetic costs decrease, sex differences in the percentage of time spent arboreal, daily energetic costs, day range and the percentage of time spent resting (i.e. activity budget) decrease. These variables are used here to summarise foraging strategy and so, as ecological costs increase, sex differences in foraging strategy also increase.

Relatively high ecological costs partly stem from a high degree of feeding competition, in response to which, the sexes appear to shift their foraging behaviour relative to one another. Although there is an overlap in activity budget, there is a divergence in habitat exploitation represented by a promotion of sex differences in ranging, activity budget and arboreal behaviour.

With an increase in sexual dimorphism, there appear to be two possible

Table 13.7. *Summary of sex differences in great ape foraging behaviour*

	High ecological cost	Low ecological cost
High degree of sexual dimorphism in body mass	Ecological separation of sexes results in sex differences in foraging behaviour, which is pronounced due to high levels of feeding competition between the sexes (e.g. orangutans, low altitude gorillas)	Although ecological separation of sexes, feeding competition is low and so sex differences in foraging behaviour are conservative and related to substrate use (where much larger males are limited to robust substrates) and are mostly due to males being much larger than females (e.g. high altitude gorillas)
Low degree of sexual dimorphism in body mass	Overlap in foraging strategy due to low sexual dimorphism in body mass combined with harsh habitat leads to high feeding competition and thus sex differences in foraging behaviour (e.g. chimpanzees)	Although overlap in feeding strategy occur because of low level of sexual dimorphism in body mass, the habitat is sufficient to support these foragers and so feeding competition is low as are sex differences in foraging behaviour (e.g. bonobos)

responses. The first of these is to increase the extent of differences between the sexes in activity budgets (e.g. orangutans). The second strategy can be proposed as an increase in the energetic (dietary) separation of males and females, but a decrease in the extent of differences in activity patterns (e.g. chimpanzees and bonobos).

Summary and discussion

A combination of habitat quality and sexual dimorphism influences sex differences in foraging behaviour (Table 13.7).

The sexes experience different selective pressures for energetic acquisition and expenditure. The males' larger body size affects resource utilisation (MacKinnon, 1971, 1974; Rodman, 1977; Galdikas, 1979), and so the degree of sexual dimorphism in body mass of a species will affect the amount of sexual segregation within a habitat.

The degree of sexual dimorphism in body mass will primarily affect the degree of overlap in daily energetic costs between the sexes and the level of intrasexual feeding competition. However, the habitat type will also affect sex differences in daily energetic costs such that a harsher habitat will promote a divergence in foraging strategy via dietary composition. This

Table 13.8. *Component pattern matrix for principal components analysis of male and female great ape foraging strategy*

Variable	Males			Females		
	C1 (40.7%)	C2 (29.5%)	C3 (12.3%)	C1 (46.7%)	C2 (24.6%)	C3 (12.2%)
DEC	0.430	−0.797	−0.288	0.354	0.483	0.538
Arboreal	−0.928	0.113	0.068	0.542	0.744	0.156
Day range	0.301	0.843	−0.040	0.608	−0.173	0.685
FPS	0.578	0.006	0.641	0.640	0.589	0.119
% Fruit	−0.579	0.689	0.167	0.822	0.042	−0.442
Rainfall	−0.851	−0.411	0.077	−0.821	−0.434	0.043
Seasonality	0.862	0.017	−0.285	−0.628	0.664	0.102
Altitude	0.360	−0.381	0.697	−0.934	0.083	0.172
Home range	0.512	0.711	−0.028	0.623	−0.642	−0.230

DEC = daily energetic costs; C = component number; FPS = foraging party size.

has the effect of reducing sex differences in daily energetic costs, causing overlap between the activity budgets for both males and females. This divergence in foraging strategy is represented by a promotion in behavioural sex differences that indicates dietary strategy such as ranging and arboreality. This divergence in foraging strategy serves to decrease feeding competition, which is increased in a harsher habitat (Table 13.8).

Therefore, sex differences in great ape foraging behaviour vary with habitat type, ecological cost (by feeding competition) and sexual dimorphism in body mass by a complex interaction of climatic variables and daily energetic costs, which form identifiable patterns for both males and for females at the local habitat level. Sex differences in foraging behaviour are attributable to local conditions and daily energetic requirements as well as to species and subspecies characteristics. Reproductive costs and sexual dimorphism in body mass are fundamental differences between the sexes. Sex differences in foraging behaviour are not solely the result of these distinctions but are closely linked to resource availability as well.

Notes

1 Daily energetic cost estimate = daily energetic costs of feeding, travelling and resting summed plus reproductive costs if female (i.e. lactation and gestation) (see Bean, 1998, for a review and Wheatley, 1982, for equation derivations).
2 Foraging index = time spent feeding/time spent travelling and indicates efficiency of foraging (see Rodman, 1977, for a review).

References

Bean, A.E. (1998). The ecology of sex differences in great ape foraging behaviour and human hunter–gatherer subsistence behaviour: the origin of sexual division in human subsistence behaviour. PhD thesis, University of Cambridge.

Bean, A.E. (in press). Habitat and foraging in great apes. In *Primate Behaviour and Evolution*, ed. C. Harcourt and R. Crompton. London: Westbury Publishing, Linnaean Society.

Caldecott, J.O. (1986a). *An Ecological and Behavioural Study of the Pig-Tailed Macaque*. Basel: Karger.

Caldecott, J.O. (1986b). Mating patterns, societies, and the ecogeography of macaques. *Animal Behaviour* **34**, 208–20.

Clutton-Brock, T.H. (1994). The costs of sex. In *The Differences Between the Sexes*, ed. R.V. Short and E. Balaban, pp. 347–62. Cambridge: Cambridge University Press.

Clutton-Brock, T.H., Alban, S.D. and Guiness, F.E. (1984). Maternal dominance, breeding success and birth sex ratios in red deer. *Nature* **308**, 358–60.

Collin's Atlas of the World, revised edition (1995). London: Harper Collins.

Doran, D.M. (1993a). Comparative locomotor behaviour of chimpanzees and bonobos: the influence of morphology on locomotion. *American Journal of Anthropology* **91**, 83–98.

Doran, D.M. (1993b). Sex differences in adult chimpanzee positional behavior: the influence of body size on locomotion and posture. *American Journal of Anthropology* **91**, 899–915.

Dunbar, R.I.M. (1988). Ecological modelling in an evolutionary context. *Folia Primatologica* **53**, 234–56.

Dunbar, R.I.M. (1992). Time: a hidden constraint on the behaviour ecology of baboons. *Behavioural Ecology and Sociobiology* **31**, 35–49.

Dunbar, R.I.M. (1994). Time budgets and other constraints. In *Primate Social Systems*, pp. 90–106. London: Chapman & Hall.

Fossey, D. (1974). Observations on the home range of one group of mountain gorillas *Gorilla gorillas beringei*. *Animal Behaviour* **20**, 36–53.

Galdikas, B.M.F. (1979). Orangutan adaptation at Tanjung Puting Reserve: mating and ecology. In *Perspectives on Human Evolution*, Vol. 5. *The Great Apes*. Menlo Park: Benjamin/Cummings.

Galdikas, B.M.F. (1988). Orangutan diet, range and activity at Tanjung Puting, Central Borneo. *International Journal of Primatology* **9**(1), 1–35.

Galdikas, B.M.F. and Teleki, G. (1981). Variations in subsistence activities of female amd male pongids: new perspectives on the origins of human labor division. *Current Anthropology* **22**(3), 241–56.

Ghiglieri, M.P. (1948). *The Chimpanzees of Kibale Forest: A Field Study of Ecology and Social Structure*. New York: Columbia University Press.

Goodall, J. (1968). The behaviour of free-living chimpanzees in the Gombe Stream Reserve. *Animal Behaviour Monographs* **1**, 161–311.

Harrison, M.J.S. (1983). Age and sex differences in the diet and feeding strategies of the green monkey, *Cercopithecus sabaeus*. *Animal Behaviour* **31**, 969–77.

Hladik, C.M. (1977). Chimpanzees of Gabon and chimpanzees of Gombe: some comparative data on the diet. In *Primate Ecology*, ed. T.H. Clutton-Brock, pp. 481–501. London: Academic Press.

Lee, P.C. and Bowman, J.E. (1995). Influences of ecology and energetics on primate mothers and infants. In *Motherhood in Human and Nonhuman Primates*, ed. C.R. Pryce, R.D. Martin and D. Skuse, pp. 47–58. Basel: Karger.

Leigh, S.R. and Shea, B.T. (1995). Ontogeny and the evolution of adult body size dimorphism in apes. *American Journal of Primatology* **36**, 37–60.

MacKinnon, J. (1971). The orang-utan in Sabah today. *Oryx* **11**(2–3), 141–91.

MacKinnon, J. (1974). The behaviour and ecology of wild orang-utans (*Pongo phymaeus*). *Animal Behaviour* **22**, 3–74.

Malenky, R.K. and Wrangham, R.W. (1994). A quantitative comparison of terrestrial herbacious food consumption by *Pan paniscus* in the Lomako Forest, Zaire, and *Pan troglodytes* in the Kibale Forest, Uganda. *American Journal of Primatology* **32**, 1–12.

McGrew, W.C., Baldwin, P.J. and Tutin, C.E.G. (1981). Chimpanzees in a hot, dry and open habitat: Mt Assirik, Senegal. *Journal of Human Evolution* 10, 227–44.

Pearce, E.A. and Smith, C.G. (1994). *World Weather Guide*, 3rd edition. Oxford: Helicon.

Pollock, J.I. (1977). The ecology and socioecology of feeding in *Indri indri*. In *Primate Ecology: Studies of Feeding and Ranging Behaviour in Lemurs, Monkeys and Apes*, ed. T. Clutton-Brock, pp. 38–68. London: Academic Press.

Pond, C.M. (1977). The significance of lactation in the evolution of mammals. *Evolution* 31, 177–99.

Portman, O.W. (1970). Nutritional requirements (NRC) of nonhuman primates. In *Feeding & Nutrition of Nonhuman Primates*, ed. R.S. Harris, pp. 87–115. New York: Academic Press.

Post, D., Hausfater, G. and McCuskey, S.A. (1980). Feeding behaviour of yellow baboons (*Papio cynocephalus*): relationship to age, gender and dominance rank. *Folia Primatologica* 34, 170–95.

Reynolds, V. and Reynolds, F. (1965). Chimpanzees of the Budongo Forest. In *Primate Behaviour*, ed. I. DeVore, pp. 368–424. New York: Holt, Rinehart & Winston.

Rijksen, H.D. (1978). *A Field Study on Sumatran Orang-utans (Pongo pygmaeus abelii. Lesson 1827): Ecology, Behaviour and Conservation*. Menlo Park: Benjamin/Cummings.

Rodman, P.S. (1977). Feeding behaviour of orang-utans of the Kutai Nature Reserve, East Kalimantan. In *Primate Ecology: Studies of Feeding and Ranging Behaviour in Lemurs, Monkeys and Apes*, ed. T.H. Clutton-Brock, pp. 383–413. London: Academic Press.

Sadleir, R.M.S. (1969). *The Ecology of Reproduction in Wild and Domestic Mammals*. London: Methuen.

Teleki, G. (1973). the omnivorous chimpanzee. *Scientific American* 228(1), 32–42.

Tutin, C.E.G. and Fernandez, M. (1984). Nationwide census of gorilla (*Gorilla g. gorilla*) and chimpanzee (*Pan t. troglodytes*) populations in Gabon. *American Journal of Primatology* 6, 313–36.

Tutin, C.E.G., McGrew, W.C. and Baldwin, P.J. (1983). Social organisation of savannah-dwelling chimpanzees, *Pan troglodytes verus*, at Mt Assirik, Senegal. *Primates* 24(2), 154–73.

Watts, D.P. (1984). Composition and variability of mountain gorilla diets in the Central Virungas. *American Journal of Primatology* 7, 323–56.

Wheatley, B.P. (1982). Energetics of foraging in *Macaca fascicularis* and *Pongo pygmaeus* and a selective advantage of large body size in the orang-utan. *Primates* 23(3), 348–63.

White, F.J. (1988). Party composition and dynamics in *Pan paniscus*. *International Journal of Primatology* 9, 179–93.

Wrangham, R.W. (1977). Feeding behaviour of chimpanzees in Gombe National Park, Tanzania. In *Primate Ecology*, ed. T.H. Clutton-Brock, pp. 503–8. London: Academic Press.

Wrangham, R.W. (1979). On the evolution of ape social systems. *Social Science Information* 18, 335–68.

Wrangham, R.W. (1987). The significance of African apes for reconstructing human evolution. In *The Evolution of Human Behaviour: Primate Models*, ed. W.G.

Kinzey, pp. 28–47. Albany: Sata University of New York Press.

Wrangham, R.W. and Smuts, B.B. (1980). Sex differences in the behavioural ecology of chimpanzees in the Gombe National Park, Tanzania. *Journal of Reproduction and Fertility* **28**, Supplement, 13–31.

Yamagiwa, J., Maruhashi, T., Yumoto, T. and Mwanza, N. (1996). Dietary and ranging overlap in sympatric gorillas and chimpanzees in Kahuzi-Biega National Park, Zaire. In *Great Ape Societies*, ed. W.C. McGrew, F. Marchant and T. Nishida, pp. 82–98. Cambridge: Cambridge University Press.

Yamagiwa, J., Mwanza, N., Yumoti, T. and Maruhashi, T. (1994). Seasonal change in the composition of the diet of eastern lowland gorillas. *Primates* **35**(1), 1–14.

14 *Hominid behavioural evolution: missing links in comparative primate socioecology*

ROBERT A. FOLEY

Introduction

Humans represent perhaps the greatest challenge to a comparative approach, for they comprise just one species, and one that is radically different from all others. However, the recognition of more and more species in the fossil record that are closer to humans than to other living primates has changed this perspective. The approximately 15 species of extinct hominid provide an excellent framework for considering hominid socioecology comparatively, and also for filling the gap between humans and chimpanzees and bonobos. The problem for a socioecological analysis is that we are entirely dependent upon the fossil record, with all that entails in terms of patchy and incomplete data, and total silence on a number of key issues relating to behaviour. Although there are technical and empirical problems is using fossil data, they do bring into focus some important issues that are often ignored in comparative analyses of extant species. Some of these will be addressed here in the context of human socioecological evolution.

Socioecology can be defined as the way in which ecological principles underlie the structure of behaviour and interactions between members of the same species, and reciprocally how social behaviour influences the exploitation of resources. That the observed ecological and social strategies of living species have thus evolved through natural selection, and represent adaptations to particular environments, is virtually axiomatic to the discipline. What is seldom made explicit, but follows from this, is that these small-scale adjustments provide the microevolutionary basis for long-term evolutionary change, and ultimately macroevolutionary patterns. Although the implications for long-term evolution are clear, for the most part socioecological analyses have not examined longer term evolutionary change. Socioecological models are essentially synchronic, comparing across the terminal twigs of evolving lineages, and treating these extant snapshots as evolutionarily stable equilibria.

This chapter explicitly considers evolution in a temporal framework, using the evolution of ecological strategies and social behaviour among the hominids as the focus. While comparative socioecology of living primates and humans takes into account the effects of selection on phylogeny, evolution is a question of inference back down the lineages to any common ancestral node, and extrapolation of intermediate and transitional states. The fossil record, which is of course available for other species as well, allows for a closer examination of these transitional states. Many fossil taxa are themselves terminal twigs that happen to have become extinct, but they at least allow for two additional elements to be incorporated into any analysis. The first element is a much fuller phylogeny. The living primates represent a partial sample of primates, and so any phylogeny derived from them is at best an incomplete approximation of the possible states (Martin, 1985). Partial sampling of a phylogeny based on what are presumably the non-random effects of extinction rates in different lineages may give a misleading perspective of a clade. An example of this might be the difference between the Cercopithecoidea and the Hominoidea (Andrews, 1992). The former number some 70 species, and this may represent up to 50% of the total number of cercopithecoid species that have ever existed. The Hominoidea, on the other hand, are represented today by less than 15 species, and these are probably less than 20% of the total lineage. The Hominoidea have been ravaged by extinction on a much larger scale than the Cercopithecoidea (McCrossin and Benefit, 1994). The contrast, therefore, between the hominoids, which have very high levels of discontinuity in socioecology between relatively few taxa, and the cercopithecoids, which show a very high degree of similarity and continuity among a much larger number of taxa, must be at least in part a function of differential patterns of extinction.

The second additional element provided by fossils is that their location in time and space permits a greater access to the actual conditions under which particular strategies may have emerged. Orangutans, for example, are now restricted to Borneo and Sumatra, but this may not be where they evolved. Fossil orangutans have a much broader Asian distribution (Ciochon and Etler, 1994) and, by inference, evolved sometime in the later Miocene when conditions were very different – more stable climates, greater expanses of forests, lower sea levels, and among a very different sympatric fauna (Foley, 1987). It is therefore important, if we ask why certain behavioural and ecological adaptations have evolved, to obtain a robust measure of the past conditions under which these events took place.

Table 14.1. *Fossil hominid taxa in time and space*

Taxon	First appearance datum	Last appearance datum	Geographical area
Ardipithecus ramidus	5.1	4.2	North eastern Africa
Australopithecus anamensis	4.2	3.9	Eastern Africa
Australopithecus afarensis[a]	3.7	2.7	North-eastern–east Africa
Australopithecus bahrelghazali	3.5	3.0	North central Africa
Australopithecus africanus[a]	3.5	2	Southern Africa
Australopithecus aethiopicus	2.6	2.3	North-eastern–east Africa
Australopithecus boisei	2.4	1.3	North-eastern–east Africa
Australopithecus robustus	2.1	1.9	Southern Africa
Australopithecus crassidens	1.6	0.65	Southern Africa
Homo rudolfensis	2.4	1.6	Eastern Africa–southern Africa
Homo habilis	1.9	1.4	Eastern Africa
Homo ergaster	1.9	1.5	Africa–Caucasus
Homo erectus	1.8	0.05	East Asia–SE Asia–Africa
Homo anteassessor	0.8	0.5?	Europe
Homo heidelbergensis	0.6	0.2	Africa and Europe (?E. Asia)
Homo helmei	0.28	?	Sub-Saharan Africa
Homo neanderthalensis	0.2	0.027	Europe–Middle East
Homo sapiens	0.15	Extant	Global

[a]It has been suggested that these taxa include material from more than one species, and therefore may be further subdivided.

The greater interest in phylogenetic inertia brought about by the development of the comparative method makes such a step even more essential. When viewed across long spans of time, evolution becomes analytically much messier, but major issues can be addressed.

The hominids and their evolutionary patterns

The term 'hominid' is used here to mean any taxon that is more closely related to humans than to *Pan*. While there is now a general consensus that the classic hominoid families are not monophyletic, for purposes of clarity the conventional definition of the Hominidae is retained here.

The data pertaining to the various hominid taxa are shown in Tables 14.1 and 14.2. Table 14.1 lists the range of taxa that have been recognised, and their chronological and geographical distribution. There has, of course, been considerable debate in the literature about both the number of hominid species and the nature of species themselves. Table 14.1 represents a 'splitting' view. Such a classification is not necessarily based on the

Table 14.2. *Adaptive grades or superspecies of hominids*

The taxa shown in Table 14.1 represent considerable splitting of the hominid fossil record, and not all authorities recognise all these species. Table 14.2 shows how the various taxa described in Table 14.1 might be combined to form superspecies. Such superspecies may or may not be monophyletic, but would perhaps also represent grades of hominid sharing relatively similar levels of organisation and adaptive patterns. A thumbnail sketch outlining the basic characteristics and known variation for each is provided.

Last common ancestor/earliest hominids

Superspecies: *Ardipithecus ramidus*   Species/subspecies: *A. ramidus*

General characteristics

This taxon is generally considered to be very close to the point of divergence from the last common ancestor with *Pan.* Poorly known, but probably in the size range of a large chimpanzee (35–45 kg), it may or may not be bipedal. Small canines, thin dental enamel and projecting face all indicate its very mixed character. Overall, it is probably a frugivorous/omnivorous ape. Brain size and sexual dimorphism are unknown, but probably similar to *Pan.* Perhaps indicates geographical origin of Hominidae in north-eastern Africa.

Variation

Not known.

Early savannah bipedal apes

Superspecies: *Australopithecus africanus*   Species/subspecies: *A. anamensis*
                                                                                      A. afarensis
                                                                                      A. africanus
                                                                                      A. bahrelghazali

General characteristics

These are the early australopithecines. Although often described as gracile, these taxa are larger than chimpanzees, and mostly fall within the range 45–60 kg. In absolute terms, brain size is between 400 and 550 cm^3. In relative terms, EQ is slightly above that of a chimpanzee (2.1) – approximately 2.3 to 2.6. Facultatively bipedal; in general terms, these australopithecines have relatively long arms, short legs, large guts and chests, suggesting a mixed locomotion/positional behaviour involving terrestrial and arboreal activities. Generally show a trend towards larger posterior teeth, with some anterior reduction. *A. africanus* is often heavily megadontic. Tooth enamel is thick. On the basis of tooth morphology and wear, most of these would have been frugivores, with elements of both coarser lower quality food and meat in the diet. Growth rates are ape-like and rapid, with age of first reproduction probably similar to *Pan.* Probably highly sexually dimorphic, these species are best considered geographical and time transgressive variants on the theme of African apes, less specialised than the later australopithecines.

Variation

A. anamensis and *A. afarensis* represent the earlier eastern forms, while *A. africanus* and *A. bahrelghazali* are slightly later southern and north-western extensions of range and thus allopatric species. They exhibit considerable body size variation within and between species (*A. anamensis* 47–55 kg, *A. afarensis* 27–45 kg. *A. africanus* 30–43 kg). Posterior tooth size and wear in *A. africanus* overlap with those of some later australopithecines.

Later savannah bipedal apes

Superspecies: *Australopithecus (Paranthropus) robustus*

Species/subspecies: *A. robustus*
A. crassidens
A. aethiopicus
A. boisei

General characteristics

These are the so-called robust australopithecines or paranthropines. Their robustness is largely cranial, although they do tend to be slightly larger than the earlier forms. Overall body size ranges from around 40 kg to over 80 kg, with an average around 50 kg. Some increase in brain size compared to other australopithecines, with an EQ between 2.2 and close to 3. Bipedal, but still with relatively long forelimbs and shorter hindlimbs, and broad thoraxes. Megadontic posterior dentition, with thick tooth enamel, highly reduced anterior dentition. All teeth show the effects of heavy wear and chewing, and have flat occlusal surfaces. Tooth wear and morphology indicate very coarse, small object foods, probably high in grit and fibre. Mostly plant foods, but likely to be eclectic on the basis of hominoid ancestry and includes some meat. Highly sexually dimorphic across all taxa where known.

Variation

The robust australopithecines are all variants on a theme. *A. boisei* is the most extreme in its megadonty, while the older *A. aethiopicus* possesses the smallest brain (410 cm³) and a projecting face. They may represent convergent evolutionary trends.

Early intelligent and opportunistic omnivores

Superspecies: *Homo habilis*

Species/subspecies: *H. habilis*
H. rudolfensis

General characteristics

Early *Homo* taxa show mixed features, in some ways similar to australopithecines but exhibiting larger brains and dental/facial reduction. Body size is very variable, but probably around 45–50 kg. Brain size exceeds 600 g, and the EQ estimates are close to 3.0. Early *Homo* is poorly known postcranially, but some specimens indicate a body structure similar to australopithecines. Problems of taxonomic assignment make estimates of sexual dimorphism problematic, but it is likely to have been considerably dimorphic. *Homo* is associated with the first stone tools, and may have been increasingly omnivorous. Growth patterns would be closer to apes than humans.

Variation

There are basically two forms – a smaller and more gracile australopithecine type (*H. habilis*), showing some brain enlargement and facial reduction, and a larger, more megadontic form with larger brain (*H. rudolfensis*).

Table 14.2. (cont.)

Superspecies: *Homo erectus*

Later intelligent and opportunistic omnivores

Species/subspecies: *H. ergaster*
H. erectus
H. heidelbergensis
H. anteassessor

General characteristics

Pleistocene *Homo* is generally larger than the Pliocene australopithecines and *Homo*, with brain sizes between 800 and 1200 cm³. EQs are greater than 3.0. Full bipedalism and linear body form are established, but with developed muscularity. Teeth are smaller, and technology extensive. Substantial hunting/meat eating may have been in place. Sexual dimorphism where known remains substantial. Growth patterns shifted towards the human condition.

Variation

Pleistocene *Homo* is very variable. There is a general temporal trend towards greater brain size, especially in African and European forms (*H. heidelbergensis*); increased robusticity in African lineage of *H. heidelbergensis*. The early African forms are much taller and more linear, with body mass above 50 kg. Later *H. erectus* are larger and more robust (> 60 kg) than *H. ergaster* (52–65 kg). *H. heidelbergensis* (55–80 kg) is often very large and robust (Bodo, Petralona). Other differences may be local geographical ones.

Superspecies: *Homo sapiens*

Technological colonisers and dominant herbivores

Species/subspecies: *H. helmei*
H. neanderthalensis
H. sapiens

General characteristics

All these forms are generally large, often greater than 60 kg, with larger cranial capacities well within the range of living humans. EQ is in excess of 5. There is full bipedalism, and a general loss of extreme cranial superstructures, facial and dental reduction. Technology is much more complex (Mode 3, 4, and 5). Sexual dimorphism is reduced, and life history parameters are likely to be close to or largely within the range of modern humans. Extreme habitat tolerance appears to be characteristic, possibly associated with high levels of omnivory (hunting).

Variation

Variability in this form is quite marked across time. *H. helmei* is the early common ancestor to the later form, and so retains more primitive characters and is robust. *H. neanderthalensis* is the most derived, with cold climate adaptation in terms of face size, body proportion, body mass (55–70 kg) and posture/locomotion. Early *H. sapiens* are large and robust, but become increasingly variable, gracile (35–70 kg) and widespread, with the most cultural and technological complexity. Sexual dimorphism is high in early forms of all taxa, but reduced in later *H. sapiens*.

biological species concept, as it is virtually impossible to determine this from fossil samples. Furthermore, the time transgressive nature of the fossil sample makes the use of most species concepts somewhat problematic. The definition of a species employed here is closer to that put forward by Simpson (1950) as the phylogenetic species concept; a species is a unit that appears to be undergoing independent evolution. The fact that we can recognise these as having distinctive morphological patterns underwrites their position as species, although it does not preclude them being able to interbreed should the situation have arisen. In using such a definition, one is also moving close to the idea of species as ecologically unique units, a distinction that is functional and adaptive rather than genetic (van Valen, 1976).

However, some of the taxa identified here are nonetheless very similar to each other, both morphologically and adaptively. Table 14.2 shows how these fine-grained taxonomic units might be united either to form super-species or else to conform to more lumping preferences in taxonomy. There is little point in maintaining a dogmatic view of the nature of hominid species in the absence of genetic data, and it is better to recognise that different species concepts may be appropriate for different evolutionary questions.

Table 14.2 also provides an overview of the basic biological and adaptive characteristics of these taxa in so far as they can be inferred from the fossil record. By their very nature, these data are patchy and prone to error, are usually second-order inferences, and do not normally compare to point estimates with confidence limits used for living species. They are sum-marised here as general indicators of overall trends rather than precise quantitative parameters. In effect, they are samples in a complex pattern of variation through time and space. Figure 14.1 summarises the data available in graphical form.

Fig. 14.1   Temporal trends in ecologically relevant parameters inferred from the hominid fossil record. Encephalisation quotient (EQ) is based on Martin (1981); body mass is estimated in kilograms from regressions based on hominid postcranial and cranial proportions; sexual dimorphism is based on inferred weight estimates; growth is a measure of deviation from chimpanzee maturation patterns (Smith, 1992, 1994); tooth size is the occlusal surface area of the lower first molar (mm^2). Hominid data from Aiello and Dean (1990); Aiello and Dunbar (1993); Bromage and Dean (1985); Foley and Lee (1991); Frayer and Wolpoff (1985); Grine (1988); Leigh (1992); McHenry (1992); Smith (1992, 1994). See also Tables 14.1 and 14.2. For abbreviations, see Figure 14.2.

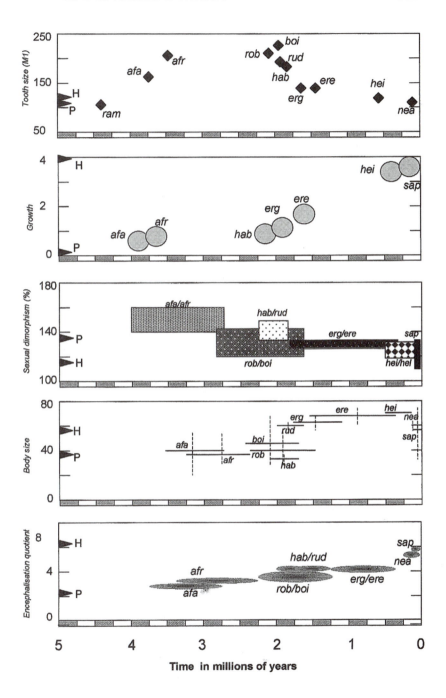

Time in millions of years

Body size

For discussion of body size see: Pilbeam and Gould (1974), Steudel (1980), Martin (1981, 1983), Foley (1987), McHenry (1988, 1992), Jungers (1988) and Trinkaus (1989). There is no simple increase in time; australopithecines are distributed around a central value of between 40 and 50 kg, which would make them somewhat larger than chimpanzees. Early *Homo* is very diverse, and *H. ergaster* is thought to be around 60 kg (Ruff, 1991; McHenry, 1992). Within *Homo*, the pattern of body size evolution is generally towards increased body mass and robusticity, with a reversal among modern humans (Brown, 1987; Clark *et al.*, 1994; Lahr and Wright, 1996).

Body shape, posture and locomotion

With the possible exception of *A. ramidus*, all hominids show bipedal adaptation, but these vary from relatively apelike proportions in australopithecines to virtually modern proportions in *H. ergaster* (Stringer, 1984; Susman, Stern and Jungers, 1984; Senut and Tardieu, 1985; McHenry, 1986; Lovejoy, 1988; Trinkaus, 1989; Schmid, 1991; Ruff, 1991; Trinkaus *et al.*, 1991).

Brain size

There is undoubtedly a major increase in brain size during the course of hominid evolution, but measuring the rate is compounded by the effect of body size (Martin, 1983; Aiello and Dean, 1990; Leigh, 1992). Australopithecines fall within the range of encephalisation quotient (EQ) of extant apes, early *Homo* is ambiguous, and there is a sharp trend of increase in the last 0.3 Myr. The encephalisation of *Homo* can be ascribed to one or more of a number of factors; increasing technological or social complexity may have acted as a selective pressure for greater intelligence (Aiello and dunbar, 1993), and life history parameters (see below) may have altered in such a way as to make larger brains sustainable (Foley and Lee, 1991).

Life history and growth

Developmental analysis of enamel histology has provided insights into changes in growth patterns during the course of hominid evolution (Bromage and Dean, 1985; Beynon and Wood, 1987; Beynon and Dean, 1988; Smith, 1989, 1992, 1994). Across the evolution of *Homo*, it is clear that there is a retardation in the rate at which teeth erupt, indicating a slower overall growth rate. Although there may be some doubt about the neander- thals, by and large it is clear that later *Homo* was much more similar to modern humans than to other primates in its growth patterns (Foley and Lee, 1991; Foley, 1992). There is a strong interaction between age of maturation and other life history parameters (Harvey, Martin and Clut- ton-Brock, 1987).

Tooth size

The marked trend in early hominids is towards posterior tooth row mega- donty (Grine, 1988). Indeed, it can be argued that this is the primary and dominant evolutionary trend for the australopithecines, occurring against a relative stability in life history, body size, brain size and post-cranial traits. The most obvious interpretation of this pattern is that there are in fact two different and independent trajectories taking place among the hominids. The first of these, seen among the early African hominids, is towards megadonty, and then the second, in *Homo*, reverses this. In dietary terms, this implies that there was, during the period 5–2 Myr, considerable selection for exploiting resources that required heavy mastication. How- ever, comparatively speaking, all hominids are likely to have been oppor- tunistic and wide ranging in diet breadth.

Sexual dimorphism

Sexual dimorphism is inextricably linked to problems of taxonomic iden- tification; one person's highly sexually dimorphic species is another per- son's two species. Two things are apparent. The first is that the early African hominids may well have been more sexually dimorphic than their closest living relative, the chimpanzee (McHenry, 1992). The higher levels of sexual dimorphism found in gorillas and orangutans may be closer to that inferred for the australopithecines. Second, there is a marked dimin- ution of sexual dimorphism during the course of *Homo* evolution (Frayer and Wolpoff, 1985). Again, this does not stop with the appearance of modern humans, but is a trend that continues through the latest evolution of our own species.

Populations, habitat and distribution

The hominids differ markedly in their extent and distribution. Some taxa are local or regional species; the robust australopithecines, for example, appear to be allopatric sister species in eastern and southern Africa. Middle Pleistocene *Homo* shows a more continental distribution, with *Homo erectus* in eastern Asia, and *Homo heidelbergensis* in Africa and southern Europe. Only very late Pleistocene/Holocene *Homo sapiens* has a fully cosmopolitan distribution (Foley and Lahr, 1997).

It can be inferred that *Homo erectus* and its descendants were more tolerant in some ways, such as being able to live at higher altitudes, and spread out of Africa. The later hominids all show a preference for more open habitats, but can also tolerate much colder climates in addition. However, the biggest contrast in hominid population distribution lies between the Pleistocene as a whole and the last 10 000 years from very small ($\sim$ 100 000) to the current vast populations.

Hominid evolutionary and behavioural ecology in comparative perspective

Key adaptive shifts

Four key adaptive shifts or grades during the course of hominid evolution can be observed (Fig. 14.2).

The last common ancestor

The fossil record is completely silent on the matter of the nature of the last common ancestor of hominids and chimpanzees, and its characteristics can be inferred only from later hominids and the living species. The nature of the early australopithecines rules out the notion that the last common ancestor was intermediate between modern humans and chimpanzees, and evolutionary change has been far more substantial in *Homo* compared to *Pan*. On this basis, it is probable that the last common ancestor was an African ape, existing in Miocene forests, hairy and quadrupedal, with an EQ of around 2.0 or less. It would have been characterised by small molars, large incisors and canines, with thin tooth enamel, and its diet is likely to have been similar to that of a chimpanzee today – fruits plus opportunistic hunting and poor season leaf-eating (Moore, 1996). On the basis of what is know about Miocene environments, African apes are likely to have been quite widely distributed, with probably a number of species and subspecies, with hominids derived from one of these situated on the eastern side of the continent. Using the socioecology of the chimpanzee as a model, we can

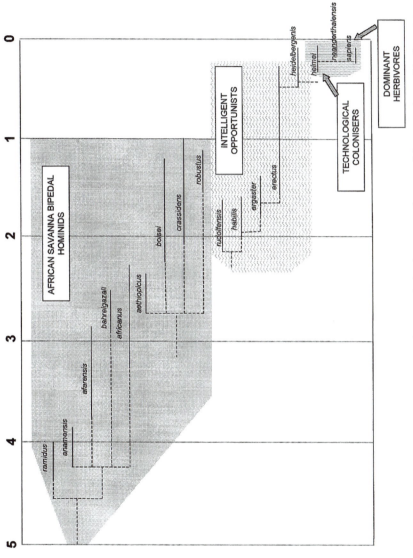

Fig. 14.2   Temporal distribution of the hominid taxa, showing general adaptive patterns in terms of the categories used in the text to describe the different evolutionary strategies of the hominids.

infer that the last common ancestor lived in relatively large communities, were male kin bonded, with female dispersal, and hostile to other communities (Wrangham, 1987; Foley, 1987, 1989).

African savannah bipedal apes

The shift to a hominid and australopithecine grade is initially associated with greater levels of bipedalism. Bipedalism is energetically efficient as a means of moving around the ground, and is likely to be associated with greater terrestriality. Time budget-based models have suggested that where approximately 60% of the feeding and travelling occurs on the ground, the advantages of bipedalism will exceed the costs (Foley, 1992; Foley and Elton, 1998). This implies that bipedalism is likely to evolve under conditions in which resources are either in trees that are far apart, or else in which the food can be reached on the ground. Bipedalism may, therefore, evolve in relatively wooded environments, although not in forests, and in particular where animals are forced into ranging more widely. The foraging ecology of the earliest hominids would thus be one in which the key determinant is extended day and home ranges, and this in turn is likely to alter other energetic parameters such as thermoregulation (Wheeler, 1985)

The primary trend among these bipedal apes over the period 5.0 to 1.5 Myr is that of megadonty, associated with some increase in body size, cranial robusticity, and perhaps high levels of sexual dimorphism. All of these suggest an essentially African ape lifestyle adapted to desiccating environments where food resources consisted of relatively hard, coarse fruits and nuts. Later australopithecines would have been more specialised forms, with an ability to survive in more open habitats, possibly even extreme grasslands. These savannah apes remained restricted to African biomes, and speciated and became extinct as local populations. The inferred socioecology of the African savannah bipeds would share many characteristics with that of the common ancestor. The move to more open environments with widely dispersed and scarce, high-quality plant foods may have led to smaller communities distributed over wider territories. Reference to living hominoids would suggest that these australopithecines could have retained male kin-bonded groups (Foley and Lee, 1989), but that a higher risk of predation would have reduced the degree of fission within communities. The nature of resource distribution is such that these hominids are likely to have had extensive day and home ranges, and this, in addition to increased heat stress, is probably the main factor underlying their key adaptive trait, bipedalism.

The diet of the megadontic australopithecines has generally been considered as a specialisation: to seeds, fibrous plant foods, and low-quality, coarse foods (Kay, 1985; Grine, 1981). While there must be an element of this in their adaptations, it should also be recognised that as large-bodied, highly intelligent apes, they would have been opportunistic frugivores, hunters and scavengers, and thus their diet would have included meat. In addition, a key to their ability to survive in these more arid environments may have been the ability to tap into underground plant resources, perhaps with the aid of simple technology.

Omnivorous intelligent opportunists

In ecological terms, the primary trend in early hominid evolution appears to be a series of increasingly specialised adaptations (bipedalism, teeth) to survival in relatively arid and open terrestrial African environments (Vrba, 1985, 1996; Reed, 1997), probably occurring in the context of an essentially African ape grade of cognitive capabilities and social strategies. The evolution of *Homo* after 2.0 million years constitutes a new trend, and one that is diametrically opposed to that of the australopithecines. Morphologically, the shift can be seen in dental and facial reduction, enlarge-ment of the brain, and loss of a more ape-like upper body as the hominid body form became more linear (Ruff, 1991). In addition, hominids became far more widespread across Africa and Eurasia, with indications of a significant technological dependence.

The socioecological basis for these trends starts with dietary change. In contrast to the australopithecines, there is less evidence for dependence upon (plant) foods requiring heavy mastication. This, in the context of archaeological evidence, may be interpreted as a shift to greater use of animal resources (Bunn and Kroll, 1986), either through hunting or scavenging. High-quality resources would have changed time budgets (higher search times, possibly reduced foraging times; e.g. Chapter 12), extended day and home ranges, increased habitat tolerance, and affected social organisation and group structure. Meat would have provided a higher quality resource, which would help fuel brain enlargement and also a reduction in gut size (thus the change in body shape) (Foley and Lee, 1991; Wheeler and Aiello, 1995). As more omnivorous opportunists, *Homo* were able to extend their species range, and although still largely confined to warmer and more open environments, they were no longer African endemics, but Old World cosmopolitans. The key social change that can be inferred from both the dietary shift and the correlated change to slower growth rates and delayed maturation is that more encephalised infants require greater parental investment. It has been argued (Foley and Lee, 1989) that it is at this point that more exclusive relationships between males and females became tacked on to an existing male kin-bonded community structure, although these are likely to have been polygynous rather than monogamous.

Technological colonisers

While early *Homo* did disperse beyond Africa, the relative stability of adaptive grades during the lower and middle Pleistocene, and the evidence for environmental limitations (e.g. exclusion from tropical forests, riverine specialisation, and inability to persist in northerly latitudes during colder phases) show that the phase of opportunistic omnivores was also evolutionarily and ecologically static. Then, however, from around 300 000 years ago, two major changes are apparent. The first of these is an acceleration in the rate of brain size evolution (Leigh, 1992), and the second is evidence for repeated dispersals and phases of population expansion (Lahr and Foley, 1994). These developments coincided with the evolution of neanderthals, the evolution of anatomically modern humans, and major changes in technology, especially the appearance of prepared core and blade production techniques of stone artefact manufacture (Foley and Lahr, 1997).

During this phase of human evolution, technology was both becoming more flexible and allowed far greater adaptive flexibility. Underlying this must almost certainly be a cognitive shift, perhaps linked to life history parameters identical to those found in modern humans. It is likely that at this stage (from around 250 Kyr) the socioecological elements would have been very much like those observable ethnographically – kin-based groups of hunter–gatherers, with cultural and linguistic traits used to mark out ethnic and population differences, elaborated culturally in many different ways. The extent to which these traits evolved over the period, and whether there were major differences between neanderthals and modern humans, is a matter of considerable dispute, but it is likely that the socioecology of this period is one in which there was great variability built around a few simple core strategies – male kin bonding, communities of culturally identifiable individuals, intergroup tensions, and high levels of parental care and within-group alliance structures – conditioned by demographic and environmental conditions. The result is the dynamic world of the later Pleistocene that runs relatively seamlessly into the ethnographic present.

Dominant herbivores

It is tempting to use the technological colonisers' phase of human ecological evolution to account for all current human variation, especially as the amount of subsequent biological change is relatively small. However, from an ecological point of view, this would be misleading, for a major change has occurred in the last few thousand years.

The evidence points to modern humans having evolved in the last 200 Kyr, but having effectively dispersed across the planet in only the last 50 Kyr. For all but the last 10 Kyr they lived as hunter–gatherers, at relatively low overall population size. During the last 10 000 years, however, human population has grown by three orders of magnitude; socioecologically, community size has become much larger, and social structures highly variable, hierarchical and complex. These changes are the result of agriculture. Whereas there has been considerable controversy about the nature of domestication, there is consensus that it involves a major control over resources and has led to both population growth and environmental change. This is the world to which most people are accustomed and in all probability adapted. The basis for this adaptation is the ability to produce large quantities of easily digestible and energy-rich plant foods – root crops and cereals, but especially the latter. Plant cultivation lies at the heart of the current socioecology of humans, despite the fact that a number of animals are also domesticated and provide an important subsistence base (see Chapter 15). The primary food resource for most people is a cereal crop, supplemented to a greater or lesser extent by meat and dairy produce. In this sense, humans are the dominant herbivores of a global ecosystem. This contrasts with much of the ecological foundation for the earlier phases of hominid evolution, in which it was the ability to widen diet breadth to incorporate meat that was critical.

The socioecological correlates of domestic plants are many, but perhaps the key ones are: they are highly predictable in space and time, although abundance can fluctuate markedly; returns can be increased very significantly by increasing effort; they are easily digestible, and thus are an excellent weaning food; access to them can be relatively easily controlled by individuals rather than shared. As a result, agriculturally based communities can have very high reproductive rates and high population densities, although these can be subject to major local fluctuations; they are likely to be territorial and experience intercommunity aggression and conflict; differential control of access to resources combined with a greater potential to coerce less well-situated individuals and communities lead to very marked hierarchical differences or despotism. Male kin-based systems predominate, as do polygynous mating patterns. Above all, the simplified ecological structure can result in marked variation from area to area, with a consequential variability in social system and cultural pattern in which membership of the community is itself a highly significant element of the adaptive process (Aunger, 1996). The socioecology of humans today and in the recent past is characterised by high levels of community membership signalling (cultural variation, language). Biologically, there is also a trend towards reduced body size, gracility of skeletal form, and reduced sexual dimorphism. Finally, the nature of the adaptation is one in which it is possible for the human population, through agriculture, to modify in very radical ways the environment itself, and in particular to homogenise it as a source of agricultural productivity.

Socioecology and the evolution of human social behaviour

There is little doubt that humans have evolved in much the same way as any other species in terms of pattern (cladogenetic radiations, extinction and adaptive trends) and process (natural selection). The three available sources of information about the socioecology of humans – living humans, living non-human primates, and the fossil record – provide a relatively coherent picture, despite the very different nature of the type of data they yield.

In terms of resource structure shaping social behaviours, the key shifts appear to have been: (1) more dispersed and poor-quality foods in the drier habitats of Pliocene Africa, with a resulting shift in locomotion that was fundamental to what followed, and an extension of day ranges and probably increased time stress; (2) greater and more efficient use of scavenged and hunted meat, providing an alternative high-quality food, resulting in the energetic conditions for major life history modifications, and greater habitat tolerance; (3) a subsequent shift to energy-rich cereals, allowing massive increases in population densities and hierarchical community structures.

The phylogenetic context for these resource shifts is important, and it can be argued that it is large, male kin-bonded groups with dispersing females that provide the thread of historical continuity to hominid socioecological evolution (Foley and Lee, 1989; Wrangham and Peterson, 1996). Changes in resource distribution, availability and returns will affect the size and degree of substructuring and the rate of fission, but will not affect the fundamental structure. As such, it is the hominid clade's capacity to maintain large groupings that will have been critical to survival either in a predator-rich environment during the Pliocene, or else in the context of antagonistic intergroup encounters. A switch to dispersing males would have been individually lethal and likely to make such groups highly vulnerable. As such, male kin bonding may represent something of an irreversible strategy in social evolution, unless there is a complete loss of sociality or communities become so large that sex-specific dispersal/residence patterns become unnecessary.

If male kin bonding is the continuity element in human social evolution, male–female relationships and the nature of parental care are the novelties. Higher quality resources which can be shared and weaning foods are the key resource structures that are likely to have changed mating and parenting strategies within the group, and led to the cognitive shifts underlying close relationships between the sexes, in association with delayed life history strategies and high levels of parental care. Finally, the very recent past has seen massive growth and diversification of the human populations, leading to a diversity of social structures, but the ethnographic record perhaps demonstrates that these are all variations on a long-term theme.

Missing links in the comparative method

In the context of comparative primate socioecology, the pattern outlined here is unusual in the extent to which it has been possible to integrate the observable patterns among extant species with the dynamics of change through time. This is only possible, perhaps, because of the amount of attention that has been focused on the hominid fossil record. However, the nature of the inferences drawn have implications for the comparative method in general. This emphasises the importance of taking phylogenetic history into account; thus, the amount of adaptive evolutionary change that has occurred is considered relative to a common ancestral node (Harvey and Pagel, 1991; Purvis, 1995; see also Chapter 3). For humans, that node would be the last common ancestor with *Pan*. An estimate of the common ancestral node for *Pan–Homo* brain size, for example, would be a midpoint between 350 and 1400 g – 875. However, the fossil record shows that this is in fact the cranial capacity of *Homo ergaster* at less than two million years, as opposed to the common ancestor at over five million years. The problem clearly arises when, of a pair of extant sister clades, all the evolutionary change in a certain character is occurring in one lineage only. This is not simply a case of selection in one lineage and phylogenetic inertia in the other. Stabilising selection is presumably the force acting in one lineage, and directional selection on the other; evolutionary forces are thus acting on each, but in different evolutionary directions. The key point about the fossil record is that it pinpoints the timing of switches in selection in a more accurate way than phylogenetic inference such as contrasts can. Returning to brain size as an example, inferred rates of encephalisation in the hominids based on phylogenetic comparison would be either 130 g/ Myr or 210 g/Myr, depending upon whether the *Pan* or midpoint ancestral estimate was used. A knowledge of the fossil record would show something very different: encephalisation rates among the australopithecines hominids over a three million-year period would maximally be 55 g/Myr; for *Homo* it would be 347 g/Myr; and in fact the evolution of *Homo sapiens* involved a rate of 800 g/Myr over the last half million years, while the Asian *Homo erectus* lineage had an encephalisation rate of 210g/Myr. While these figures should be corrected for body size, the basic message would be unchanged: evolutionary change is not distributed equally among lineages, but is highly variable. The variability of the nature and direction of change, or the balance between stabilising and directional selection, is the heart of evolutionary issues. Comparisons between species and higher taxa can provide some insight into that variability; adding the phylogeny of living taxa can refine that, but it is still incomplete without a knowledge of the actual distribution of events through time. To put this all another way, our understanding of primate socioecology is strongly influenced by the effects of differential extinction.

Conclusions

The key points relating to hominid socioecology can be summarised as follows. First, when a lineage such as the African ape/human clade is considered with the fossil record incorporated, and thus with a greater emphasis on time, the overall pattern becomes far more complex. The pruning of an evolutionary tree by extinction may well remove entire evolutionary trends, as is the case with the australopithecines. Second, the fossil record shows that human behavioural traits did not evolve as a package, but were accumulated during the course of separate transitions over a number of events. However, the full consequences of these adaptive shifts, as observable in the high-density living of post-neolithic populations, occurred only in the last 10 000 years and subsequent to any major biological changes in the human population. Third, the scale and pattern of human evolution are consistent with microevolution, with a balance of stabilising and directional selection operating on fluctuating populations being a sufficient mechanism. Fourth, if there are key points in the evolution of humans and their ancestors, these are most likely to have been: during the terminal Miocene when novel locomotor features evolved; at the base of the Pleistocene when changes in foraging strategy (meat eating) allowed dispersal into multiple habitats and beyond Africa; around 300 000 years ago when it appears the basis for modern human life history, cognitive and behavioural traits were established; and during the last 20 000 years when, for the first time in hominid history, population densities really became a significant global factor, and the socioecology of the human species went beyond the normal expectations of the comparative method.

References

Aiello, L.C. and Dean, C. (1990). *An Introduction to Human Evolutionary Anatomy.* London: Academic Press.

Aiello, L.C. and Dunbar, R.I.M. (1993). Neocortex size, group size and the evolution of language. *Current Anthropology* **34**, 184–93.

Andrews, P.J. (1992). Evolution and environment in the Hominoidea. *Nature* **360**, 641–6.

Aunger, R. (1996). Acculturation and the persistence of indigenous food avoidances in the Ituri Forest, Zaire. *Human Organisation* **55**, 206–18.

Beynon, A.D. and Dean, M.C. (1988). Distinct dental development patterns in early fossil hominids. *Nature* **335**, 509–14.

Beynon, A.D. and Wood, B.A. (1987). Patterns and rates of molar crown formation times in East African hominids. *Nature* **326**, 493–6.

Bromage, T.G. and Dean, M.C. (1985). Re-evaluation of the age at death of Plio-Pleistocene fossil hominids. *Nature* **317**, 525–8.

Brown, P. (1987). Pleistocene homogeneity and Holocene size reduction: the Australian human skeletal evidence. *Archaeology in Oceania* **22**, 41–67.

Bunn, H.T. and Kroll, E. (1986). Systematic butchery by Plio-Pleistocene hominids at Olduvai Gorge, Tanzania. *Current Anthropology* **27**, 431–52.

Ciochon, R. and Etler, D. (1994). Reinterpreting past primate diversity. In *Integrative Paths to the Past*, ed. R. Corruccini and R. Ciochon, pp. 37–68. Englewood Cliffs, NJ: Prentice Hall.

Clark, J.D., de Heinzelin, J., Schick, K.D. *et al.* (1994). African *Homo erectus*. Old radiometric ages and young Oldowan assemblages in the middle Awash Valley, Ethiopia. *Science* **264**, 1907–10.

Foley, R. (1987). *Another Unique Species: Patterns in Human Evolutionary Ecology.* Harlow: Longman.

Foley, R.A. (1989). The evolution of hominid social behaviour. In *Comparative Socioecology*, ed. V. Standen and R.A. Foley, pp. 473–94. Oxford: Blackwell Scientific Publications.

Foley, R.A. (1992). Evolutionary ecology of fossil hominids. In: *Evolutionary Ecology and Human Behavior*, ed. E.A. Smith and B. Winterhalder, pp. 131–64. Chicago: Aldine de Gruyter.

Foley, R.A. and Elton, S.E. (1998). Time and energy: the ecological context for the evolution of bipedalism. In *Primate Locomotion: Recent Advances*, ed. E. Strasser, A.L. Rossenberger, J.M. Fleagle and H.M. McHenry, pp. 523–31. New York: Alan Liss.

Foley, R.A. and Lahr, M.M. (1997). Mode 3 technologies and the evolution of modern humans. *Cambridge Archaeological Journal* **7**, 3–32.

Foley, R.A. and Lee, P.C. (1989). Finite social space, evolutionary pathways and reconstructing hominid behavior. *Science* **243**, 901–6.

Foley, R.A. and Lee, P.C. (1991). Ecology and energetics of encephalization in hominid evolution. *Philosophical Transactions of the Royal Society, London Series B* **334**, 223–32.

Frayer, D. and Wolpoff, M.H. (1985). Sexual dimorphism. *Annual Review of Anthropology* **14**, 343–428

Grine, F.E. (1981). Trophic differences between 'gracile' and 'robust' australopithecines: a scanning electron microscope analysis of occlusal events. *South African Journal of Science* **77**, 203–30.

Grine, F.E., ed. (1988). *Evolutionary History of the 'Robust' Australopithecines*. New York: Aldine de Gruyter.

Harvey, P.H., Martin, R.D. and Clutton-Brock, T.H. (1987). Life histories in comparative perspective. In *Primate Societies*, ed. B.B. Smuts, D.L. Cheney, R.M. Seyfarth, R.W. Wrangham, and T.T. Struhsaker, pp. 181–96. Chicago: University of Chicago Press.

Harvey, P. and Pagel, M. (1991). *The Comparative Method in Evolutionary Biology*. Oxford: Oxford University Press.

Jungers, W.L. (1988). New estimates of body size in australopithecines. In *Evolutionary History of the 'Robust' Australopithecines*, ed. F. Grine, pp. 115–25. New York: Aldine de Gruyter.

Kay, R.F. (1985). Dental evidence for the diet of *Australopithecus*. *Annual Review of Anthropology* **14**, 315–42.

Lahr, M.M. and Foley, R.A. (1994). Multiple dispersals and the origins of modern humans. *Evolutionary Anthropology* **3**(2), 48–60.

Lahr, M.M. and Wright, R. (1996). The question of robusticity and the relationship between cranial size and shape in *Homo sapiens*. *Journal of Human Evolution* **31**, 157–91.

Leigh, S.R. (1992). Cranial capacity in *Homo erectus* and early *Homo sapiens*. *American Journal of Physical Anthropology* **87**, 1–14.

Lovejoy, C.O. (1988). The evolution of human walking. *Scientific American* **259**, 82–9.

Martin, R.D. (1981). Relative brain size in terrestrial vertebrates. *Nature* **293**, 57–60.

Martin, R.D. (1983). *Human Brain Evolution in an Ecological Context*. 52nd James Arthur Lecture on the Evolution of the Brain. American Museum of Natural History.

Martin, R.D. (1985). Primates: a definition. In *Major Topics in Primate and Human Evolution*, ed. B.A. Wood, L. Martin and P.J. Andrews, pp. 1–31. Cambridge: Cambridge University Press.

McCrossin, M. and Benefit, B. (1994). Moboko Island and the evolutionary history of Old World monkeys and apes. In *Integrative Paths to the Past*, ed. R. Corruccini and R. Ciochon, pp. 95–122. Englewood Cliffs, NJ: Prentice Hall.

McHenry, H.M. (1986). The first bipeds: a comparison of *Australopithecus afarensis* and *Australopithecus africanus* postcranium and implications for the origins of bipedalism. *Journal of Human Evolution* **15**, 177–91.

McHenry, H.M. (1988). New estimates of body weight in early hominids and their significance to encephalization and megadontia in 'robust' australopithecines. In *Evolutionary History of the 'Robust' Australopithecines*, ed. F.E. Grine, pp. 133–48. Chicago: Aldine de Gruyter.

McHenry, H.M. (1992). How big were the early hominids? *Evolutionary Anthropology* **1**, 15–20.

Moore, J. (1996). Savanna chimpanzees, referential models and the last common ancestor. In *Great Ape Societies*, ed. W.C. McGrew, L.F. Marchant and T. Nishida, pp. 00–00. Cambridge: Cambridge University Press.

Pilbeam, D. and Gould, S.J. (1974). Size and scaling in human evolution. *Science* **186**, 892–901.

Purvis, A. (1995). A composite estimate of primate phylogeny. *Philosophical Transactions of the Royal Society of London Series B* **348**, 405–21.

Reed, K. (1997). Early hominid evolution and ecological change through the African Plio-Pleistocene. *Journal of Human Evolution* **32**, 289–322.

Ruff, C. (1991). Climate and body shape in human evolution. *Journal of Human Evolution* **21**, 81–105.

Schmid, P. (1991). The trunk of the australopithecines. In *Origines de la Bipede chez les Hominides*, ed. Y. Coppens and B. Senud, pp. 187–98. Paris: CNRS.

Senut, B. and Tardieu, C. (1985). Functional aspects of Plio-Pleistocene hominid limb bones: implications for taxonomy and phylogeny. In *Ancestors: the Hard Evidence*, ed. E. Delson, pp. 193–201. New York: Alan R. Liss.

Smith, B.H. (1989). Dental development as a measure of life history in primates. *Evolution* **43**, 683–8.

Smith, B.H. (1992). Life history and the evolution of human maturation. *Evolutionary Anthropology* **1**(4), 134–42.

Smith, B.H. (1994). Patterns of dental development of *Homo, Australopithecus, Pan* and *Gorilla*. *American Journal of Physical Anthropology* **94**, 307–25.

Steudel, K. (1980). New estimates of early hominid body size. *American Journal of Physical Anthropology* **52**, 63–70.

Stringer, C. (1984). Human evolution and biological adaptation in the Pleistocene. In *Hominid Evolution and Community Ecology: Prehistoric Human Adaptation in Biological Perspective*, ed. R.A. Foley, pp. 55–84. London: Academic Press.

Susman, R.L., Stern, J.T. and Jungers, W.L. (1984). Arboreality and bipedality in the Hadar hominids. *Folia Primatologica* **43**, 113–56.

Trinkaus, E. (1989). The Upper Pleistocene transition. In *The Emergence of Modern Humans*, ed. E. Trinkaus, pp. 42–66. Cambridge: Cambridge University Press.

Trinkaus, E., Churchill, S.E., Villemeur, I., Riley, K.G., Heller, J.A. and Ruff, C.B. (1991). Robusticity versus shape: the functional significance of neanderthal appendicular morphology. *Journal of Anthropological Society of Nippon* **99**, 257–78.

Vrba, E. (1985). Ecological and adaptive changes associated with early hominid evolution. In *Ancestors: the Hard Evidence*, ed. E. Delson, pp. 63–71. New York: Alan Liss.

Vrba, E. (1996). *Palaeoclimate and Neogene Evolution*. New Haven, CT: Yale University Press.

Wheeler, P. (1985). The evolution of bipedalism and the loss of functional body hair in hominids. *Journal of Human Evolution* **14**, 23–8.

Wheeler, P.E. and Aiello, L.C. (1995). The expensive tissue hypothesis. *Current Anthropology* **36**, 199–222.

Wrangham, R.W. (1987). The significance of African apes for reconstructing human evolution. In *The Evolution of Human Behavior: Primate Models*, ed. W.G. Kinzey, pp. 28–47. Albany: SUNY Press.

Wrangham, R. and Peterson, D. (1996). *Demonic Males*. London: Bloomsbury.

15 Evolutionary ecology and cross-cultural comparison: the case of matrilineal descent in sub-Saharan Africa

RUTH MACE AND CLARE HOLDEN

Introduction

The comparative method will always be an important tool when testing adaptive hypotheses about human behaviour, particularly as experimentation is not usually an option. Cross-cultural comparison has a long history in anthropology, but in recent decades, formal statistical comparison has fallen into decline. This was partly due to broad philosophical and ideational changes in the field as whole. The specificity and uniqueness of individual cultures are the favoured theme of most anthropological research, which is not quantitative. Even amongst cultural ecologists or sociobiologists, who were not opposed in principle to quantitative approaches, the appearance of cross-cultural studies reporting highly statistically significant associations between traits (without concern for the non-independence of cultures), that were sometimes accompanied by fantastical hypotheses, sowed suspicion of a statistical approach in the minds of those reading them.

Evolutionary studies of human behaviour of any kind can also be controversial, whether employing cross-cultural comparison or not. In social anthropology, the term 'evolutionary' is generally taken to imply the erroneous paradigm of a linear progression of stages that cultures move through from a more primitive to a less primitive state. Few social anthropologists today are even aware that the views of modern evolutionary ecologists could not be further from this position, and that it is the variation observed in phenotypes in different environments that is the central focus of research in evolutionary ecology. Amongst evolutionary biologists, the status of human sociobiology was really little better than in the social sciences, until very recently.

The field was not helped by an obsession with debating the extent to which any particular trait could be considered to be genetic or cultural in origin. Virtually all behaviour and even physiology have a large environmental component and, in humans, a large part of that environmental component will be a property of the culture in which the individual lives. Behaviour is almost always plastic in the face of environmental variation, and selection favours mechanisms by which a range of responses will arise when faced with a range of environmental circumstances. But the adaptiveness of a phenotype (or range of phenotypes) can be studied without reference to the precise mechanisms by which that variation arose. Life-history theory, optimality and game theory provide examples of thriving fields of evolutionary biology where this is done. There is a growing field of human evolutionary ecology in which cultural traits are now being successfully investigated, within an adaptive framework. There is now a substantial number of studies finding that cultural traits are mechanisms by which reproductive success is enhanced, and very few indicating that cultural transmission leads to long-term maladaptation. There is now as much or more reason to see selection on genes and on culture as forces generally pulling in the same, not opposite, directions (Betzig, 1997).

The aim in this chapter is to investigate the hypothesis that the occurrence of matrilineal descent systems in sub-Saharan Africa can be understood in adaptive terms. Matrilineal descent, as described in anthropological texts, is the inheritance of titles and property rights down the maternal line. In the case of males, this means that they will inherit from their uncles (their mother's brothers) rather than from their father. This system of inheritance is fairly common in parts of sub-Saharan Africa, although patrilineal descent (in which fathers pass titles and wealth to their children, usually their sons) is the most common societal norm. (Biologists will recognise matrilineal descent as being somewhat similar to the descent of a mitochondrial gene, whereas patrilineal descent is more analogous to that of a Y chromosome.) Africa shows particularly strong lineality, whereas elsewhere it is quite common for descent to be mixed, with more emphasis on the nuclear family rather than on the lineage.

There is a range of different systems of matrilineal descent (Schneider and Gough, 1961). The classic texts emphasise the power of the maternal uncle and other senior males over the other members of the matriline. Men were guardians of all resources (thus inheritance is often considered male biased), and these guardians frequently played a key role in protecting women from having their resources taken from them. However, the key resources (usually fields) were mainly the property of the matriline, and men would die with little or nothing in the way of individual possessions. It should be noted that a man passing the fields used by his sister and her children (for whom he is the legal guardian) to the custody of his sister's sons (who are guardians for their own sisters' children) is operationally the same as a mother passing her fields to her daughters. In some cases, women may have more property rights than in others. For example, during our own recent fieldwork among the matrilineal Chewa in Malawi, informants indicated that land was owned by women, who usully passed it directly from mother to daughter.

Matrilineality has puzzled evolutionary anthropologists because it is not clear that male fitness is enhanced by passing resources on to a sister's son in preference to the man's own son. Tensions between men wishing to invest in their own children rather than in the mamtrilineal heirs are commonly reported. In the first descriptions of matrilineal societies by Western anthropologists (who were struck by the differences from their own culture), marriage and marital fidelity in matrilineal societies appeared relatively weak. If paternity uncertainty were very high, then fitness might be better enhanced by investing in your uterine sister's sons than in your wife's sons. But Hartung (1985) has demonstrated that paternity uncertainty would have to be at unrealistically high levels for this condition to hold. Hartung shows that matrilineality, whilst not adaptive for males, is adaptive for females under any level of paternity certainty below 1. This is because a grandmother's matrilineal heirs will always be more closely related to her than her patrilineal heirs. Thus, matrilineality may be a female strategy. Even if political power rests with males, on a proximate level, it is the female gene-line that benefits most from the inherited resources.

Why a female strategy should predominant over a male strategy is not known. One possibility is that this is a system that arises when the key resources are largely produced by females. Anthropologists in Africa noted that matrilineal descent was most common in farming and horticultural societies, which are usually heavily dependent on women's labour. It is also interesting to note that matrifocal (if not matrilineal) family structures, with weak or non-existent marriage, are frequently observed outside Africa where male earning power is limited. Some Caribbean cultures provide examples of this. A former Minister of State for Wales (John Redwood) was famously horrified when observing a similar family structure during a visit to a Welsh housing estate where male unemployment was very high and state benefits to mothers were one of the main sources of income.

In the African context, matrilineal systems are frequently described as being under threat, as new economic opportunities arise for men. Pastoralism may be one factor that enables men to build up personal wealth, in

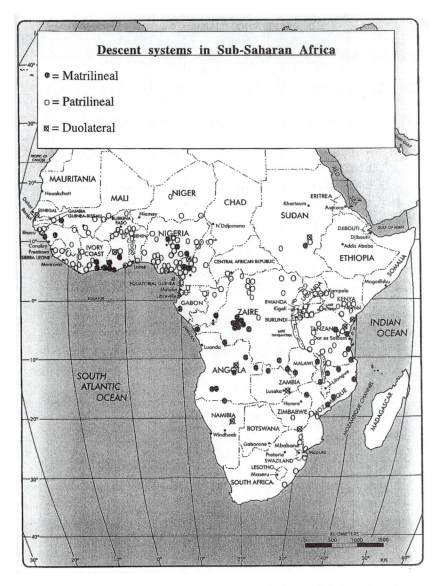

Fig. 15.1   A map of the main descent system of NigerKordofanian cultures in sub-Saharan Africa.

Table 15.1. *Sub-Saharan African cultures listed in Murdock's The Ethnographic Atlas, grouped by main system of descent and dependence on pastoralism, excluding those where descent is unclassified*

Dependence on pastoralism (%)	Main descent system[a]			
	Patrilineal	Duolateral	Matrilineal	Others
0	11 (50)	1 (5)	10 (45)	5
10	114 (70)	6 (4)	43 (26)	4
20	86 (86)	5 (5)	9 (9)	7
30	48 (81)	4 (7)	7 (11)	2
40	16 (88)	1 (6)	1 (6)	1
50+	14 (93)	1 (7)	0 (0)	0

[a]Percentages given in parentheses.
$n = 289$.

the form of cattle herds, which they can pass on to their own children. It has been noted that matrilineal systems are generally not pastoralist, and tend to cluster in a 'matrilineal belt' (Fig. 15.1) across Africa. It has even been suggested that this belt corresponds with the tsetse-infested areas of Africa where it was not possible to keep cattle (Aberle, 1961). Here, that hypothesis is tested by seeking to see if matrilineality is negatively associated with pastoralism in sub-Saharan Africa.

Phylogenetic approaches in cross-cultural studies

Table 15.1 shows all the sub-Saharan cultures listed in Murdoch's *World Ethnographic Atlas* (computer version). Murdock codes dependence on various modes of subsistence quantitatively, including pastoralism, and the main system of descent. Duolateral cultures are those that trace descent through both the male and the female line, i.e. both systems co-exist and sometimes one resource (such as cattle) may be inherited differently from another (such as fields down the matriline). There is a decrease in the proportion of cultures that are matrilineal as dependence on pastoralism increases. But each culture cannot be considered an independent data point, for statistical purposes. Cultures descend from common ancestors and are thus related hierarchically. Cultural traits are likely to be more similar between related cultures, so both the traits of interest and a range of background variables will not be independently distributed amongst cultures. If a particular group of related cultures happened to inhabit a cattle-free region, and happened to have a matrilineal descent system for some other reason, any statistics based on counting each individual culture within that group as an independent unit of statistical information could suggest a highly significant association between these two traits that would be spurious.

The difficulties that the hierarchical relatedness of groups pose for the statistical interpretation of cross-cultural data have long been appreciated – they were first pointed out by Francis Galton in 1889. 'Galton's problem' introduced debate and controversy into the issue of the analysis of cross-cultural data long before most zoologists were aware of the problem in their field of enquiry. Felsenstein (1985) argued that comparison between species, which are all hierarchically related, has to be investigated by counting independent changes along the branches of the phylogeny of the species concerned. Pagel and Harvey (1988) and Harvey and Pagel (1991) developed this phylogenetic comparative approach so that it can be used to study a range of quantitative and categorical variables. Mace and Pagel (1994) argue that similar approaches can be used to account for Galton's problem in cross-cultural comparison.

The diversity that we are interested in here is between populations rather than species, and there are differences between these two cases. These differences arise because it is possible for both genes and culture to pass horizontally from one population to the next. It is more difficult to establish a phylogeny of cultures. A branching phylogeny without anastomoses (joining of branches) may not be a full representation of the ancestry of cultures. The Fst trees produced by Cavalli-Sforza, Menozzi and Piazza (1994) are a representation of genetic similarity between populations around the world. An alternative method of categorising human cultures is by linguistic similarity; and comprehensive classifications exist, such as that of Ruhlen (1991). Cavalli-Sforza *et al.* (1994) have pointed out that trees of human cultures based on language similarity and on genes show broad similarities, which they attribute to a similar mode of evolutionary change: modification by descent, which diverges as groups are separated. The trees produced by either method correspond broadly with a consensus view of human history based on archaeological, linguistic and genetic evidence, although particular phylogenies generated by different methods will have many differences in the particulars.

Those differences are partly due to the differing power (statistical or otherwise) of the different methods used to construct the trees. They are also due to differences in how the two trees would be influenced by the mixing of populations. In a genetic Fst tree, of the type produced by Cavalli-Sforza *et al.* (1994), interbreeding between two groups would make the branches shorter and two groups appear to share a recent common ancestor. In a linguistic classification, languages rarely mix in way similar to genetic mixing. Even if vocabulary is borrowed, the basic structure remains clearly that of the ancestral group. It is thought to be more common for one language to supplant the other (Renfrew, 1987); and in that case the culture that lost its language would simply be lost from the tree as a unit of information (Mace and Pagel, 1994). Deep ancestral nodes are less resolved in language trees, as language evolution is faster than genetic evolution and cannot give us much information about the separation between groups in the distant past.

Ultimately, a tree will only be of use if it is a reasonable model of the history of the population concerned. If one were interested in a genetic trait (such as lactose digestion capacity), a genetic tree might be most informative; for cultural traits (such as matrilineality), a cultural tree based on linguistic similarity might be better. We stress *reasonable* model becuse the model will never be perfect; but standard statistical regressions effectively assume a 'star' phylogeny (equal relatedness between all cultures), and if the tree is a better representation of population history than that, then constructing a phylogeny, and making use of it in statistical tests, is likely to be more informative. When it is not clear which of a number of possible trees best reflects population history (which is commonly the case in within-species trees), hypotheses can be tested using each of them, to see whether any of the results found to be significant are reliant on a particular model of history, or whether they hold up more generally.

Cultural diffusion

Diffusion is a term used to describe the spread of a trait from one culture to another, by virtue of geographic proximity. This is a problem that does not have to be considered in cross-species comparisons. Guglielmino *et al.* (1995) contrast three mechanisms of the transmission of cultural traits, which they call cultural diffusion, demic diffusion (by which they mean inheritance down a phylogeny), and ecological adaptation, all of which are important. Figure 15.1 shows that matrilineality and patrilineality do appear to be geographically clustered. Clearly, spatial clustering could result from cultural diffusion or demic diffusion (inheritance down a phylogeny of a group that is not very migratory) and/or from ecological adaptation to a geographically distinct region (the last of which is a functional explanation and is compatible with either of the two mechanistic explanations). Guglielmino *et al.* (1995) point out that demic diffusion (down a phylogeny) appears to be particularly widespread in traits relating to kinship and economy: in other words, these traits appear to leave strong phylogenetic signatures. Thus, the use of phylogenetic comparative methods is very important.

The formal comparative methods used here are designed to identify ecological adaptation. Some statistical methods of cross-cultural comparative analysis have considered methods by which the effects of diffusion due to geographic proximity can be excluded (e.g. Dow *et al.*, 1984). Whether or not this is considered important depends on your model of cultural evolution, and whether or not the adoption of a trait from a neighbour is considered by be of any functional significance. If cultural traits are thought to be picked up from neighbours simply because of proximity (rather like the flu), then it would be necessary to control for proximity when testing an adaptive hypothesis. The methods used here do not do this: the adoption of a trait from a neighbouring culture is here counted as a unit of information that can be used to test a functional hypothesis. The rationale for this is the assumption that cultures do not take up everything their neighbours do, and remain separate cultures. Therefore, instances of the adoption of particular traits by individual cultures (or groups of phylogenetically related cultures) are compared statistically in the same way as if the trait had originated de novo. If the adoption of a trait, from a neighbour or from elsewhere, occurs repeatedly, on different parts of the tree, in the presence of another trait, then these statistical methods will consider the two traits to be correlated (Mace and Pagel, 1994).

The authors have used the same approaches to investigate the evolution of the ability to digest lactose as an adult, which is a genetic polymorphism (Holden and Mace, 1997). The frequency of adults with lactose digestion capacity in cultures (other than those formed by recent mixing) followed a bimodal distribution; cultures could thus be categorised as having either high or low lactose digestion capacity. We found high lactose digestion capacity tended to arise after the adoption of pastoralism, but was not associated with some other variables with which it has sometimes been linked using non-phylogenetic comparative tests, such as high latitude. In each population, the capacity to digest lactose may not have arisen due to a unique mutation; it may have entered a population through interbreeding with a lactose-tolerant group. Thus, as with cultural traits, horizontal transmission may be occurring. However, if it has reached reasonably high levels, our assumption is that it has been selected for in that population. Thus, the spread of traits between populations does not invalidate a phylogenetic approach, unless that spread is for so many traits that populations can no longer be meaningfully identified as distinct units.

Three phylogenetic comparative tests of an association between pastoralism and descent

Comparisons between sister groups on a genetic tree

Figure 15.2 shows a genetic phylogeny for sub-Saharan African agriculturalists and pastoralists. Three outlying groups have been excluded – the Khoisan cultures, the Hadza and the Mbuti – as all are predominantly hunter–gatherer groups without strongly lineal systems. They also all cluster outside the clade shown. Two individual cultures within this clade have also been excluded from the analysis as they are reported as not showing clear lineality of any kind. Only those cultures that could be identified in Murdock's (1967) *The Ethnographic Atlas* (updated computer version) and could be placed on the genetic tree of Cavalli-Sforza *et al.* (1994) were used. This left 89 cultures, with many of the nodes unresolved. Shading on branches in Figure 15.2 indicates the most parsimonious estimate of ancestral character states, derived from the program MacClade (Maddison and Maddison, 1992). White branches indicate matrilineal cultures, and black shading indicates patrilineal cultures. Grey indicates duolateral cultures, in which both matrilineal and patrilineal descent occur simultaneously. For example, Pennington and Harpending (1993) describe some herds of cattle being in the control of matrilines, along with houses and gardens, whilst other herds were owned by patrlines, in the Herero. Duolateral groups are in the minority, but do constitute a significant proportion of those cultures showing some matrilineality. The pattern of matrilineality on the trees suggests that matrilineality is not confined to a single subgroup, but appears to have been adopted by several different groups that are not closely related.

The tree was used to construct phylogenetic contrasts, so that matrilineal groups (or clusters of groups) can be compared with sister groups that differ in the relevant variable (Felsenstein, 1985; Pagel and Harvey, 1988). Whether or not a group is matrilineal, and the extent to which it is dependent on cattle pastoralism, were taken from the updated computer version of Murdock (1967). This reveals eight contrasts (see Fig. 15.2) between sister groups with and without matrilineality. The computer program CAIC (Purvis and Rambaut, 1995) can be used to calculate the value of these contrasts. The magnitude of the difference from 0 (predicted

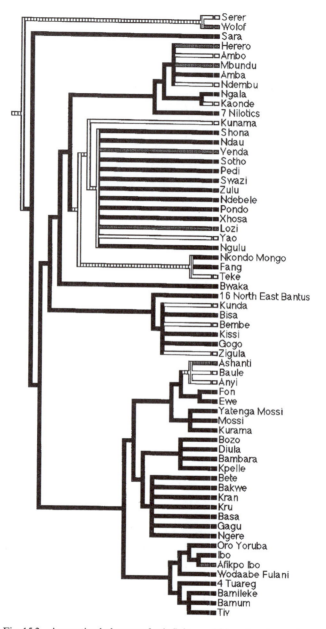

Fig. 15.2   A genetic phylogeny of sub-Saharan African farming and pastoralist groups, based on Cavalli-Sforza *et al.* (1994). Shading indicates the most parsimonious reconstruction of descent in ancestral nodes (after MacClade: Maddison and Maddison, 1992). Black indicates patrilineality, grey is duolateral descent, and white is matrilineality.

Table 15.2. *Associations with descent and mode of subsistence, using phylogenetic contrasts*

	Matrilineality			Patrilineality		
	Mean	df	Significance	Mean	df	Significance
Genetic tree Dependence on pastoralism	− 0.227	8	$p < 0.351$	0.409	7	$p < 0.091$
Language tree Dependence on pastoralism	− 0.192	25	$p < 0.062$	0.277	26	$p < 0.028$

by the null hypothesis that there is no association with matrilineality) can be tested by a one sample t-test. There was no statistically significant tendency for the matrilineal half of the contrast to be less dependent on cattle than the patrilineal half (Table 15.2). Table 15.2 also shows the same test for patrilineality. There are nine contrasts, some of which are different from those used in the matrilineal comparison because the duolateral cultures do show patrilineality (in addition to matrilineality) and are thus coded positive in both tests. In this case, there was a stronger, positive association between the presence of patrilineality and dependence on pastoralism, but it was not statistically significant. Relying only on data from those cultures for which the necessary genetic data have been collected to place them on a genetic phylogeny means that a large number of cultures has to be excluded. The number of contrasts is further reduced by the lack of resolution on the tree, leaving small sample sizes in this case.

Comparisons between sister groups on a linguistic tree

The same procedure was repeated using a linguistic tree based on Ruhlen's (1991) classification, for which it is possible to include a much greater number of cultures listed in Murdock's atlas. The genetic tree shown in Figure 15.1 is dominated by NigerKordofanians, but does include some other language groups (such as some Nilo-Saharan language speakers). On a language tree, these two groups would cluster separately. In this case, the authors constructed a tree only from the NigerKordofanian language group (Nilotics do not show much matrilineality). They created a tree of 227 cultures (the tree is not shown, but these are the cultures plotted on the map in Fig. 15.1), which resulted in 28 contrasts. The results (shown in Table 15.2) were in a similar direction to those found using the genetic tree. Patrilineality was significantly associated with pastoralism; the association between matrilineality and pastoralism was negative but no significantly so.

A model of the co-evolution of categorical characters based on maximum likelihood

Another comparative method, that does not use contrasts, was used to investigate the same effect. The method used was that of Pagel (1994), implemented by the program DISCRETE, which seeks evidence for the co-evolution of two categorical traits using a maximum likelihood approach. the authors use this method to test whether the evolution of matrilineality and the keeping of cattle (and also whether the evolution of patrilineality and the keeping of cattle) was dependent in any way. The model explores all possible means by which different rates of change in the characters concerned could produce the outcome observed, weighting them by their likelihood, if they are evolving independently. The procedure is repeated for a model in which either character may influence the rate of evolution of the other. The significance of the difference in the likelihood of the dependent model and the likelihood of the independent model can be tested. This method has some advantages over contrast methods. One is that it does not rely on contrasts that are based on a single reconstruction of events on the tree (the one that is most parsimonious). Cultural traits, in particular, might evolve rapidly, and therefore a number of reconstructions of history are all possibilities. The hatched branches on Figure 15.3b show that a single, most parismonious history of cattle adoption cannot even be identified in several areas of the phylogeny in this case. Further, the method has the potential to make use of far more of the information that can be inferred from the distribution of characters on the tip of a tree than does a simple correlation coefficient: the rate and likely direction of all poossible evolutionary transitions between states can be estimated (Pagel, 1994, 1997).

At present, DISCRETE requires a fully resolved (bifurcating) tree, so the authors used a composite tree based on the genetic tree in which multiple nodes had been resolved on the basis of linguistic similarity. Where nodes could not be resolved, and the cultures showed no variation in the relevant characters, they were clumped into a single tip. As only one tree was tested (Fig. 15.3), the authors consider this a preliminary test. Table 15.3 summarises the state of the cultures: all possible combinations of the presence or absence of cattle keeping and patrilineality are found, but with different frequencies. The correlational results were consistent with those found using the contrast method. The model of co-evolution of matrilineality and cattle keeping was not significantly different from the model of the two traits changing independently of each other. But there was an association between patrilineality and cattle keeping: the model of dependent evolution was far more likely ($p < 0.02$). The most likely rates of transition from one state to another are shown in Figure 15.4. It should be noted that the magnitude of a transition rate (q value) does not necessarily indicate its significance. Transitions between all states are likely to occur, but some are more likely than others. Whether or not a transition rate is significantly different from zero, or from any other rate, can be tested by fixing q values (either to zero, or equal to other q values), and comparing the likelihood of that model with that of dependent evolution in which all characters are unfixed. We tested all transitions and have marked those transition rates that are highly significantly different from zero with black bars. The flow diagram (Fig. 15.4) illustrates that cultures without cattle may lose or gain patrilineality. The state combining patrilineality and cattle appears to be quite stable. Looking at all the transition together, it appears that if a matrilineal culture without cattle were to become a culture with both cattle and patrilineality, the most likely direction of transition would be first the acquisition of cattle, followed by the acquisition of patrilineality. If the transition from no patrilineality to patrilineality either without cattle ($q31$) or with cattle ($q42$) is fixed to be equal, then the likelihood decreases significantly, indicating that $q42$ is significantly larger than $q31$; this indicates that matrilineal cultures are much more likely to gain patrilineality if they have first acquired cattle. This is a test of causation rather than simple correlation.

Discussion

Taken together, all these phylogenetic comparative methods suggest that patrilineality and pastoralism are positively associated. The adoption of cattle increases the likelihood that patrilineal descent will arise, but pastoralism does not necessarily threaten matrilineal descent per se. Several cattle-keeping societies combine matrilineal and patrilineal descent. These results are consistent with Hartung's idea that matrilineal inheritance is adaptive for women, and governs the resources over which they have most control, whereas patrilineal inheritance is adaptive for men, and is found when men have resources that they can use to accumulate their own wealth, in this case cattle.

The striking, spatial clumping of descent system, which spans a range of habitats, suggests that there may be functional reasons why patrilineal or

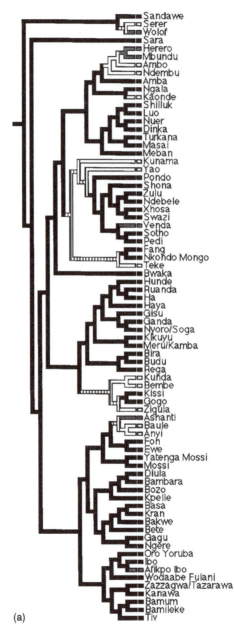

(a)

Fig. 15.3   (a) A composite phylogeny of sub-Saharan African farming and
pastoralist cultures, with patrilineality marked in black, duolateral descent in
grey, and matrilineality in white. Shading on branches shows the most
parsimonious reconstruction of ancestral states, for illustrative purposes only
(using MacClade: Maddison and Maddison, 1992). This is only one of the many
possible reconstructions considered in the maximum likelihood model.

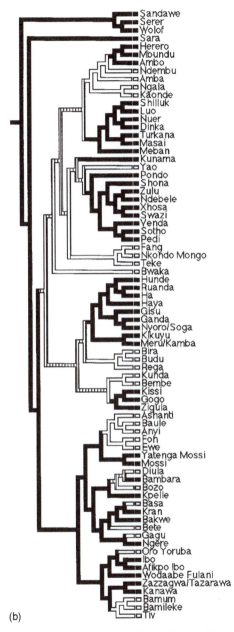

(b) The same phylogeny is as (a), showing cattle keeping (cattle keepers are marked in black). Hatched shading on a branch indicates that no single most parsimonious reconstruction of ancestral states can be established.

Table 15.3. *The states of that sample of cultures examined using* DISCRETE[a]

	− Patrilineality	+ Patrilineality
+ Cattle	4	44
− Cattle	8	20

[a]Also shown in Figure 15.3.

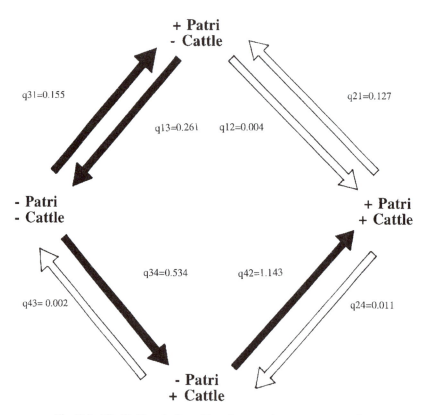

Fig. 15.4 The likelihood of transitions between character states. *q* values indicate the rate of transition between two states, and the significance of that rate is indicated by the shading of the arrow (black = $p < 0.01$), estimated by the program DISCRETE (Pagel, 1994).

matrilineal cultures might do better when neighbouring cultures are of the same descent system as themselves. It is interesting to consider what might occur when individuals from neighbouring groups of different descent intermarry. A matrilineal man would have difficulty acquiring the resources to marry and support a patrilineal woman, neither of whom would be likely to own or inherit any substantial resources. This situation would change if the man managed to acquire some independent source of income. A man from a patrilineal descent group might marry a woman of matrilineal descent, but she would remain in her matriline, as would their children. The patrilineal man would not gain any claim to the land owned by the matrilineal group into which he was marrying; thus, a patrilineal group migrating into a matrilineal area might have difficulty in gaining a foot-hold. As a genetic strategy, marriage with a matrilineal woman could be very successful, but the genetic success would not necessarily correlate with the spread of the cultural trait of patrilineality. A patriline may be reluctant to give resources to those who have married into a matrilineal group for fear of losing control of those resources. However, if the offspring of such unions were lucky enough to inherit resources from both their matriline and patriline, they might do rather well, and it is tempting to speculate that duolateral groups may have arisen in this way.

Whilst the authors have provided both empirical support and theoretical justification for an association between pastoralism and patrilineality, this does not constitute an explanation of how matrilineal systems ever arose and are maintained. Matrilineal descent is what is predicted to occur if females are in control of inherited resources, but neither these authors, nor any others, have yet provided a very convincing explanation of why females would have such control. The ethnographic literature suggests that males still wield most of the political power in matrilineal societies and, even without that, their physical strength would render them the usual winner in a dispute between man and wife. Whereas farming is an arena in which females work hard, men also work on farms and benefit from their produce. Men in matrilineal societies are frequently expected to work hard on their wives' fields.

We suggest an adaptive hypothesis here for why matrilineal systems arise and persist. Matrilineality is a strategy that favours the female line, but it arises not because females have 'won' in battle to control resources, but because men do not want to control those resources. Men are choosing an alternative strategy, in which they make use of mating and marriage opportunities whenever they arise. Sticking with one wife, guarding the land and raising the children may not be as beneficial to a man's reproductive success as moving on to another wife or lover, should the opportunity arise to father more children elsewhere. The costs of desertion of the first wife, when the economy is based on a resource that women can utilise independently of their husband if necessary, may be small relative to the benefits to be found by helping a new mate raise a new child. This may not apply in the case of cattle pastoralism, as herds are in need of constant protection from theft (which women cannot provide). Cattle are mobile and can produce resources immediately they are acquired (unlike fields, which may have little intrinsic value, deriving much of their worth from the hard work necessary to produce each harvest). Livestock can be used by men to enhance marriage opportunities directly, such as by raising a brideprice (e.g. BorgerhoffMulder, 1987; Mace, 1996). Where husbands are transient partners, matrilineal systems will arise as women seek support from their matriline, and it is to the advantage of matrilineal relatives to provide that help.

A prediction from this theory would be that if inherited land became more valuable, for example due to increasing population density and land shortage, men may then wish to gain more control. In modern matrilineal systems, there are many cases of women signing their fields over to their husbands, who may be more literate and capable of dealing with modern bureaucracy, in order to protect them from being claimed by others as land rights are formalised by government officials. Further, if opportunities for paternal investment arose that would make a greater contribution to the success of children, then, again, matrilineal systems would come under pressure as the nuclear family became a better strategy for both men and women. In modern Africa, fathers are generally expected to pay school fees. All these modern pressures could account for why a matrilineal 'roving male' strategy may no longer be the best in some societies. Similarly, in places where men are finding fewer opportunities to make useful investments in their offspring, due to unemployment at home or the promise of better opportunities from long or frequent migration to earn money elsewhere, matrilineality may be as strong as ever, or even appearing in places where it was formerly absent.

Acknowledgements

This research was funded by the Royal Society, NERC and the Leverhulme Trust.

References

Aberle, D.F. (1961). Matrilineal descent in cross-cultural comparison. In *Matrilineal Kinship*, ed. D. Schneider and K. Gough, pp. 655–730. Berkeley: University of California Press.

Betzig, L., ed. (1997). *Human Nature: A Critical Reader*. New York: Oxford University Press.

Borgerhoff Mulder, M. (1987). On cultural and biological success: Kipsigis evidence. *American Antoropologist* **89**, 619–34.

Cavalli-Sforza, L.L., Menozzi, P. and Piazza, A. (1994). *The History and Geography of Human Genes*. Princeton: Princeton University Press.

Dow, M., Burton, M., White, D. and Reitz, K. (1984). Galton's problem as a network autocorrelation. *American Ethnologist* **11**, 754–70.

Felsenstein, J. (1985). Phylogenies and the comparative method. *American Naturalist* **125**, 1–15.

Guglielmino, C.R., Viganotti, C., Hewlett, B. and Cavalli-Sforza, L.L. (1995). Cultural variation in Africa: role of mechanisms of transmission and adaptation. *Proceedings of the National Academy of Sciences, USA* **92**, 7585–9.

Hartung, J. (1985). Matrilineal inheritance: new theory and analysis. *Behavioural and Brain Sciences* **8**, 661–88.

Harvey, P. and Pagel, M. (1991). *The Comparative Method in Evolutionary Biology*. Oxford: Oxford University Press.

Holden, C. and Mace, R. (1997). A phylogenetic analysis of the evolution of lactose digestion in adults. *Human Biology* **69**, 605–28.

Mace, R. (1996). Biased parental investment and reproductive success in Gabbra pastoralists. *Behavioural Ecology and Sociobiology* **38**, 75–81.

Mace, R. and Pagel, M. (1994). The comparative method in anthropology. *Current Anthropology* **35**, 549–64.

Maddison, W.P. and Maddison, D.R. (1992). *MacCalde: Analysis of Phylogeny and Character Evolution. Version 3.0*. Sunderland, MA: Sinauer.

Murdock, G.P. (1967). *The ethnographic Atlas*. Pittsburg: Pittsburg University Press.

Pagel, M. and Harvey, P. (1988). Recent developments in the analysis of comparative data. *Quarterly Review of Biology* **63**, 413–40.

Pagel, M. (1994). Detecting correlated evolution on phylogenies: a new method for the analysis of discrete categorical data. *Proceedings of the Royal Society London Series B* **255**, 37–45.

Pagel, M. (1997). The inference of evolutionary processes from phylogenies. *Zoologica Scripta* **26**, 331–48.

Pennington, R. and Harpending, H. (1993). *The Structure of an African Pastoralist Community: Demography, History and Ecology of the Ngamiland Herero*. New York: Oxford University Press.

Purvis, A. and Rambaut, A. (1995). Comparative analysis by independent contrasts. *Computer Applications in Biosciences* **11**, 247–51.

Renfrew, C. (1987). *Archaeology and Linguistics*. Cambridge: Cambridge University Press.

Ruhlen, M. (1991). *A Guide to the World's Languages*. London: Edward Arnold.

Schneider, D.M. and Gough, K. (1961). *Matrilineal Kinship*. Berkeley: University of California Press.

Editor's conclusion. Socioecology and social evolution

Each of the chapters in this book aims to address a problem of major theoretical importance in relation to primates, their ecology, evolution and social diversity. The central thrust is comparative, either at the gross level of interspecific variation or at the more fine-grained intraspecific level. Some issues will remain unresolved until further information on the less well-known species is available; others may simply be too deeply rooted in the phylogenetic history of the primates to be explored effectively at the kind of levels emphasised in this book. Furthermore, the more we know about how different populations, or even social units, within a species vary, the more difficult it becomes to construct satisfactory generalised causal explanations for social system evolution.

What are the major issues in primate socioecology, and how have the chapters in this book helped to address them? We now have a better feel for the interaction between group size and social system, an issue that has long been conflated. If group size represents solutions to problems of intergroup competition within an ecologically tolerable range of sizes, then we can predict what range of group sizes should be expected (Williamson and Dunbar, Chapter 12). However, there is a further issue not explored here in depth, that of predation and its independent effects on group size. Predation can also be an extrinsic pressure on group size, as has long been noted. But neither predation nor ecologically sensitive group sizes fundamentally determines internal group structures; these are a consequence of the needs of females to ensure resources in order to maximise fertility and of the needs of males to maximise access to mates. Out of these principles of female and male distributions through time and space emerges the social system in all its complexity and variation (Fig. 1). This model is simply a heuristic device for structuring social options, and there are thus two further questions that can be raised. The first of these concerns an exploration of the range of variation observed in patterns of female and male distributions; the primates offer exceptionally rich examples of the extent of variation possible, particularly well illustrated by lemuroids (Chapter 10), neotropical monkeys (Chapter 11), and apes (Chapter 13). Sociality is not unique to the primates, but having arisen, the primates have co-opted their

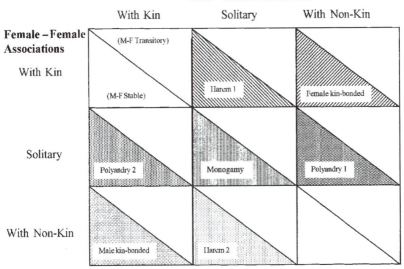

Fig. 1 Theoretical distributions of males and females based on associations
between same-sexed individuals and opposite-sexed individuals. This model
describes the limits to possible 'social space'.

sociality as a means for ecological adaptation. This is their specialisation,
but, again, it is not common to all forms. A schematic phylogeny of social
traits in primates (Fig. 2) suggests two important considerations: first, that
maintained sociality, based on repeated interactions and relationships
between the same individuals over time, appears to be basal among the
anthropoid primates; and second, that maintained sociality can arise independently (as in the lemurs), or indeed be lost (as in the orang-utan).
Furthermore, as Strier (Chapter 11) notes, the radiation of female-kin
structures is specific to the lineage of Old World monkeys, excluding the
colobids. While highly successful within this lineage, it is actually surprisingly rare amongst other primate groups.

The next question, and one of fundamental interest in this book on
comparative socioecology, concerns causality in the sex-specific distributions posed in Figure 1. Are these distributions a function of the need
to co-operate in reproduction (Ross and Jones, Chapter 4), in ensuring
infant growth and survival (Lee, Chapter 5), in food defence (Kappeler,
Chapter 10) and food sharing (Blurton Jones *et al.*, Chapter 6), or in the
prevention of infanticide (van Schaik *et al.*, Chapter 8). Indeed, each of
these may be important for different lineages, or have played a role in
shaping observed social systems at different times in a group's evolutionary

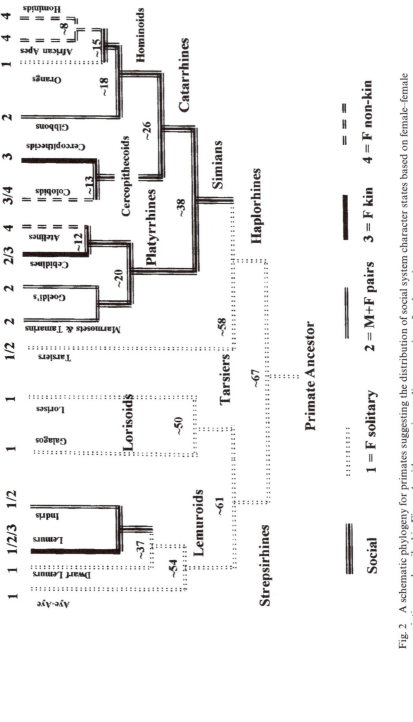

Fig. 2 A schematic phylogeny for primates suggesting the distribution of social system character states based on female–female associations as described in Figure 1, with approximate divergence times for the major taxa.

history. Unresolved questions as to ecological causality are specifically highlighted by the limited application of resource defence models to the neotropical primates (Strier, Chapter 11). Finally, we can examine the relationship between social systems and biological constraints. What are the roles of digestive, reproductive or cognitive morphology in setting the conditions for social evolution? Here, we need far greater exploration of causality as opposed to consequence in order to separate the effects of biology from outcomes in social evolution. If, as Barton (Chapter 7) argues, large primate brains are the product of visual system specialisation both for food perception and facial discrimination, then the evolution of the primate brain plays a central role both in foraging ecology and social strategies.

If there is any final message in the book, it is an appeal to continue the attempt to explore primate sociality in all its variation: to attempt to synthesise the mechanisms, be they physiological or psychological, with behavioural outcomes; to understand the evolutionary patterns within as well as between primate lineages; and to seek novel explanations where hypotheses are simply no longer adequate for the data.

Furthermore, the extent of intraspecific variation in social structure should point the way towards the development of a new generation of models. These models may need to be either specific to the lineage of interest, or general across primates from the strepsirhines to humans. It is only by asking the kinds of comparative questions addressed in this book that we can seek patterns, understand variation and explore causality for social evolution. It is hoped that the chapters will generate new approaches, and in turn stimulate new comparative analyses of primate socioecology.

Index

Species names are not listed, since species comparative data are presented throughout the chapters.